33.00
80E

OBSERVATIONAL AND THEORETICAL ASPECTS OF
RELATIVISTIC ASTROPHYSICS AND COSMOLOGY

OBSERVATIONAL AND THEORETICAL ASPECTS OF RELATIVISTIC ASTROPHYSICS AND COSMOLOGY

Proceedings of the International Course
held at Santander, Spain, September 3-7, 1984

edited by
J. L. Sanz and L. J. Goicoechea

U.I.M.P.
Universidad Internacional Menéndez Pelayo

World Scientific

Published by

World Scientific Publishing Co. Pte. Ltd.
P. O. Box 128, Farrer Road, Singapore 9128

OBSERVATIONAL AND THEORETICAL ASPECTS OF RELATIVISTIC ASTROPHYSICS AND COSMOLOGY

Copyright © 1985 by World Scientific Publishing Co Pte Ltd.

All rights reserved. This book, or parts thereof, may not be reproduced in any form or by any means, electronic or mechanical, including photocopying, recording or any information storage and retrieval system now known or to be invented, without written permission from the Publisher.

ISBN 9971-978-19-9

Printed in Singapore by Kyodo-Shing Loong Printing Industries Pte Ltd.

OBSERVATIONAL AND THEORETICAL ASPECTS OF RELATIVISTIC ASTROPHYSICS AND COSMOLOGY

SPONSOR: U. I. M. P.

with the collaboration of:
Consejería de Cultura del Gobierno de Cantabria
G. I. F. T.
Facultad de Ciencias de la Universidad de Santander

ORGANIZING COMMITTEE:
J. L. Sanz and L. J. Goicoechea
(Universidad de Santander, Spain)

FOREWORD

The specialized Course "Observational and Theoretical Aspects of Relativistic Astrophysics and Cosmology" was held in Santander (Spain) during 3-7 September 1984 at the International University Menéndez Pelayo (U. I. M. P.)

The speakers were Professors B. J. Carr (Inst. of Astronomy, Cambridge), R. Dominguez (Univ. Autónoma de Madrid), A. C. Fabian (Inst. of Astronomy, Cambridge), B. J. Jones (Nordita, Copenhagen), R. Lapiedra (Univ. de Valencia), M. A. H. MacCallum (Queen Mary College, London), J. Marcaide (Max-Planck, Bonn), P. J. E. Peebles (Princeton Univ.), J. Uson (Princeton Univ.) and E. Verdaguer (Univ. Autónoma de Barcelona). All of them covered different theoretical and observational aspects of Astrophysics and Cosmology such as: the microwave and X-ray backgrounds, the large-scale mass distribution, relativistic cosmological models, the generation of new solutions, the structure and evolution of galaxies, primordial black holes, the dark matter problem, primordial multheosynthesis, relativistic effects in binaries, radioastronomical observations, etc.

Many of these topics have been rapidly evolving during the last decade, therefore our interest to promote active research and propagate ideas in those areas, have been the main objectives of the Course. We hope this meeting, which follows the same line initiated by X. Fustero and E. Verdaguer with another International Course on Astrophysics and Cosmology held in Sant Felíu de Guixols during June 1984, to be continued with more Courses on different topics on Astrophysics and Cosmology in the near future in Spain

The organizers express their cordial thanks to all participants especially to our speakers who kindly accepted our invitation since the first moment, gave agreeable lectures directed to a wide audience of theoretical and observational physicists and did a remarkable effort to give us their manuscripts promptly.

We are also deeply indebted to the sponsor Institutions: the U. I. M. P. and as collaborators: the Consejería de Cultura del Gobierno de Cantabria, the G. I. F. T. and the Facultad de Ciencias de la Universidad de Santander.

Finally, we specially thank Professor A. Cendrero and the Spanish Relativity Group that encouraged us with the organization of this Course from the beginning.

Santander, April 1985   J. L. Sanz and L. J. Goicoechea

CONTENTS

Foreword	vii
Black holes, pregalactic stars, and the dark matter problem B. J. Carr	1
Primordial nucleosynthesis and the gravitino problem R. Dominguez	79
The X-ray background A. C. Fabian	103
Cosmology and galaxy formation — An introduction to some recent results B. J. T. Jones and E. Martinez	123
Clean relativistic binaries: Some observational consequences of the CM velocity R. Lapiedra	165
Relativistic cosmological models M. A. H. MacCallum	183
Present and future of high resolution radio observations of active galactic nuclei J. M. Marcaide	229
The halo puzzle P. J. E. Peebles	253
The microwave background radiation J. M. Uson	269
Solitons and the generation of new cosmological solutions E. Verdaguer	311
List of participants	351

BLACK HOLES, PREGALACTIC STARS, AND THE DARK MATTER PROBLEM

B.J.Carr

School of Mathematical Sciences, Queen Mary College,
University of London, England

Astrophysics Group, Fermilab, Batavia,
Illinois, U.S.A.

ABSTRACT

We review the different ways in which black holes might form and discuss their various astrophysical and cosmological consequences. We then consider the various constraints on the form of the dark matter and conclude that black holes could have a significant cosmological density only if they are of primordial origin or remnants of a population of pregalactic stars. This leads us to discuss the other cosmological effects of primordial black holes and pregalactic stars.

1. BLACK HOLES IN COSMOLOGY

1.1 Introduction

One of the most exciting predictions of general relativity theory is that there can exist regions of space-time in which gravity is so strong that nothing, not even light, can ever escape. As shown by the spherically symmetric Schwarzschild solution, this happens whenever a mass M is concentrated within a radius

$$R_S = \frac{2GM}{c^2} \approx 3\left(\frac{M}{M_\odot}\right) \text{ km}. \tag{1}$$

Of course, even though a black hole may exist mathematically, it is not obvious that the enormous compression required to form it can arise in nature: eqn (1) implies that the density of a region which has entered its Schwarzschild radius is

$$\rho_S = \frac{3c^6}{32\pi G^3 M^2} \simeq 10^{18} \left(\frac{M}{M_\odot}\right)^{-2} \text{ g cm}^{-3} \qquad (2)$$

and, for a solar mass object, this is a thousand times greater than nuclear density. Nevertheless, the discovery of neutron stars in 1967 (with radius only ten times larger than that required for collapse) forced astrophysicists to realize that such extreme conditions are not necessarily implausible[1]. Indeed our understanding of stellar evolution now suggests that sufficiently massive stars almost inevitably leave black hole remnants.

The realization that black holes could exist in the real Universe prompted a renewed interest in their mathematical properties and the last twenty years have seen some remarkable developments in this respect. Spherically symmetric gravitational collapse is well understood[2] and exhibits two crucial features. Firstly, an _event horizon_ forms when the radius of the object falls below R_S; this is the boundary of the black hole and events inside it can never be seen by an outside observer. Secondly, having formed such an event horizon, the infalling matter collapses to a point of infinite density called a _singularity_; at such a point all known laws of physics break down. It is not obvious that these same features would be exhibited in a more general (non-spherical) collapse. However, powerful theorems show that a singularity must always occur somewhere once a body gets sufficiently compressed[3]. It would be embarrassing if this singularity (with its associated impredictability) could influence the outside world, so the _cosmic censorship_ hypothesis (still not rigorously proved) states that an outside observer will always be shielded from its effects by an event horizon[4].

Another remarkable theorem shows that, however messy the collapse, the black hole will always settle down to a stationary state which depends only on the mass, angular momentum, and charge of the original object[5-7]. All other information about the object is lost and any irregularities are radiated away as gravitational and electromagnetic radiation[8]. This is called the No Hair Theorem and it makes the study of black holes remarkably simple: unlike all other astrophysical objects (which display a wide variety of properties), black holes are described by only three parameters. The stationary solution to which black holes evolve is called the Kerr-Newman solution[9] and, in the limit in which the angular momentum and charge are zero, it just becomes the Schwarzschild solution.

A third theorem shows that the surface area of a black hole never deceases[10]. This Area Theorem implies that black holes can never bifurcate, even though two of them can merge. However, the proof of this theorem applies only in classical theory and it can be violated by quantum effects. This was demonstrated by Hawking's discovery in 1974 that black holes are not black at all but radiate due to quantum effects with a temperature[11]

$$T_{BH} = \frac{hc^3}{8\pi GkM} = 10^{-7} \left(\frac{M}{M_\odot}\right)^{-1} K. \tag{3}$$

Since the holes lose energy in this way, they must shrink and eventually disappear altogether, even though this contradicts the Area Theorem. For holes of stellar origin, however, the temperature given by eqn (3) is tiny and the evaporation timescale is much longer than the age of the Universe, so the classical laws are still effectively valid.

1.2 How Black Holes Form

1.2.1 Stellar remnants. The most plausible mechanism for black hole formation invokes the collapse of stars which have completed their nuclear burning. However, this can only happen for sufficiently massive ones. Stars smaller than $4 M_\odot$ are supposed to leave white dwarfs because the collapse of their remnants can be halted by electron degeneracy pressure[12], while stars in the mass range $4-8 M_\odot$ probably explode due to degenerate carbon ignition[13]. Stars larger than $8 M_\odot$ but smaller than about $10^2 M_\odot$ are supposed to burn stably until they form an iron/nickel core[14]. At this stage no more energy can be released by nuclear reactions and so the core collapses. If the collapse can be halted by neutron degeneracy pressure, a neutron star will form and a reflected hydrodynamic shock then ejects the envelope of the star, giving rise to a type II supernova. If the core is too large, however, it necessarily collapses to a black hole, in which case it is not clear whether envelope ejection occurs. We do not know for certain what circumstances give rise to the formation of black holes rather than neutron stars, but it is probably reasonable to assume that a black hole will result if the initial stellar mass exceeds some critical value M_*.

It is difficult to predict the value of M_* theoretically. Indeed one could not be sure from numerical calculations that <u>any</u> stars in the $4 - 10^2 M_\odot$ range undergo collapse. However, there is good evidence that at least a few stellar black holes exist. For even though black holes can never be seen, one can still see their effects on surrounding objects. In particular, one can infer their presence in binary systems, especially when they are able to accrete material from the companion star and thereby generate X-rays. The first candidate for such an object was Cygnus X1 [15]. It was discovered by the X-ray satellite UHURU in 1972 and is still the best case of its kind, even though there are now several other ones (Circinus X-1 [16], LMCX-3 [17], GX339-4 [18]). The existence of stellar black holes is therefore likely on both theoretical and observational grounds, and M_* probably lies between 20 $M_\odot$ and 50 $M_\odot$.

1.2.2 <u>VMO remnants</u>. Stars larger than $10^2 M_\odot$ are radiation-dominated and therefore unstable to nuclear-energized pulsations during their hydrogen and helium burning phases [19]. It used to be thought that the resulting mass loss would be so rapid as to preclude the existence of such Very Massive Objects (VMOs). However, it is now thought that the pulsations will be dissipated as a result of shock formation [20] and this could reduce the mass loss enough for VMOs to survive for at least their main-sequence time [21]; this is just a few million years, independent of the VMO's mass. In fact, there is evidence for the existence of a few VMOs even at the present epoch (in particular, 30 Doradus [22], η Carina [23], and SN1961 [24]). However, VMOs encounter the much more serious "pair instability" as soon as they commence oxygen core burning; this is because the temperatures attained in this phase are enough to generate electron-positron pairs[25]. (This applies only for an oxygen core mass exceeding about 40 $M_\odot$, which is why ordinary stars are able to burn stably until they form an iron/nickel core.) This instability has two effects: sufficiently large cores collapse to black holes, while smaller ones explode[26]. Semi-analytical calculations[27], as well as numerical results[28], indicate that the critical dividing mass is M_{oc}=100 $M_\odot$ to an accuracy of 10% if there is no rotation, though the figure could rise to 500 $M_\odot$ if rotation is as large as possible[29]. The critical mass (M_c) for the initial hydrogen stars depends upon the amount of mass loss in the hydrogen and helium burning phase, but it would have to be at least 200 $M_\odot$. There is no direct evidence for any VMO holes in the real Universe, but all the tentative VMO candidates would appear to exceed the critical mass.

1.2.3 <u>SMO remnants</u>. Stars in the mass range above $10^5 M_\odot$ are unstable to general relativistic instabilities[30]. Such Supermassive Objects (SMOs) may collapse directly to black holes without any nuclear burning at all, at least if they have zero metallicity and no angular momentum. The presence of either metals or rotation may permit SMOs to explode in some mass range above $10^5 M_\odot$ but sufficiently massive ones will still collapse[31]. Although there is no definite evidence for the existence of SMOs, one could plausibly envisage their formation through relaxation at the centres of dense star clusters: the stars would be disrupted through collisions and a single supermassive star could then form from the newly released gas[32]. Indeed SMOs were originally invoked as an explanation for the violent activity associated with quasars, although accretion by their black hole remnants is now regarded as a more plausible explanation[33]. The holes would need to have a mass of about $10^8 M_\odot$, although it should be noted that supermassive holes would not necessarily derive from supermassive stars; they might also derive from the coalescence of smaller holes[34] or from accretion onto a single smaller hole[35].

Since quasars are probably precursors of galaxies, one might expect many galaxies to contain giant black holes in their nuclei even today[36]. In the last ten years considerable evidence has accumulated to support this view. Velocity dispersion and light curve measurements at the centre of M87 suggest[37] that it could contain a hole of $2 \times 10^9 M_\odot$ and rotation curve measurements at the centre of our own galaxy indicate that it may house a $3 \times 10^6 M_\odot$ hole[38]. The violence associated with some galactic nuclei could also result from the presence of a giant black hole: for example, the activity and X-ray emission of Centaurus A may be accounted for by a $10^8 M_\odot$ hole[39].

1.2.4 <u>Primordial remnants</u>. Equation (2) shows that the formation of giant black holes does not involve the same extreme conditions as arise in stellar collapse: an object of $10^9 M_\odot$ would only have the density of water on falling inside its event horizon. On the other hand, the formation of black holes smaller than a solar mass would require even more compression. Such conditions are unlikely to arise at the present epoch. They may, however, have arisen naturally in the first few moments of the Big Bang and this has led to the suggestion[40] that primordial black holes may have formed with mass much less than 1 $M_\odot$. Such primordial holes could have formed from initial inhomegeneities if the Universe started off "semi-chaotic" or they could have formed spontaneously at a cosmological phase transition. We will be discussing these formation mechanisms in detail in Section (3). For the present it is sufficient to point out that primordial holes are expected to have a mass of order that of the particle horizon at their formation epoch. They could thus span an enormous mass range: from 10^{-5} g for those forming at the Planck time to $10^5 M_\odot$ for those forming at 1 s. Although there is no conclusive evidence that primordial holes ever formed, they are of great theoretical interest since they are the only holes small enough for quantum effects to be important.

1.3 When Black Holes Form

A population of black holes could form at a variety of cosmological epochs, as indicated in Fig.1. The epoch of formation is not necessarily related to the mechanism of formation, although primordial holes form during the very early Universe by definition. Fig.1 also associates a "probability" with each scenario. The P estimates are necessarily subjective since it is difficult to assess the likelihood of any scenario in a field as prone to changing fashions as cosmology. Nevertheless, the probabilities are supposed to be a fair reflection of current trends, i.e., if one took a poll, the probabilities quoted might correspond to the fraction of cosmologists who would have credence in each scenario!

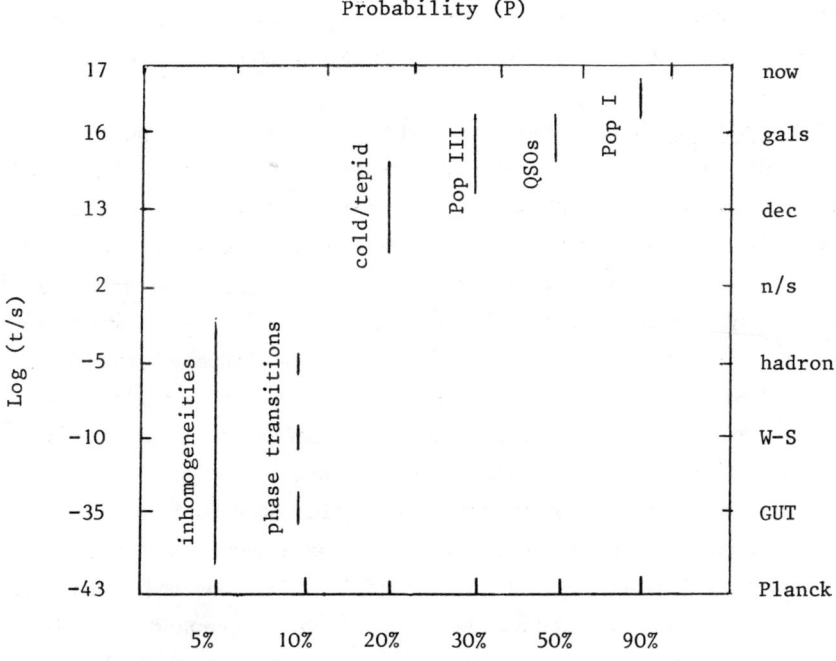

Figure (1): Probability of black holes forming at various epochs

1.3.1 Protogalactic holes. The holes most likely to exist are those which derive from ordinary Population I stars (i.e. stars which form in the discs of spiral galaxies). The number of such holes depends on the uncertain mass M_* but it would be very surprising if there were none of them at all. The probability of 90% is really an estimate of the likelihood that the prime candidate, Cygnus X-1, is a black hole. Since galactic discs probably do not form until a fairly recent epoch ($z \sim 1$), Population I holes are assumed to form at around 10^{10}y.

A probability of 50% has been assigned to the hypothesis that supermassive black holes power quasars and active galaxies. If the holes form through relaxation processes at the centres of protogalaxies, then they would arise shortly after the protogalaxies themselves. Since protogalaxies are expected to bind at about z=10, and since quasars are observed back to z=4, this probably corresponds to a time of order 10^9y. However, one cannot exclude the alternative hypothesis that the holes actually formed before the galaxies.

1.3.2 Pregalactic holes. As discussed in Section 4.2, Population III stars would be expected to form before galaxies providing the density fluctuations surviving at decoupling extended down to subgalactic scales. This is expected if the initial fluctuations were isothermal[41] and it may also apply with adiabatic fluctuations if the Universe is dominated by "cold" particles like axions[42]. The time at which such pregalactic stars would form depends on the amplitude of the density fluctuations: it would necessarily exceed the time of decoupling (10^6y) in a hot Universe and it would have to be before 10^9y. On the other hand, even if galaxies are the first objects to form, one could still postulate that a protogalactic generation of Population III stars formed. Indeed, in the sense that they have zero metallicity, the first stars to form are necessarily Population III. Thus the probability of the Population III scenario per se might be regarded as 100%. However, what is not certain is the characteristic mass of the first stars. The probability of 30% indicated in Fig.1 is supposed to specify the likelihood that some of these stars will be large enough to leave black hole remnants. If some of the Population III stars were actually SMOs, they could even produce the holes required to power quasars.

1.3.3 **Primordial holes.** The probability of primordial black hole formation has been estimated as only 5%. This is because, if the holes derive from initial inhomogeneities, then the amplitude of those inhomogeneities has to be very finely tuned if primordial holes are to be produced in a sufficient number to be interesting without being overproduced[43]. If the holes form at a phase transition, the situation is somewhat better because one does not need to depend on the prior existence of density fluctuations.

This conclusion assumes that the early Universe has a radiation equations of state (i.e. it presupposes the conventional hot Big Bang scenario). If the Universe started off "cold" (without any radiation content), the situation would be very different because the equation of state would go soft after the hadron era at 10^{-4}s. In this case, bound objects could form very prolifically on scales larger than 10 $M_\odot$ at very early times [44]. The same would apply if the Universe started off "tepid"[45] (with a photon-to-baryon ratio S much less than its present value of 10^9) because, in this case, the equation of state would go soft after $10^{-5} S^2$s. However, in both the cold and tepid scenarios the first objects would not be expected to form black holes directly; unless they were very large, they would produce primordial stars, in which case the black holes themselves would not arise until a much later cosmological epoch.

The consequences of this scenario are therefore rather similar to those of the Population III scenario, at least as far as black hole formation is concerned. However, there is an important difference in that one still has to generate the 3K background. The most natural way to accomplish this is through the stars themselves[46]. If most of the Universe is processed through VMOs, then one can shown that the photon-to-baryon ratio generated by their nuclear burning should be

$$S \approx \left(\frac{Gm_p^2}{\hbar c}\right)^{-1/4} \approx 10^{10}. \tag{4}$$

Thus the observed value of S is explained rather naturally. However, as discussed in Section 4.4.2, it turns out to be very difficult to thermalize starlight efficiently unless one invokes rather exotic grains [47]. An alternative scheme is to generate the radiation by black hole accretion at such an early time ($t < 10^6$ y) that light can be thermalized by free-free processes[48]. However, only SMOs larger than $10^6 M_\odot$ collapse on a timescale less than $10^6 y$. Fig.1 assigns a low probability of 20% to the cold and tepid scenarios, not only because of the problems involved in thermalizing the 3K background, but also because of the problems in generating the light elements whose abundances are so naturally explained by cosmological nucleosynthesis in the hot Big Bang picture[49].

1.4 How Many Black Holes

The sorts of holes which are most likely to exist would not be expected to contribute appreciably to the cosmological density. There could be a billion stellar holes in our own galactic disc but they would still only comprise a small fraction of the mass of the visible galaxy. For example, if one assumes that the mass distribution of large stars is described by the Scalo-Miller[50] spectrum and that a fraction ϕ_B of the mass of all stars larger than M_* ends up as a black hole, then the density of these holes (in units of the critical density) is only

$$\Omega_B \simeq \Omega_* \phi_B \left(\frac{M_{min}}{M_*}\right)^{1.2} \simeq 10^{-5} \left(\frac{M_*}{20 M_\odot}\right)^{-1.2} \left(\frac{\phi_B}{0.1}\right) \tag{5}$$

Similarly, even if every galaxy contains a $10^8 M_\odot$ black hole in its nucleus, the associated cosmological density would only be

$$\Omega_B \simeq \Omega_{gal} \left(\frac{M_{BH}}{M_{gal}}\right) \simeq 10^{-6} \tag{6}$$

The question therefore arises of whether the other sorts of holes discussed above might not make a more significant contribution to the density, even though the probability of their existence may be a priori smaller. Could they make up in Ω for what they lack in P?

As we discuss in Section 2.2, there is now considerable evidence that a large fraction of the Universe's mass is dark[51]. There seem to be several dark components and, since darkness is such a pronounced property of objects which have undergone gravitational collapse, one naturally wonders whether black holes could provide one of them. This is certainly not an inevitable conclusion since there are several other viable candidates (in particular, some kind of elementary particle). Nevertheless, we will argue that black holes could at least be a plausible explanation for the dark matter in galactic halos. If this is indeed the case, one naturally wonders how such a large fraction of the Universe could have gone into black holes in the first place. There are probably only two possible answers: either they are the remnants of a first generation of Population III stars or they formed primordially but with a mass sufficiently large to have avoided evaporation by the present epoch.

We will be examining these options in turn in Sections (3) and (4). However, our discussion will not only focus on the dark matter issue; it turns out that primordial black holes and Population III stars can have important cosmological consequences even if they are not numerous enough to explain galactic halos. The general theme of these lectures might therefore be regarded as the cosmological consequences of black holes or their precursors. However, before scrutinizing primordial and Population III holes in particular, it may be useful to summarize some of the more general cosmological effects of black holes.

1.5 What Black Holes Do

Black holes could have a variety of astrophysical and cosmological consequences, even if their density is rather modest. The nature of these consequences will depend on the mass of the holes, as indicated in Table (1). Some of the effects will be discussed in greater detail in later sections but we summarize them below for completeness.

Table (1): Formation mechanisms for black holes and their consequences

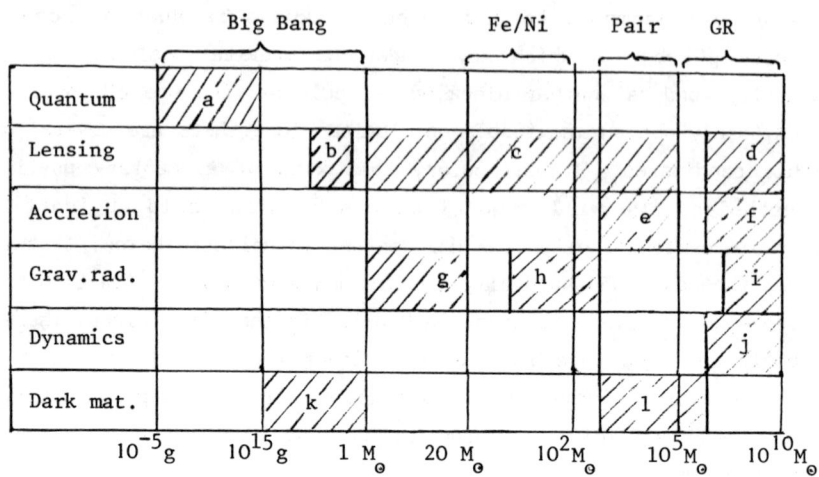

(a) See Table (3)   (e) IR background?   (i) Doppler detection
(b) Intensity change (f) X-ray background? (j) Disc heating
(c) Line/continuum (g) Bar detection   (k) Encounters?
(d) Image-doubling (h) Laser detection (l) See Table (4)

1.5.1 <u>Dynamical effects</u>. We will find that black holes could provide a lot of dark matter only if they are primordial with a mass between 10^{16}g and 1 $M_\odot$ or of Population III origin with a mass between $10^2 M_\odot$ and $10^6 M_\odot$. In the second case, the holes could have important dynamical effects. For example, if they reside in galactic halos, they could puff up the galactic disc, disrupt star clusters, and sink into the galactic nucleus through dynamical friction [52]. In general, these effects are significant only for holes larger than $10^5 M_\odot$ (i.e. only at the upper end of the permitted mass range). In the first case, however, the holes would be virtually undetectable by dynamical means. Even a minihole encountering the Earth would have no appreciable effect unless it were bigger than 10^{25} g and such an encounter could occur at most once every 10^5y. Thus the suggestion that the Tunguska event was induced by a small black hole [53] is rather implausible.

1.5.2 <u>Lensing effects</u>. Since the gravitational field associated with a black hole will bend light, anomalous effects are likely to occur whenever it traverses the line of sight of any light source. Both the shape and intensity of the source may change. In order for this traversal to occur with a reasonable probability, the source usually has to be an object like a quasar at a cosmogicial distance. Of course, this lensing effect is not specific to black holes (any object would do it) and so we will discuss the effect in Section 2.3.4 in the general context of the dark matter problem. However, in anticipation of that discussion, we note that gravitational lensing could permit the detection of black holes over the entire mass range from $10^{-4} M_\odot$ to $10^6 M_\odot$. It is thus the only effect which can probe both of the black hole scenarios for the dark matter mentioned above.

1.5.3 <u>Quantum effects</u>. The cosmological consequences of the evaporation of holes smaller than 10^{15} g will be discussed in detail in Section (3). We will argue that these holes could never have made a significant contribution to the cosmological density. Their evaporations may nevertheless have produced observable consequences. These are summarized in Table (2). In particular, they may have generated a detectable background of gamma-rays, positrons, and antiprotons; they may even have produced the cosmic deuterium abundance. The final explosive phase of evaporation could be particularly dramatic, tho energy of a billion megaton bomb being released from a region whose size is only one thousandth of a fermi!

1.5.4 <u>Gravitational radiation</u>. Black holes should be the most efficient generators of gravity waves in nature. Most of the radiation would appear as a burst at the time of the original collapse [54]. The efficiency ϵ with which radiation energy is generated from the original rest mass depends on the asymmetry of the collapse. To ensure a high value of ϵ, the collapsing object has to undergo bounces or fragment when rotational effects become important: in the optimal case ϵ might be as high as 0.1 but, in general, it would be much less[55]. A variety of types of gravitational wave detectors should be operative within the next few years and, between them, they could seek for bursts from almost all the types of hole discussed in Section 1.2: bars could detect ordinary stellar collapses, laser interferometers could detect VMO collapses, and doppler tracking of interplanetary spacecraft could detect SMO collapses. If a large number of holes formed at some epoch, one would expect their bursts to overlap and form a background of gravitational waves. The characteristics of this background are discussed further in Section 4.4.8.

1.5.5 **The clustering effect**. The formation of black holes could generate large-scale density fluctuations. This is because one would expect there to be statistical ($\sqrt{N}$) fluctuations in their number density, at least over scales sufficiently large for their formation to be uncorrelated [56]. This number density fluctuation will not necessarily produce a growing fluctuation in the total density (because each hole would be expected to be initially surrounded by an underdense region), but it may do so in two circumstances: (1) if the holes form primordially when the Universe has a hard equation of state [57]; and (2) if the holes are born with large peculiar velocities [58]. In the first case, the density fluctuations are of the $\sqrt{N}$ form on all scales but only begin to grow at decoupling or when the holes dominate the density at $t_B = 10^{10} \Omega_B^{-2}$ s (whichever is earlier). This effect could conceivably generate the fluctuations required to make galaxies, providing the holes have a mass of at least $10^5 M_\odot$. In the second case, the $\sqrt{N}$ fluctuations are only set up on the scale over which the black holes' distibution can be randomized by their peculiar velocities. In order for this effect to explain galaxies, the holes must have a mass of at least $10^6 M_\odot$ and they need velocities of order 10^3 km s^{-1}. Such velocities might conceivably be produced by the gravitational radiation recoil effect[59-61].

1.5.6 **Black hole accretion**. Black holes may generate radiation through accretion [62]. At the present epoch, this may be the chief hallmark of their existence, since the resulting luminosity can be very large for black holes in binary systems or galactic nuclei. However, it should be emphasized that the luminosity generated by black hole accretion at pregalactic epochs could also be significant (providing, of course, the holes exist then). If we assume that the holes accrete gas at the Bondi rate [63] and that the accreted material is converted into radiation with an efficiency η, then this will exceed the Eddington limit for some period after decoupling providing the hole mass exceeds $10^3 \eta^{-1} M_\odot$. Calculating how long the Eddington phase persists is complicated because the radiation generated will heat the background gas and thus suppress the accretion [64]. In general, however, one expects the accretion to have important consequences: besides affecting the thermal history of the Universe, it will also generate a significant background radiation density. If the holes are sufficiently large ($M \simeq 10^8 M_\odot$), it has been proposed that their pregalactic accretion could explain the hard X-ray background [65,66]. On the other hand, the accretion of somewhat smaller holes ($M \simeq 10^6 M_\odot$) has been invoked to produce an infrared background [67]. Neither of these proposals should be taken too seriously (since they both presume specific accretion models) but it is certainly plausible that pregalactic black holes may have generated a lot of radiation in some waveband.

In concluding this section, it must be emphasized that we cannot be sure that any of the black holes whose formation mechanisms are indicated in Table (1) actually exist. They could exist in theory, but that is no guarantee that the conditions for their formation ever arose in practice (cf. the probabilities indicated in Fig.1). Accordingly, all the cosmological consequences discussed above are equally speculative. Nevertheless, even the possibility that black holes could have such a variety of effects emphasizes how important it is to study them.

2. THE DARK MATTER PROBLEM

In this section we will first review the evidence for the existence of different types of dark matter and we will then discuss the various candidates for explaining it. By collecting together all the different constraints on the mass of the dark objects, we will argue that it is unreasonable to expect all the dark matter problems to have a single solution. We suggest that black holes would be a plausible explanation for at least the dark matter in galactic halos.

2.1 Evidence For Four Types Of Dark Matter

2.1.1 <u>The galactic disc</u>. It has been known for many years that the local density of material in the galactic disc, as inferred from its velocity dispersion, exceeds the density observed in gas and visible stars. The most recent calculations[68] indicate that 50% of the disc mass is dark and this corresponds to a density parameter $\Omega \simeq 0.01$. Although this is a fairly modest density (compared to that associated with the other dark matter problems), it is the dark component for which the evidence is most unambiguous. The observations also indicate that the disc dark matter must have a velocity dispersion of less than 50 km s^{-1}. Thus it must itself be confined to the disc and cannot be associated with any of the other dark components discussed below.

2.1.2 <u>Galactic halos</u>. In several dozen spiral galaxies, the rotation curve measurements indicate that the rotation velocity is constant as far as the visible stars extend[69]. This corresponds to a density $\rho(R) \propto R^{-2}$, and hence to a mass $M(R) \propto R$, whereas the density of visible stars falls off as R^{-3}. In many cases the rotation velocity can be measured out to distances well beyond the visible stars (eg. by making 21 cm observations of neutral hydrogen) and yet still remains constant[70]. This indicates that spirals have a dark component which extends further than the visible material and contains considerably more mass. For our own galaxy, the rotation velocities of giant molecular clouds[71] and the velocity dispersion of globular clusters[72] suggest that the dark material extends to at least 30 kpc, while the dynamics of our satellites [73] may increase this to 60 kpc. Independent evidence for the existence of galactic halos may come from the persistence of warps in discs [74] and from the fact that an extended halo may be required to stabilize discs against bar formation [75]. Both these features would also require that the halo have a spheroidal distribution [76]. The mass-to-light ratio and mass of a typical spiral halo depend on the radius R_H to which it extends. If $R_H \simeq 50$ kpc, we would require M/L $\simeq$ 100 and $M_H \simeq 10^{12} M_\odot$; this corresponds to a density parameter $\Omega \simeq 0.1$. It is unclear whether elliptical galaxies have halos [77], though there is some evidence that dwarf spheroidals do [78].

2.1.3 <u>Clusters of galaxies</u>. Measurements of the velocity dispersion in rich clusters of galaxies indicate [79] that their total mass exceeds the mass in their visible galaxies by at least a factor of 10. This corresponds to a typical mass to light ratio M/L $\simeq$ 300 and a density parameter $\Omega \sim 0.3$. The dark matter in clusters cannot be gas since, having a virial temperature of 10^8 K, it would produce far more X-ray emission than is observed. However, some X-rays are seen and this suggests that the gas density is at least comparable to that in galaxies [80]. The dark matter in clusters could in principle be the the same as that in galactic halos. Indeed, in the hierarchical clustering picture [81], one would expect all the galaxies inside a cluster to be stripped of their individual halos, thus forming a collective dark component. However, this would explain the amount of dark matter in clusters only if the original galactic halos had a sufficiently large value of R_H.

2.1.4 <u>The closure density</u>. There are various theoretical reasons for expecting that that the Universe should have at least the critical density required for it to eventually stop expanding. For example, if the early Universe underwent an inflationary phase (thereby explaining the isotropy of the 3K background and the flatness problem [82]), one would expect the total density parameter to be 1 to at least 60 places of decimal! The isotropy of the Universe may also indicate that $\Omega = 1$, even if inflation never occurred.[83] Finally there may be aesthetic reasons for wanting the Universe to be closed (eg. to satisfy Mach's principle[84]). According to all these arguments, there must be a dark component with a mass-to-light ratio of at least $M/L \simeq 10^3$, corresponding to $\Omega \simeq 0.99$. This would not necessarily contravene dynamical observations, providing the component is distributed much more smoothly than galaxies themselves[85]; to avoid clustering, it would need to have a velocity dispersion exceeding 10^3 km s^{-1}. Thus, if the closure density dark matter does exist, it must be distinct from the preceding types of dark matter.

2.2 Candidates For The Dark Matter

There are many possible explanations for the various forms of dark matter discussed above. This is hardly surprising since most objects in the Universe are dark; for this reason, several of the explanations may turn out to be correct. The candidates may be grouped into two categories: <u>non-baryonic</u> types (in which the dark object is some sort of elementary particle) and <u>baryonic</u> types (in which it is something astrophysical). In the first case, the existence of the dark object (which may degenerately be termed an "ino") goes back to the very early Universe. In the second case, the dark object forms out of the background gas at a relatively late stage (viz. 10^6-10^9 y after the Big Bang); this may be termed the "Population III" scenario. The two possibilities are illustrated qualitatively by the first two diagrams in Fig. 2 and the candidates are listed explicitly in Table (2). Note that primordial black holes will be included in the non-baryonic category (even though the holes may be large enough to be regarded as astrophysical) since they form at a time when the baryons only comprise a tiny fraction of the Universe's total density.

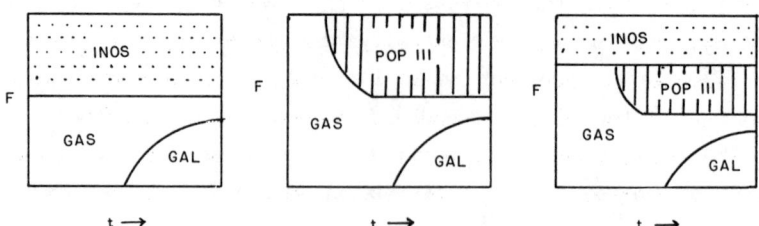

<u>Figure (2)</u>: Three scenarios for dark matter and galaxy formation

2.2.1 <u>Elementary particle candidates.</u> In the conventional hot Big Bang picture any particle should exist in thermal equilibrium with all other particles at times sufficiently early that the various interaction rates exceed the cosmological expansion rate. When this condition first fails (at some temperature T_F), the particle concerned will "freeze out"; providing it is relativistic at this time (i.e. providing its rest mass m_x is less than kT_F) and providing it survives until the present epoch (i.e. providing it does not annihilate or decay), its present number density should be [86]

$$n_x \simeq g\, n_\gamma \simeq 10^2 g\, cm^{-3}, \quad g = \left(\frac{T_x}{T_\gamma}\right)^3. \tag{7}$$

Here n_γ is the number density of the 3K photons and the factor g arises because the annihilation of other particle species after the freeze-out time will increase the relative photon density. Thus, as T_F increases, $g(T_F)$ decreases in a series of steps, each step corresponding to the rest mass of some particle. Since the particles are assumed to be non-relativistic today, eqn (7) implies that their present density is

$$\Omega_x \simeq g \left(\frac{m_x}{100 eV}\right) \tag{8}$$

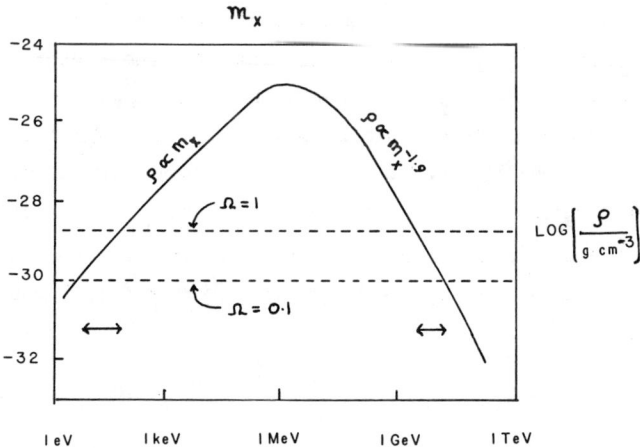

Figure (3): Density of "relict" inos as a function of their mass

If the particle is non-relativistic at freeze-out, the value of n_x is reduced by a Boltzmann factor [87] and this implies that the Ω_x decreases as a power of m_x for $m_x > kT_x$. The overall dependence of Ω_x on m_x is thus as indicated in Fig. 3. In order to explain the halo or cluster dark matter problems without having more density than would be consistent with observations of the cosmological deceleration parameter, m_x must lie in one of two narrow bands, as shown by the arrows.

The original elementary particle candidate was the neutrino [88-91]. This freezes out when the weak interaction rate falls below the expansion rate at $T_F = 1$ MeV and $g = 3/11$. Thus eqn (8) shows that the neutrino could be of cosmological significance if its rest mass exceeds about 10 eV. Attention was focussed on this possibility as a result of tritium decay experiments, which appeared to indicate [92] a neutrino rest mass in the range 14 eV $< m_\nu < 46$ eV. Independent evidence may come through the detection of neutrino oscillations: electron and muon neutrinos of energy E_ν should transform into each other over a distance [93]

$$L \simeq \frac{E_\nu}{m_{\nu e}^2 - m_{\nu \mu}^2} \simeq 10 \left(\frac{E_\nu}{10 \text{MeV}}\right)\left(\frac{m_\nu}{10 \text{eV}}\right)^{-2} \text{m}. \qquad (9)$$

At one stage there were claims to have found such an effect over a distance of 10 metres, though later experiments have not confirmed this. Indirect evidence for neutrino oscillations may come from the solar neutrino experiment [94]: the observed flux of electron neutrinos appears to be about a third that expected from nuclear reactions within the core of the Sun and this may be explained rather naturally if there are three neutrino species which oscillate into each other. However, eqn (9) implies that the oscillation lengthscale is less than the distance to the Sun providing m_ν exceeds 10^{-6} eV, so the solar neutrino problem does not in itself imply that m_ν is large enough to be of cosmological significance.

2.2.2 <u>Warm and cold elementary particles</u>. Another possibility is to invoke a particle whose mass is larger than 10 eV but which decouples earlier (reducing g) so that Ω_x is not too large. Candidates for such a particle arise in the supersymmetry theories (for which $T_F \simeq 10^6$ GeV and $g \simeq 0.01$) and include the gravitino[95], the right-handed neutrino[97], the photino[98,99], and sneutrino[100]. The mass of these particles ranges from 1 keV to 1 GeV and, if m_x is very large, one may need to invoke annihilations as well as an decrease in g to avoid an excessive value of Ω_x. Since the present temperature of these particles should should just be $g^{1/3}$ times the temperature of the microwave background, their velocity dispersion - prior to any clustering - should be

$$\langle v^2 \rangle^{\frac{1}{2}} \simeq \frac{kT_x}{m_x c} \simeq 20 \, (\frac{m_x}{10eV})^{-1} \, g^{1/3} \, (1+z) \, \text{kms}^{-1} \tag{10}$$

This decreases with increasing m_x and so particles much heavier than 10 eV are termed "warm" to distinguish them from particles like the neutrino which are termed "hot".

It should be stressed that the free-streaming of the particles will erase any density fluctuations on scales less than they can traverse by the time they go non-relativistic. This corresponds to a mass-scale[90]

$$M \simeq \frac{m_{planck}^3}{m_x^2} \simeq 10^{15} \, (\frac{m_x}{10eV})^{-2} \, M_\odot \tag{11}$$

For hot particles like neutrinos, this scale is very large and so it may be very difficult to explain how the observed large-scale structure could have evolved by the present epoch. For warm particles like the gravitino, the damping scale could be as small as a galaxy and the problem is less pronounced. However, a currently more popular solution is to invoke a "cold" particle like the axion which is not subject to free-streaming at all[101-104]. The axion is associated with an extra symmetry introduced so that QCD does not exhibit CP violation. This symmetry is spontaneously broken at an energy scale $f_a \simeq 10^{12}$ GeV. At a lower temperature, $T_I \simeq 1$ GeV, the axion develops a mass $m_a \simeq 10^{-5}$ eV due to QCD instanton effects and the associated density is

$$\rho_a \simeq \left[\frac{f_a^2}{M_p T_I}\right] m_a T^3 \tag{12}$$

The value of m_a is constrained by various astrophysical observations but Ω_a could certainly be large enough to be of cosmological significance.

Doubtless theorists will concoct many more types of elementary particles in the years to come, many of which might in principle have survived as relics of the Big Bang, so it is to difficult at this stage to assess the front runner. The latest candidate, for example, is "shadow matter". This is matter which resembles ordinary matter but only interacts with it gravitationally[105]; its existence may be predicted by superstring theory. It is probably premature to lay bets on any particular candidate. However, with such a large zoo of "inos", it is not implausible that at least one of them will turn out to be cosmologically significant. Note that certain types of elementary particles can already be excluded: for example, magnetic monopoles could not have a significant cosmological density without contravening the Parker limit[106]. The only remaining non-baryonic candidate, primordial black holes, will be disussed in detail in Section (3). Suffice it to say that only non-evaporating ones in the mass range 10^{16}g to 1 $M_\odot$ could possibly be relevant to any of the dark matter problems discussed above.

2.2.3 <u>Population III candidates.</u> It is very difficult to predict a priori the mass of the Population III objects since it depends on the mass-scale at which fragmentation of the first bound clouds ceases. (This is discussed in detail in Section 4.2.) The suggestions range from objects as small as snowballs to objects as large as supermassive stars. We will discuss each of the possibilities in turn, although we will find later that many of them can be rejected on empirical grounds.

Snowballs of condensed hydrogen have been proposed but can be excluded immediately[107]. In order to have avoided collisions within the age of the Universe, they must have a size of at least 1 cm but they would then have been evaporated by the 3K background radiation. In any case, it is rather unlikely that any fragmentation scenario in the hot Big Bang picture could produce fragments as small as this. One might, on the other hand, envisage the fragments being as small as Jupiters (i.e. objects in the mass range $M < 0.08\ M_\odot$ which are too small to ignite their nuclear fuel). We will see later that such objects could only be detectable by their gravitational lensing effects.

There would be a better chance of detecting Population III objects which derive from nuclear-burning stars. Stars smaller than $1 M_\odot$ would still be burning but they would have to be at least as small as $0.1 M_\odot$ in order to have a mass-to-light ratio large enough to explain any of the dark matter problems[51]. Stars larger than $1 M_\odot$ would no longer exist but they could still have produced dark remnants. For example, we have seen that white dwarfs and neutron stars could derive from stars in the mass ranges $1 - 4 M_\odot$ and $8 - 100 M_\odot$, respectively, and that black holes could possibly derive from stars at the upper end of this mass range. However, as discussed in Section 1.1, only VMOs in the mass range above $M_c \simeq 200 M_\odot$ and SMOs in the mass range above $10^5 M_\odot$ could collapse to black holes without ejecting a large fraction of their initial mass first. In any case, a priori, all Population III objects could produce dark matter except those in the mass ranges $0.1 - 1 M_\odot$ (which are too bright), $4 - 8 M_\odot$ (which explode due to degenerate carbon ignition), and $100 - 200 M_\odot$ (which explode due to the pair instability).

2.3 Constraints On The Dark Matter

2.3.1 <u>Cosmological nucleosynthesis</u>. In the standard hot Big Bang picture, the neutron-proton ratio freezes out at a temperature $T_F \simeq 10^{10} K$ (i.e. at a time $t_F \simeq 1$ s), when the rate for the for the weak interactions $p + e^- \to n + \nu$, $n + e^+ \to p + \bar{\nu}$ falls below the expansion rate. At this point the ratio has a value

$$\left(\frac{n}{p}\right)_F \simeq \exp\left[-\left(\frac{m_n - m_p}{kT_F}\right)\right] \simeq \frac{1}{8} \tag{13}$$

Since all neutrons (except the small fraction which are lost through β-decay) burn first into deuterium and then into helium at about 10^2 s, the resulting helium abundance is[49]

$$Y \simeq 2\left(\frac{n}{p}\right)_F \Big/ \left[1 - \left(\frac{n}{p}\right)_F\right] \simeq \frac{1}{4} \tag{14}$$

There are also small residual abundances of deuterium, helium-3, and lithium-7. These depend very sensitively on the total baryon density Ω_b, but it is a remarkable triumph of the standard model that the predicted abundances of all these elements[108] is consistent with observation providing $\Omega_b \simeq 0.1$. In particular, the observed deuterium abundance of 10^{-5} can only be explained by cosmological nucleosynthesis if $\Omega_b \lesssim 0.1(H_0/50)^{-2}$. Thus the dark matter in galactic halos (and conceivably even clusters) could be of baryonic origin, but a critical density certainly could not be unless one sacrifices the hot picture altogether[109]. We infer that Population III candidates could not have $\Omega \simeq 1$. This conclusion could be circumvented only for primordial black holes that formed before the neutron-proton freeze-out time; such holes would necessarily be smaller than $10^5 M_\odot$.

2.3.2 <u>Enrichment constraints</u>. The existence of Population I stars with metallicity as low as 10^{-3} excludes any of the dark matter problems being solved by stars with produce an appreciable metal yield. If the fraction of the initial stellar mass which is is left as a remnant is ϕ_r and the fraction returned as metals is Z_{ej}, the maximum density of the remnants is

$$\Omega_r = 10^{-3} \phi_r Z_{ej}^{-1} \Omega_g, \qquad (14)$$

where Ω_g is the gas density before the stars form. This constraint is disussed in more detail in Section 4.3.3. In the present context, it is sufficient to point out that it already excludes neutron stars as an explanation of anything except the local dark matter problem. Since these can only derive from stars in the mass range 8 - 60 $M_\odot$, for which $Z_{ej} > 0.2$ and $\phi_r < 0.2$, Ω_r could be at most $10^{-3}\Omega_g$ for neutron stars. A similar argument does not work for white dwarfs since these derive from stars in the range 1 - 4 $M_\odot$ and these return helium (for which the constraints are rather weak) rather than metals. However, even in this case Ω_r could be at most $\phi_r \Omega_g$ and this could only be large enough to explain the local dark matter.

2.3.3 <u>Source count constraints</u>. We have seen that stars with mass around 0.1 $M_\odot$ might in principle have a sufficiently high mass-to-light ratio to explain any of the dark matter problems. However, such stars could still be detectable as high velocity infrared sources[110] and searches already indicate[111] that their number density near the Sun can be at most

0.01 pc^{-3}. This is a hundred times too small to explain the local dark mattter problem and ten times too small to explain the halo problem, so M-dwarfs would seem to be excluded. A similar conclusion is indicated by infrared observations of some other spiral galaxies; these suggest that the mass of the halo objects must be less than 0.08 $M_\odot$, which would preclude any main-sequence stars[112]. Indirect arguments may even exclude Jupiters[107]: for it would be surprising if any fragmentation scenario could lead to fragments smaller than (say) 0.004 $M_\odot$ and, unless the fragments have a very steep spectrum, this implies there would still be too many light-producing stars above 0.1 $M_\odot$.

2.3.4 Lensing constraints.

Both high and low mass Population III objects could produce interesting gravitational lens effects. If one has a population of objects with mass M and density Ω_r, then the probability that one of them will lie close enough to the line of sight of a quasar to image-double it is about Ω_r and the separation between the images is[113]

$$\theta \approx 10^{-6} \left(\frac{M}{M_\odot}\right)^{\frac{1}{2}} \text{ arcsec.} \tag{15}$$

Thus the VLA - with a resolution of 0.1 arcsec - could search for lenses as small as $10^{10} M_\odot$ (indeed it has already found them), and the VLBI - with a resolution of 10^{-3} arcsec - could search for ones as small as $10^6 M_\odot$. This effect is important only for very large Population III objects. However, a galaxy itself can act as a lens and, if one is suitably positioned to image-double a quasar, then it can be shown that there is also a high probability than an individual halo object will traverse one of the lines of sight. This will give appreciable intensity fluctuations in one but not both images.[114] This effect would be observable for stars larger than $10^{-4} M_\odot$ but the timescale of the fluctuations, being or order $40(M/M_\odot)^{1/2}$ y, would only be detectable over a reasonable period for M < 0.1 $M_\odot$. However, yet another kind of lensing effect could permit the detection of objects with M > 0.1 $M_\odot$. This derives from the fact that such objects could modify the ratio of the line to continuum output of the quasar[115]; the fluxes are affected differently because they come from regions which act as extended and pointlike sources, respectively, unless M is very large. This already excludes a critical density of objects with $0.1 < M/M_\odot < 10^5$, though not necessarily the tenth critical density required for halos. SMOs with $M > 10^5 M_\odot$ are not excluded by this effect because, for them, the whole quasar nucleus would act as a point source. However, we have seen that holes only slightly larger than this might be detected by their direct image-doubling. Thus lensing constrains the Population III mass spectrum over nearly the entire range above $10^{-4} M_\odot$.

2.3.5 <u>Dynamical constraints</u>. A variety of dynamical effects constrains the masses of all four types of dark matter. For example, the survival of binaries in the galactic disc already requires [116] that the objects which comprise the local dark matter be smaller than $2 M_\odot$. On the other hand, if the dark matter is dissipationless, it must necessarily form in the disc and this immediately excludes any ino solution. The requirement that the disc should not be puffed up too much by the heating effect of traversing halo objects implies [117] that the halo objects must be no larger than $10^6 M_\odot$, an effect which is discussed in detail in Section 4.4.8. Even if the dark matter in clusters is different from the halo dark matter, the absence of unexplained tidal distortions of visible galaxies in (for example) the Virgo cluster implies [118] that the dark objects must still be smaller than $10^9 M_\odot$. The fact that any dark matter which contains the critical density must avoid clustering like galaxies implies that it must have a velocity dispersion of at least 10^3 km s^{-1}. This would seem to exclude any Population III candidates: even black holes could not be expected to born with recoil velocities this large. However, this argument would not exclude inos from having the closure density. In fact, phase space considerations - together with eqn (10) - imply that inos can form clusters of velocity dispersion σ and radius R only if [119]

$$M_x > 30 \left(\frac{R}{10\text{kpc}}\right)^{-1/2} \left(\frac{\sigma}{200\text{kms}^{-1}}\right)^{-1/4} \text{eV} \qquad (16)$$

Thus inos can explain the dark matter in clusters only if $m_x > 4$ eV and they can explain dark halos only if $m_x > 20$ eV; in the latter case, eqn (8) implies $\Omega_x > 0.5$, which is obviously too large. Thus the same feature

which makes inos an attractive explanation of the closure dark matter
detracts from their attractiveness as an explanation of the halo dark
matter.

2.4 Conclusion

The various constraints discussed above are brought together in
Table (2), which indicates which sorts of dark matter could explain each
of the four dark matter problems. The shaded regions in this figure are
excluded by at least one of the arguments given above. Whether the
dotted region is excluded depends on whether the cosmological
nucleosynthesis constraint demands that the dark matter in clusters be
non-baryonic; this is marginal.

Table (2): Constraints on types of dark matter

		LOCAL	HALO	CLUSTER	CLOSURE
POPULATION III	SMO			. . . . .	
	VMO			. . . . .	
	MO				
	NS				
	WD				
	LMO			. . . . .	
INOS	PBH				
	cold				
	warm				
	hot				

$\uparrow$ M $\downarrow$

The prime message of Table (2) is that one could not expect any single dark component to resolve all four dark matter problems. This should be of little surprise since most things in the Universe are dark. On the other hand, the figure does give some indications of what the best solutions might be: (i) the best candidate for the local dark matter would seem to be white dwarfs or Jupiters; (2) a possible solution for the halo dark matter would seem to be the black hole remnants of VMOs or low mass SMOs, though one cannot exclude primordial black holes or warm or cold inos; (3) the dark matter in clusters would need to be primordial black holes or inos if one adopts the cosmological nucleosynthesis limit in its strongest form but the other halo candidates would be viable if one adopts the weaker form; (4) the closure dark matter could only be hot inos.

Our analysis does not provide a unique answer to the four dark matter problems but it at least narrows down the range of possibilities. After all, a priori, there was an uncertainly of 10^{80} in the mass scale of dark matter, ranging from the 10^{-5} eV axion to the $10^{10} M_\odot$ SMO. Lacking a unique answer, each of us will doubtless assess the likelihood of the various candidates according to our own individual prejudices. Thus presumably particle physicists will prefer ino solutions, while astrophysicists will prefer Population III solutions. In a spirit of compromise, however, it is perhaps worth stressing that both ino and Population III solutions may be relevant: inos could provide the closure density and perhaps the dark mass in clusters, while black holes may provide the dark matter in halos. It is in this spirit of compromise that the third picture in Fig.3 is offered!

3. PRIMORDIAL BLACK HOLES

In this section we study the different ways in which primordial black holes (PBHs) may have formed in the early Universe and we derive their likely mass spectrum. We then consider their cosmological consequences, with particular emphasis on the evaporation of very small PBHs. Such holes are unlikely to have had an appreciable cosmological density but they may nevertheless have observable consequences.

3.1 The Formation of PBHs From Initial Inhomogeneities.

It was first pointed out by Hawking[40] that black holes could have formed in the first few moments of the Big Bang if the early Universe contained density inhomogeneities. Overdense regions would stop expanding with the background and undergo collapse providing they were larger than the Jeans length at maximum expansion, the Jeans length being $\sqrt{\gamma}$ times the horizon size if the background equation of state is $p = \gamma\rho$. The condition for this can be derived as follows. Consider a region with mass M which is overdense by a factor δ_o at some initial time t_o. When that region falls within the particle horizon at time t_H, the density fluctuation will have grown to

$$\delta_H = \delta_o \left(\frac{M}{M_o}\right)^{2/3} \tag{17}$$

(independent of γ) where M_o is the horizon mass at t_o. In this equation δ_H and δ_o represent the gauge-invariant energy density perturbations measured with respect to the comoving spatial hypersurface. After t_H the scale of the region and its overdensity evolve as

$$R \propto t^{\frac{2}{3(1+\gamma)}}, \quad \delta \propto t^{\frac{2(1+3\gamma)}{3(1+\gamma)}} \tag{18}$$

Thus the region binds (i.e. δ has grown to about 1) at a time

$$t_B = t_H \delta_H^{-\frac{3(1+\gamma)}{3(1+3\gamma)}} \tag{19}$$

and its scale is then

$$R_B = ct_B \delta_H^{1/2} \tag{20}$$

(i.e. it is smaller than the particle horizon by a factor $\sqrt{\delta_H}$). In order to collapse against the pressure at t_B we therefore need

$$\delta_H \gtrsim \gamma, \quad \delta_o \gtrsim \gamma \left(\frac{M}{M_o}\right)^{-2/3}. \qquad (21)$$

On the other hand, a region cannot be larger than the particle horizon at maximum expansion without forming a closed Universe, topologically disconnected from our own.[43] One way to see this is to consider the spatial hypersurface containing the region at t_B: its 3-curvature is $(G\rho) \sim (ct_B)^{-2}$, so it will close up on itself if the overdensity extends beyond a scale $\sim ct_B$. Thus we require $\delta_H < 1$. Note that δ_H is a measure of the metric perturbation at all times prior to t_H. Thus a region with $\delta_H > 1$ is <u>always</u> disconnected from our Universe; it does not evolve to that state.

Equation (20) implies that, unless the equation of state is soft ($\gamma = 0$), PBHs must have of order the horizon mass M_H at formation. More precisely, holes forming at time t should have an initial mass

$$M(t) \simeq \gamma^{3/2} M_H(t) \simeq \gamma^{3/2} \left(\frac{c^3 t}{G}\right) \simeq 10^{38} \gamma^{3/2} \left(\frac{t}{s}\right) g. \qquad (22)$$

Thus holes forming at the Planck time ($t_p \sim 10^{-43}$s) should have the Planck mass ($M_p \sim 10^{-5}$g), those forming at 10^{-23}s should have a mass of 10^{15} g, and those forming at 1 s should have a mass of $10^5 M_\odot$. This means that PBHs could span an enormous mass range, encompassing both the ones which are small enough to evaporate and the ones which are large enough to have significant astrophysical effects.

Equation (22) is roughly confirmed by the detailed hydrodynamical calculations of Nadejin <u>et al</u>.[120], who model PBH formation by a patching part of a k = +1 Friedmann universe onto a k = 0 Friedmann universe via a vacuum transition region. The evolution is found to depend on two parameters: the ratio of the size of the region (R) to the size of a k = +1 universe with the same density (R_{max}) and the ratio of the width of the transition region (Δ) to R. The first parameter indicates how close the overdense region is to being a separate universe; the second parameter determines the pressure gradient. As shown in Fig. 4, one needs a minimum value for R/R_{max} if a black hole is to form and, as the ratio Δ/R decreases (so that the pressure gradient

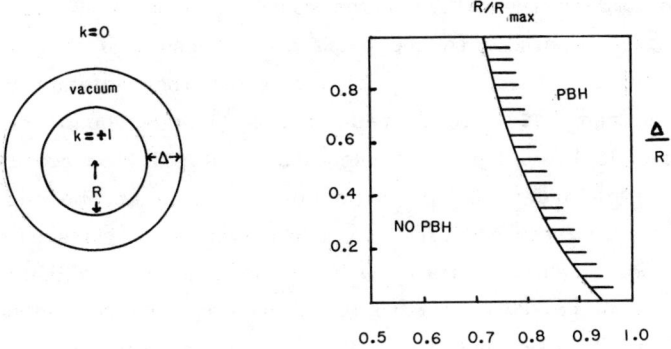

Figure (4): Condition for PBH formation from primordial inhomogeneities

increases, thereby making black hole formation more difficult), the value of R/R_{max} required increases. For example, as Δ/R goes from 1.0 to 0.1, the value of R/R_{max} required goes from 0.8 to 0.9. Perturbations with R/R_{max} less than the critical value turn into sound waves when they fall inside the Jeans length and are dissipated; only perturbations with R/R_{max} larger than the critical value grow large enough for gravitational collapse to ensue. The mass of the resulting black hole turns out to lie in the range 0.01 to 0.06 times the horizon mass at the formation epoch, which is somewhat smaller than the naive estimate given by eqn (22). The precise value depends on the equation of state.

3.2 The Growth and Mass Spectrum of PBHs

One issue which has attracted a lot of attention is the question of how much a PBH, once formed, can grow through accretion. A simple Newtonian argument of Zeldovich and Novikov[121] suggests that the black hole mass should evolve according to

$$M \simeq M_H(t) \left[1 + \frac{t}{t_1} \left(\frac{M_H(t_1)}{M_1} - 1 \right) \right]^{-1} \qquad (23)$$

where M_1 is the mass when the hole forms at time t_1. This implies that holes much smaller than the horizon cannot grow much at all, whereas holes of size comparable to the horizon could continue to grow at the same rate as the horizon ($M \propto t$) throughout the radiation-dominated era. Since eqn (22) indicates that a PBH must be of order the horizon size at formation, this suggests that all PBHs could grow to have a mass of order $10^{15} M_\odot$ (the horizon mass at the end of the radiation era). There are very strong observational limits on how many such giant holes could exist in the Universe[52], so the implication would be that very few PBHs ever formed. However, the Zeldovich-Novikov argument is clearly questionable since it neglects cosmological expansion effects and these are presumably going to hinder a black hole's growth. Indeed the notion that PBHs can grow at the horizon rate was disproved by Carr and Hawking[122], who showed that there is no spherically symmetric similarity solution which contains a black hole attached to a k = 0 Friedmann background via a pressure-wave. Since it must therefore soon become smaller than the horizon, at which stage cosmological effects become unimportant and the Zeldovich-Novikov argument does pertain, one concludes that a PBH cannot grow much at all.

An interesting special case arises if the equation of state is stiff ($p = \rho$) when the holes form. This possibility cannot be excluded, especially for holes smaller than 10^{15}g which form in the first 10^{-23}s after the big bang. Lin et al.[123] have argued that a similarity solution containing a black hole attached to a k = 0 Friedmann universe does exist in this situation, basically because the sound-wave propagates out at the same rate as the particle horizon. However, Bicknell and Henriksen[124] show that this solution contains a rather unphysical feature at the sound surface: one needs the incoming matter to be transformed into an ingoing null fluid. This feature actually permits growth at the horizon rate even if $p = \rho/3$[125]. Unless one grants this possibility, growth will be limited even in the stiff situation. Indeed the most likely consequence of a stiff equation of state would be the suppression of PBH formation.

Another important question concerns the mass spectrum of PBHs. This is related to the form of the density fluctuations fairly straightforwardly [43]. If the size of a region is to exceed the Jeans length at maximum expansion, we have seen that the amplitude of the density fluctuation on entering the horizon must exceed γ. If one assumes that the density fluctuations on a mass-scale M are spherically symmetric and have a Gaussian distribution with a root-mean-square value $<\delta_H(M)^2>^{1/2} \equiv \varepsilon(M)$, then the probability of a region of mass M forming a black hole is

$$\beta(M) \sim \varepsilon(M) \exp\left[-\frac{\gamma^2}{2\varepsilon(M)^2}\right] \tag{24}$$

Providing $\beta \ll 1$, so that the probability of the hole being in a region which collapses later is small, this is directly related to the fraction of the Universe's mass which ends up in holes of mass M. Although one might question the necessity of a Gaussian distribution for the density fluctuations, none of our qualitative conclusions will depend on this except the exponential sensitivity of β on ε. Lindley has presented a more sophisticated treatment, allowing for deviations from a Gaussian distribution and for the effects of holes being swallowed by larger holes,[126] while Marochnik et al.[127] and Barrow and Carr [128] have discussed the effects of deviations from spherical symmetry.

For "constant curvature" fluctuations (in which δ_0 scales as $M^{-2/3}$), ε is scale-independent and the present number density of PBHs in the mass range M to M+dM can be shown to be $\Pi(M)dM$ with

$$\Pi(M) \propto \beta M^{-\alpha}, \quad \alpha = \left(\frac{1+3\gamma}{1+\gamma}\right) + 1. \tag{25}$$

The integrated mass density goes as $M^{2-\alpha}$. The dependence of the exponent α on the equation of state parameter γ stems from the fact that a region's mass is reduced by redshift effects before it undergoes collapse but not thereafter. For $\gamma = 1$, $\alpha = 3$; for $\gamma = 1/3$ (the most likely value), $\alpha = 5/2$; and as $\gamma \to 0$, $\alpha \to 2$. In fact, equations (24) and (25) do not apply if γ is very small because β may then be of order 1; in this case lots of holes are forming all the time and the effective value of α is 1 as a result of swallowing [44]. In any other situation, α exceeds 2, so most of the mass in PBHs will be in the smallest ones.

If the initial fluctuations go as $\delta_0 \propto M^{-n}$ with $n > 2/3$ (i.e. if they fall off faster than constant curvature fluctuations), then $\varepsilon(M)$ decreases with M and

$$\pi(M) \propto \exp\left[-\frac{\gamma^2}{2\varepsilon_0^2}\left(\frac{M}{M_0}\right)^{\frac{(n-2/3)(1+3\gamma)}{(1+\gamma)}}\right] \qquad (26)$$

where δ_0 is the value of δ on the scale M_0. Thus the PBH spectrum is exponentially cut off above a mass

$$M_{max} \simeq 10^2 \varepsilon_0^{\left[\frac{2(1+\gamma)}{(1+3\gamma)(n-2/3)}\right]} M_0. \qquad (27)$$

For example, if $M_0 \simeq 10^{-5}$ g and $\gamma = 1/3$, this gives $M_{max} \simeq 10^{-3} \varepsilon^{4/(3n-2)}$ g. If $n < 2/3$, so that the initial fluctuations fall off less steeply than $M^{-2/3}$, ε increases with scale. However, this seems rather implausible since one necessarily gets separate closed universes on a sufficiently large scale. Thus only constant curvature fluctuations can generate PBHs with an extended mass spectrum.

3.3   The Formation of PBHs at a Phase Transition

Even if ε is very small, black holes might still form prolifically at any epoch when the Universe goes pressureless ($\gamma << 1$). For example, this might occur if the Universe's mass is ever channeled into particles[130] which are massive enough to be non-relativistic or if the equation of state softens[129] at some sort of cosmological phase transition. The Universe might also go pressureless after the nuclear density epoch (t $\sim 10^{-4}$s) if it starts off "cold" (i.e. without any background radiation). In all these situations the important parameter determining the collapse probability is not ε but the asymmetry of the collapsing region[132]. For example, even if p = 0, turbulent effects might still prevent the formation of PBHs much smaller than the particle horizon. In this case, the mass spectrum would be given by eqn (25) with $\gamma = 0$.

So far we have assumed that the PBHs form from initial inhomogeneities. However, it is possible that black holes could form spontaneously at a cosmological phase transition even if the Universe starts off perfectly smooth. For example, bubbles of broken symmetry might arise at a spontaneously broken symmetry epoch and it has been suggested that black holes [133] could form as a result of bubble collisions. In fact, this only happens if the bubble formation rate is finely tuned: if it is too large, the entire Universe undergoes the phase transition immediately; if it is too small, the bubbles never collide. In consequence, the black holes should have of order the horizon mass at formation. Thus PBHs forming at the GUT epoch (10^{-35}s) would have a mass of order 10^3g, whereas those forming at the Weinberg-Salam epoch (10^{-10}s) would have a mass of order $10^{-5} M_\odot$.

Another possibility, suggested by Crawford and Schramm, is that PBHs could form spontaneously at the quark soup to hadron phase transition (10^{-6}s) as a result of the fact that the potential between two quarks increases with their separation.[135] The idea is that, when a hadron forms, neighbouring quarks will be able to feel the colour charge of quarks farther away across the hadron "gap" than in other directions (where colour screening reduces the interaction range). This means that hadron formation is more likely where a hadron already exists, resulting in spontaneous density fluctuations and black hole formation. The PBHs should have a mass of up to $0.1 M_\odot$ in this case. It is hard to envisage a phase transition occurring later than this in a hot Universe, so one could not expect PBHs larger than $0.1 M_\odot$ to form in the absence of initial inhomogeneities.

3.4 The Cosmological Effects of PBHs

Even if PBHs have a significant density today, they can only have comprised a tiny fraction of the cosmological density at early times. This is because the 3K background radiation density ($\Omega_R \approx 10^{-4}$ in units of the critical density), though small today, increases as $(1+z)^{-4}$ as one goes back in time, whereas the PBH density increases only as $(1+z)^{-3}$. Thus the fraction of the Universe in PBHs at time t is

$$\beta(t) = \left(\frac{\Omega_{PBH}}{\Omega_R}\right)(1+z)^{-1} \simeq 10^{-6} \Omega_{PBH} \left(\frac{t}{s}\right)^{1/2} \tag{28}$$

where the t dependence applies before the time $t_{eq} \simeq 10^{10}\Omega^{-2}$ s at which the Universe is matter dominated. We assume $\Omega_{PBH} \lesssim 1$ since this is the maximum value consistent with observations of the cosmological deceleration parameter. Using eqn (22), we thus have

$$\beta(M) < 10^{-8} \left(\frac{M}{M_\odot}\right)^{1/2} = 10^{-17} \left(\frac{M}{10^{15}g}\right)^{1/2} \qquad (29)$$

where 10^{15}g is the mass above which $\Omega_{PBH}(M)$ will have been unaffected by evaporations. Equation (24) therefore implies $\varepsilon(M) < 0.05$ on all scales above 10^{15}g. This has the important consequence that the early Universe cannot have been "chaotic" (with $\varepsilon \sim 1$) after 10^{-23}s.

This conclusion must be qualified. If the early Universe were truly chaotic, one would expect it to exhibit large anisotropies as well as inhomogeneities; in this case the assumption of spherical symmetry, on which eqn (24) is based, would fail. Barrow and Carr argue that PBH formation would be suppressed in an anisotropy-dominated Universe because the shear provides an effectively "stiff" equation of state.[128] In this case, one would not necessarily contravene limit (29), although one would still anticipate too many PBHs forming once the anisotropy became dynamically insignificant if $\varepsilon \sim 1$. However, the Barrow-Carr argument is rather simplistic; in another paper they argue for the opposite conclusion - that PBHs will be produced *more* abundantly in an shear-dominated Universe.[136] Therefore it seems likely that the exclusion of chaotic cosmologies is justified.

Another situation could invalidate eqn (28): the limit assumes that the Universe always has a radiation equation of state before t_{eq} but this assumpion would fail if the Universe started off cold (without any background photons). For example, if the 3K background were generated at some time t_R, eqn (29) would be replaced by

$$\beta(t) \simeq \left(\frac{\Omega_{PBH}}{\Omega}\right) \min\left[1, \left(\frac{t_R}{t_{eq}}\right)^{1/2}\right] \qquad (30)$$

for $t_R > t > 10^{-4}$s. Thus, if t_R exceeds t_{eq}, one could in principle permit most of the Universe to go into PBHs in the mass range above 1 $M_\odot$. On the other hand, the equation of state should still be hard before 10^{-4}s because of strong interactions, so one still has a strong limit on the fraction of the Universe going into evaporating PBHs:

$$\beta(M) < 10^{-10} \left(\frac{M}{10^{15}g}\right)^{\frac{1}{2}} \min\left[1, \left(\frac{t_R}{t_{eq}}\right)^{\frac{1}{2}}\right]. \quad (31)$$

Cold scenarios are now rather out of vogue because of the problems they face in generating and thermalizing the 3K background; one also has to give up the cosmological nucleosynthesis explanation for the light element abundances. In any case, one expects the first bound regions to form primordial stars and so, even though these stars may eventually give rise to black holes, the resulting scenario resembles the Population III picture more than the usual PBH picture. For the rest of this section we will therefore confine attention to the standard hot Big Bang scenario, reverting to a discussion of the cold scenario in Section (4).

The fact that $\beta(M)$ must be small in the hot scenario should occasion no surprise in view of the exponential sensitivity of β on ε. Indeed the striking feature of eqn (28) is that Ω_{PBH} could be significant even if $\beta(M)$ is tiny. Thus PBHs could be associated with all of the sorts of cosmological effects indicated in Table (1). Of course, it requires very fine tuning of ε if Ω_{PBH} is to have an interesting value: if ε is slightly larger than 0.05, the PBHs are grossly overproduced; whereas, if ε is slightly less than 0.05, their density is negligible. (For example if ε has the sort of value ~ 0.01 required to explain galaxy formation, Ω_{PBH} would only be 10^{-220}!) Therefore it might be regarded as a priori unlikely that PBHs would form prolifically enough to be interesting. This at least applies if the PBHs derive from primordial inhomogeneities. If they derive from a phase transition, there would still be a danger of violating eqn (29) but the exponential dependence of β upon ε would no longer apply, so an interesting value of Ω_{PBH} would not be a priori so unlikely.

The cosmological consequences of PBHs larger than 10^{15}g are discussed in Sections (1) and (2), so for the rest of this section we will confine attention to the consequences of the quantum mechanical evaporation of those smaller than 10^{15}g. Note that these holes form before 10^{-23}s, when the density of the Universe exceeds 10^{54}g cm^{-3}. At such extreme densities, we cannot necesssarily assume that the simple gravity-dominated scheme for PBH formation discussed in Section 3.1 is realistic, so the spectrum predicted by eqn (25) should only be adopted with caution.

3.5 The Evaporation of PBHs

The crucial feature of sufficiently small PBHs is that they can shrink through quantum effects[11,137]. In general a black hole emits particles with energy in the range (E, E+dE) at a rate given by

$$\dot{dN} = \frac{\Gamma dE}{2\pi\hbar} \left[\exp\left(\frac{E - nh\Omega - e\phi}{h\kappa/2\pi c} \right) \pm 1 \right]^{-1}. \quad (32)$$

Here Ω, ϕ and κ are the angular velocity, electric potential and surface gravity of the hole; Γ is the absorption probability for the species of particle involved (in general a function of its spin); and the + and − signs apply for fermions and bosons respectively. One would expect $\phi = 0$ since eqn (32) implies that a black hole discharges on a much shorter timescale that it evaporates, at least for $M < 10^5 M_\odot$[138]. Although it is not so clear why angular momentum should be lost on a shorter timescale, unless there exists a massless scalar particle[139], we will assume $\Omega = 0$ also. In this case eqn (32) implies that the hole emits approximately as a black body with temperature

$$\theta = \frac{h\kappa}{2\pi c} = \frac{hc^3}{8\pi GkM} \approx 10^{26} M^{-1} K \quad (33)$$

where here and throughout this section M is in grams. The radiation is not exactly black-body because the Γ factor in eqn (32) is frequency-dependent. At high frequencies the effective cross-section is $27\pi G^2 M^2/c^4$ for all particle species, but at low frequencies the cross-section is reduced in a way which depends on the spin of the particle. The overall emission rate tends to decrease with increasing spin.

A black hole should emit any particle species whose rest mass is less than its emission temperature. Thus all holes should emit zero-rest-mass particles like photons, neutrinos, and gravitons, and

Table (3): Cosmological consequences of PBH evaporations

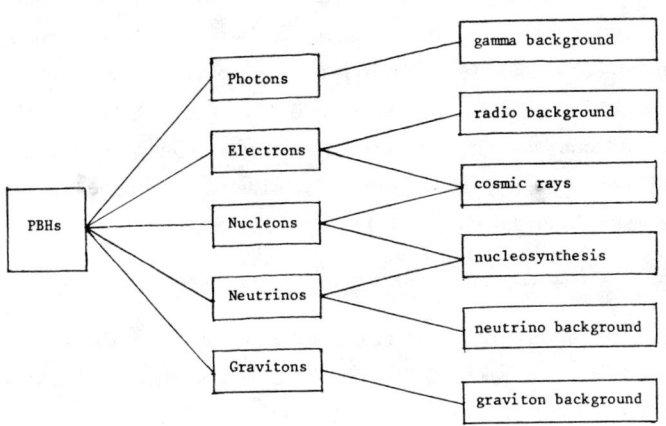

sufficiently small ones should also emit electrons and nucleons. Table (3) shows that interesting cosmological effects may be associated with each of these particle species. However it must be stressed that there is no necessity to invoke PBHs to explain any of the cosmological features alluded to in Table (3), since each of them has other possible explanations. The table should merely be regarded as exemplifying the wide range of cosmological effects which may be associated with PBH evaporations. As we shall see, it also indicates the various ways in which one can infer upper limits on the number of PBHs which may have formed in various mass ranges.

Page [139] finds that, for holes with $M > 10^{17}$ g (which can emit only zero-rest-mass particles), the fractions of the initial mass emitted in gravitons, photons, and neutrinos are $\varepsilon_g = 0.02$, $\varepsilon_\phi = 0.17$, and $\varepsilon_\nu =$

0.81, respectively. This assumes there are three neutrino species. For holes with 10^{15} g $< M <10^{17}$ g (which are hot enough to emit electrons and positrons), the values are $\varepsilon_g = 0.01$, $\varepsilon_\phi = 0.09$, $\varepsilon_\nu = 0.45$, and $\varepsilon_e = 0.45$. For holes with 10^{14} g $< M < 10^{15}$ g (which are hot enough to emit muons, these subsequently decaying into electrons and neutrinos), $\varepsilon_g = 0.01$, $\varepsilon_\phi = 0.07$, $\varepsilon_\nu = 0.51$, and $\varepsilon_e = 0.41$. For $M < 10^{14}$ g, the hole can also emit hadrons and the emission fractions depend on highly uncertain details of particle physics. These considerations suggest that one may write the mass loss rate as

$$\frac{dM}{dt} = -3 \times 10^{25} M^{-2} f(M) \text{ g s}^{-1} \tag{34}$$

where $f(M)$ depends on the number of particle species which can be emitted and is normalized to be 1 for holes which emit only massless particles ($M > 10^{17}$ g). The associated lifetime is

$$t_{evap} = 9 \times 10^{-27} f(M)^{-1} M^3 \text{ s} \tag{35}$$

and this implies that a PBH evaporates within the age of the Universe (10^{10} y) if its mass is less than

$$M_* = 5 \times 10^{14} \text{ g}. \tag{36}$$

Page has made more refined calculations of M_*, accounting for the effects of the black hole having spin[140] or charge[141]; these modify eqn (34) but only by about 10%.

The nature of the final explosive phase of a black hole's evaporation depends crucially on the form of $f(M)$ for $M < 10^{14}$ g. In the "Elementary Particle" picture, all hadrons are supposed to be made up of a finite number of fundamental particles like quarks and gluons. In this case only these fundamental particles are emitted directly and $f(M)$ never exceeds ~ 100. The most natural picture would be one in which the black hole emits quark and gluon jets, these subsequently fragmenting into hadrons and leptons. If we regard the explosive phase as beginning when $t_{evap} \simeq 10$ s, it occurs when M reaches about 10^{10} g. In the "Composite Particle" picture, the hadrons can be regarded as being made up of each other[142]. In this case all hadrons are equally fundamental and all can be emitted directly. This means that f increases exponentially when T reaches the "ultimate" temperature of 160 MeV or, equivalently, when M falls to 6×10^{13}g. However, this picture is not now regarded as being very plausible.

In discussing the cosmological effects of PBH evaporations, it is useful to note that the total number of particles of a given species emitted by a PBH of initial mass M is of order $10^{11} \varepsilon_i M^2$, where ε_i is the fraction of mass that goes into that species. Most of these particles will have an energy of order $10^{22} M^{-1}$eV but there will also be an E^{-3} tail of higher energy particles emitted in the later phases of evaporation. Note that the fraction of the Universe which is in PBHs of initial mass M at their evaporation epoch, $\alpha(M)$, is related to the corresponding fraction at the formation epoch, $\beta(M)$, by [143]

$$\alpha(M) = \beta(M) \left(\frac{t_{evap}}{t_{form}} \right)^{\frac{2\gamma}{1+\gamma}} = \beta(M) \left(\frac{M}{M_p} \right)^{\frac{4\gamma}{1+\gamma}}$$

where γ specifies the equation of state between t_{form} and t_{evap}. Thus, if $\gamma = 1/3$, the conversion factor is just M/M_p. The effect of evaporations on the PBH mass spectrum is to change eqn (25) to

$$\Pi(M) \propto M^{-\alpha} \left[1 + \frac{t_{now}}{t_{evap}(M)} \right]^{-(\frac{\alpha+2}{3})} \propto \begin{cases} M^{-\alpha} & (M > M_*) \\ M^2 & (M < M_*) \end{cases} \quad (37)$$

Thus the spectrum is unchanged above M_*; its form below M_* is associated with the E^{-3} tail effect.

3.6 The Contribution to the Photon Background

Particles emitted by PBHs after some redshift z_{free} will not have interacted with the background Universe; they will therefore preserve their original spectrum apart from being redshifted. Their present background spectrum should thus have the form indicated in Fig.5, α being the exponent of the PBH mass spectrum [144]. The E^{-3} part for $E > 100$ MeV comes from the tail of the PBHs of mass M_* which explode today and the contribution at lower values of E comes from PBHs with $M < M_*$ which evaporated earlier. There are changes of slope at two points; these derive from the different relationships between redshift and time during the free-expansion, matter-dominated, and radiation-dominated eras. Page and Hawking show that the photon spectrum drops off below

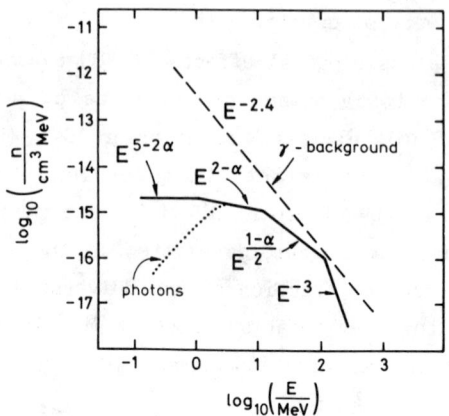

Figure (5): Spectrum of particles from evaporating PBHs

about 1 MeV (corresponding to a value for z_{free} of order 100) primarily due to pair-production off background nuclei[145]. Comparison with the observed γ-ray spectrum[146], which goes like $E^{-2.4}$, shows that the best limit on the number of PBHs derives from those of mass M_* providing $\alpha <$ 5.8 (as expected). Since the observed background γ-ray density at 10^2 MeV is about 10^{-14} cm^{-3}, or $\Omega_\gamma \sim 10^{-9}$ in units of the critical density, and since the fraction of mass coming out in γ-rays is of order 0.1, we infer an upper limit[147] on $\Omega_{PBH}(M_*)$ of 10^{-8}. The corresponding limit on $\beta(M_*)$ is

$$\beta(M_*) = \frac{\Omega_{PBH}(M_*)}{\Omega_R}\left[1+z_{form}(M_*)\right]^{-1} < 10^{-26} \tag{38}$$

where Ω_R denotes the background radiation density and we assume $\gamma = 1/3$ before t_{eq}. This shows that the fraction of the Universe going into black holes at 10^{-23} s must have been even tinier than indicated by eqn (**29**) and the upper limit on ε drops from 0.05 to 0.03. Since eqn (25) implies that the mass density in holes larger than M_* should be smaller than $\Omega_{PBH}(M_*)$ by a factor $(M/M_*)^{-\frac{1}{2}}$, one might conclude that the <u>total</u> PBH density is also less than 10^{-8}. However, this conclusion would fail if PBH formation were inhibited on scales smaller than M_* (e.g. by the equation of state being stiff[73] or by the Universe being highly anisotropic[128] before 10^{-23} s) or if the PBHs formed at a phase transition after 10^{-23}s.

Photons which are emitted sufficiently early, before a redshift $z_{therm} \simeq 10^6 \Omega_i^{-4}$ where Ω_i specifies the ionized gas density in units of the critical density[148], will be completely thermalized, and so PBHs smaller than

$$M_{therm} \simeq 10^{11} \Omega_i^{8/3} \text{ g}$$

will merely boost the primordial photon-to-baryon ratio. As shown by Zeldovich and Starobinskii[149], an initial ratio of S_o would by today have been increased to

$$S = (1 + S_o) \, \beta(M) \, \left(\frac{M}{M_P}\right) . \tag{39}$$

If the PBHs exist over an extended mass range, we expect $\beta(M)$ to be constant, so the largest contribution to S should come from the holes with mass M_{therm}. These PBHs can generate all of the 3K background ($S \simeq 10^8 \Omega^{-1}$, $S_o = 0$) providing $\beta \simeq 10^{-8} \Omega_i^{-11/3}$. This is compatible with the γ-ray limit only if the PBH spectrum is cut off before M_* or if β falls off faster than M^{-2} rather than being constant. In any case, eqn (39) yields an upper limit

$$\beta(M) < 10^3 M^{-1} \Omega^{-1} \quad (M < M_{therm}). \tag{40}$$

For initial fluctuations which fall off faster than $M^{-2/3}$, the PBH mass spectrum declines exponentially above 10^{-5}g, so one cannot generate an appreciable value of S.

Photons emitted at epochs intermediate between z_{therm} and z_{free}, rather than being thermalized or propagating freely, will merely distort the 3K background spectrum. Calculations of Naselskii[150] indicate that PBHs in the range 10^{11}g $< M < 10^{13}$g must have

$$\beta(M) < 10^{-18} \left(\frac{M}{10^{11}\text{g}}\right)^{-1} \tag{41}$$

if the distortion is not to exceed the observational limits.

3.7 PBH Explosions Today

We have seen that the γ-ray background observations require $\Omega_{PBH}(M_*) < 10^{-8}$. This implies that the mean number density of such holes can be at most $10^4 pc^{-3}$, although, if the holes are clustered inside galactic halos, the local density could be as high as $10^{10} pc^{-3}$. The corresponding explosion rate is at most $10^{-6} pc^{-3} y^{-1}$ for unclustered holes or $1 pc^{-3} y^{-1}$ for clustered holes. We now discuss the prospect of detecting these explosions.

In the Composite Particle picture, we have seen that the evaporation becomes catastrophic at a mass $M_{crit} \approx 6 \times 10^{13} g$. This should generate a "hadron fireball"[151], releasing around 10^{34} ergs in 250 MeV γ-rays over a period of about $10^{-7} s$ [145]. To detect one explosion per month, one would need a detecting area of at least $4 \times 10^5 cm^2$ for unclustered holes or $40 cm^2$ for clustered holes. Observations of COSMOS 561[152], with an area of $300 cm^2$, give a very weak upper limit of 75 explosions $pc^{-3} y^{-1}$; this is nearly two orders of magnitude larger than that permitted by the γ-background observations. On the other hand, Porter and Weekes[153] show that atmospheric Cerenkov techniques, with an effective area of $10^9 cm^2$, give a limit of 0.04 explosions $pc^{-3} y^{-1}$, which is better than the background limit providing the holes are clustered inside halos.

In the Elementary Particle picture, evaporation becomes explosive at a mass $M_{crit} \approx 10^{10} g$, some 10^{30} ergs being released as 5×10^6 MeV γ-rays in around $10 s$[144]. The number flux of photons is now so small that only atmospheric Cerenkov techniques can be used. However, all the limits presently available are weaker than the γ-background limit: Porter and Weekes[154] get an upper limit of 3×10^4 explosions $pc^{-3} y^{-1}$ and Fegan et al.[155] get a limit of 6×10^3 explosions $pc^{-3} y^{-1}$. This suggests that there is little prospect of detecting the photons emitted from PBH explosions in the Elementary Particle picture. However, this conclusion may be overly pessimistic. Recent calculations by MacGibbon suggest that most of the explosive energy may be emitted as quark jets in the Elementary Particle picture and each of these may generate a low energy tail of photons with energy down to 100 MeV.[173] The prospect of detecting these soft photons may be much better.

Rees[156] has pointed out that the situation could also be improved if the PBHs explode in a region where there is an appreciable magnetic field. In the interstellar medium, for example, $B = 5 \times 10^{-6}$ G. In this case, the interactions with this field of the shell of electrons and positrons emitted will generate a burst of radiation. The wavelength at which the burst appears is $r_{max} \gamma^{-2}$ where $\gamma \simeq (M_{crit}/10^{16} g)^{-1}$ is the Lorentz factor of the electrons and r_{max} is either the radius where the $e^{\pm}$ shell is braked,

$$r \simeq 10^{16} \gamma^{-1} \left(\frac{B}{5\times 10^{-6} G}\right)^{-2/3} \text{ cm} \qquad (42)$$

or the radius at which its conductivity breaks down,

$$r \simeq 4 \times 10^{19} \gamma^{-3/2} \left(\frac{B}{5\times 10^{-6} G}\right)^{1/2} \text{ cm} \qquad (43)$$

whichever is smaller. In the interstellar medium, the braking radius is smaller (so that most of the $e^{\pm}$ energy goes into electromagnetic waves) providing $\gamma < 10^7$ or $M_{crit} > 10^9$ g. Thus, if $M_{crit} \simeq 10^9$ g, one gets about 10^{30} ergs released at a wavelength of 10^4 Å; such an optical burst would be detectable out to 1 kpc. If $M_{crit} \simeq 10^{11}$ g, one gets about 10^{32} ergs being released at a wavelength of 10 cm. Arecibo could detect such a radio burst as far away as Andromeda. More detailed calculations of the radio burst characteristics have been presented by Blandford[157].

In fact, Rees' mechanism does not work in all circumstances. A pulse is produced only if the explosion timescale is less than the characteristic period of the generated radiation and this applies only if there exist many extra particle species which can be emitted above 10^2 GeV. On the other hand, one needs $\gamma > 10^3$ ($M_{crit} < 10^{13}$ g) to avoid most of the energy going into swept-up plasma and $\gamma > 10^5$ ($M_{crit} < 10^{11}$ g) to avoid electrons and positrons annihilating too quickly. One therefore requires a picture for the black hole explosion intermediate between the Composite Particle picture and usual Elementary Particle picture. If radio bursts *are* produced, one can already infer very strong limits on the number of PBH explosions. Observations at 400 Hz[158] give an upper limit 2×10^{-9} pc^{-3} y^{-1}; and observations at 10^2 MHz and 10^3 MHz [159] give limits of 5×10^{-7} pc^{-3} y^{-1} and 4×10^{-5} pc^{-3} y^{-1}, respectively. These are much better than the γ-ray background limits. However, the optical burst limit[160-162] is 0.03 pc^{-3} y^{-1}, which is considerably weaker.

3.8 The Contribution to the Positron Background

The electrons and positrons emitted by black hole evaporations could be of interest in their own right, even if the Rees mechanism is not operative. For PBHs in the mass range 10^{14} g $< M < 10^{17}$ g, both ε_{e^-} and ε_{e^+} should be about 0.2. Since observations of the 100 MeV positron background[163] show that n_{e^+}(100 MeV) $\sim 10^{-11}$ cm^{-3}, this implies a limit $\Omega_{PBH}(M_*) < 10^{-6}$ if the PBHs are unclustered or $\Omega_{PBH}(M_*) < 10^{-10}(t_{leak}/10^8 y)^{-1}$ if they are clustered inside galactic halos with the positrons escaping in a time t_{leak}. Since t_{leak} would probably be about 10^8 y,[166] the positron limit on $\Omega_{PBH}(M_*)$, and hence $\beta(M_*)$, may be better than the γ-ray background limit by two orders of magnitude.[144] The background electron density at 100 MeV is larger than the positron density by a factor of 10 and is therefore less interesting. If one considers the spectrum of cosmic ray electrons expected from PBHs evaporating at previous epochs, as well as today, one gets a form similar to that shown in Fig.5 except that it is scaled by a factor $\varepsilon_e/\varepsilon_\phi$ and falls off below 10 MeV (i.e. for $t_{evap} < 10^{15}$ s) on account of the electrons being degraded by inverse Compton scattering off the 3K background photons. The predicted spectrum is conceivably compatible with that observed if $\Omega_{PBH}(M_*) \sim 10^{-10}$, although solar modulation effects make a direct comparison difficult.

More refined calculations of the background of positrons expected from PBHs evaporating in the <u>present</u> epoch, allowing for their diffusion and degradation within the galactic halo, have been presented by Nazel'skii and Pelikhov[167]. These authors also calculate an indirect limit for PBHs exploding in the interstellar medium associated with the fact that both electrons and positrons there will generate synchrotron radio emission via interaction with the interstellar magnetic field. From observations of the 300 MHz background, they infer $\Omega_{PBH} < 10^{-9}$.

Another interesting effect could derive from the positrons generated by PBHs exploding near the galactic centre, since the PBH density should be higher there. One would expect some of these positrons to annihilate, producing a 0.511 MeV line, so it is relevant that such a line has indeed been detected from the galactic centre[164]. The intensity of the line corresponds to about 8×10^{42} annihilations s^{-1}. Okeke and Rees[165] have shown that any positrons from PBHs will be slowed by ionization losses, thus permitting their annihilation, providing their energy is less than $E_{slow} \simeq (50 - 100)$ MeV, i.e. providing they come from PBHs larger than $M_e \simeq 10^{14}$ g. Given the form of the PBH mass spectrum [viz.eqn (37)], the associated annihilation rate goes like $M^{-\alpha-1}$ for $10^{17} g > M > M_*$, like M for $M_* > M > M_e$ and like M^4 for $M < M_e$. The biggest contribution therefore comes from PBHs with $M \sim M_*$ and one would need about 10^{20} of them (i.e. about $10^2 M_\odot$ worth) within the central kpc of the galaxy to produce the observed 0.511 MeV line. If one asumes that the number density of PBHs in the halo falls as R^{-2} with galactocentric distance, like the rest of the halo material[51], one infers a limit $\Omega_{PBH}(M_*) < 10^{-9}$ which is about one order of magnitude stronger than the γ-ray background limit, though possibly weaker than the positron limit itself.

3.9 The Contribution to Cosmic Ray Antiprotons

PBHs smaller than about 10^{14} g would be hot enough to emit protons and antiprotons; those emitted in the tail of the PBH explosions occurring today would contribute to the cosmic ray background. However, the cosmic ray proton flux falls off as $E^{-2.6}$ in the energy range 1 GeV to 10^{11} GeV (compared to the E^{-3} expected from PBHs) and the integrated energy density is $\Omega_{CR} \sim 10^{-4}$ in units of the critical density[168]. Both features would seem to preclude cosmic ray protons deriving from evaporating black holes; indeed, in view of the γ-ray limit [$\Omega_{PBH}(M_*) < 10^{-8}$], one would infer that at most 10^{-4} of the energy in cosmic ray protons could so derive. The situation with antiprotons is much more interesting since observations[169] suggest that, in the energy range 130-320 MeV, $n_{\bar{p}}/n_p = (2.2 \pm 0.6) \times 10^{-4}$. The $\bar{p}$ flux is therefore comparable to that which could have been generated by PBH evaporations. This possibility is accentuated by the fact that more conventional explanations for the antiproton flux seem to be unsatisfactory. It is usually assumed that antiproton cosmic rays are secondary particles, produced by the spallation of the interstellar medium by primary cosmic rays. However, the observed $\bar{p}$ flux at 130-320 MeV exceeds the predicted secondary flux by a factor of 10^2; even at energies around 10 GeV, the observed $\bar{p}$ flux still exceeds the predicted value by a factor of 3^{170}.

These considerations have prompted Kiraly et al.[171] to examine whether PBH evaporations could produce the antiprotons. If one normalizes the expected flux to that observed at 10 GeV, one expects a spectrum of the form

$$\frac{dN}{dE} \simeq 10^{-2} \left(\frac{E}{\text{GeV}}\right)^{-3} \text{cm}^{-2} \text{s}^{-1} \text{GeV}^{-1}. \tag{44}$$

This accords with observation providing $\Omega_{PBH}(M_*)$ is of order 10^{-9}, which is a factor of 10 smaller than the maximum permitted by the γ-ray background limit. Turner[172] has suggested a similar scheme, allowing for an extended spectrum of PBHs. He finds that the dominant contribution derives from those PBHs with $M \sim 10^{13}$g which evaporate at about 10^{15} s, antiprotons produced before then annihilating with protons in the background Universe. Both the Kiraly et al. and Turner models are rather simplistic since they assume that the fraction of mass emitted as antiprotons is $\varepsilon_{\bar{p}} \simeq 0.1$ (independent of M). More detailed calculations[173], based on the jet picture allow one to predict the value of $\varepsilon_{\bar{p}}$ and its energy dependence more precisely. These calculations also allow one to relate the production of antiprotons by PBHs with that of positrons and gamma rays. Thus there is a close connection between the considerations of this subsection and the last. If cosmic ray positrons and antiprotons really do derive from PBHs, their observed spectra could give vital information about particle physics.

3.10 The Effect on Cosmological Nucleosynthesis

Several limits can be placed on the fraction of the early Universe which goes into evaporating PBHs by considering ways in which the evaporations would mar the standard picture of cosmological nucleosynthesis (discussed in Section 2.4.1). Firstly, if the number of photons generated by the PBHs in the period after nucleosynthesis is large enough to change the primordial photon-to-baryon ratio [cf. eqn (39)], the value of S at nucleosynthesis will be less than its present value of $10^8 \Omega^{-1}$. This will increase the helium abundance and decrease the deuterium abundance. Detailed calculations by Miyama and Sato [174] show that this imposes a limit

$$\beta(M) < 10^{-15} \Omega \left(\frac{M}{10^9 g} \right)^{-1} \qquad (10^9 g < M < 10^{13} g). \qquad (45)$$

This limit has also been obtained by Vainer and Nasel'skii [175]. Such limits do not apply for PBHs smaller than 10^9g (which evaporate before $t_F \sim 1$ s) but, from eqn (40), one also has a limit in this mass range from the requirement that one does not generate too many thermalized photons.

A more subtle effect of the photons emitted from PBHs after cosmological nucleosynthesis is that they could photodissociate the small amount of deuterium produced previously. Lindley shows that the survival of deuterium requires [176]

$$\beta(M) < 10^{-21} \Omega \left(\frac{M}{10^{10} g} \right)^{1/2} \qquad (M > 10^{10} g) \qquad (46)$$

which is stronger than the Miyama and Sato limit. It should be noted, however, that Lindley does not consider the photodissociation of helium, which could conceivably <u>increase</u> the final deuterium abundance.

Vainer and Nasel'skii [177] point out that neutrinos from PBHs with $10^9 g < M < 10^{11} g$, which evaporate in the period $1 s < t < 10^5 s$, could also effect nucleosynthesis by modifying the n_N/n_P freeze-out ratio. However, Zeldovich et al. [178] show that the effect of the neutrinos on nucleosynthesis is much less important than that of the nucleons emitted. While protons or antiprotons may be confined near the holes [179], the neutrons and antineutrons will not be so confined and they will modify nucleosynthesis both by altering the n_N/n_P freeze-out value for $10^9 g < M < 10^{10} g$ and by spallation of already formed helium for $10^{10} g < M < 10^{13} g$. The limit associated with the first effect, which is also discussed by Rothman and Matzner[180], is

$$\beta(M) < 10^{-16} \Omega \left(\frac{M}{10^9 g}\right)^{-1/2} \quad (10^9 g < M < 10^{10} g) \quad (47)$$

The helium spallation limit derives from the requirement that the deuterium abundance produced by the spallation not exceed that observed. Since any deuterium produced after around 10^3 s will survive, whereas the spallation will continue to operate until arund 10^5 s, one requires

$$\beta(M) < 10^{-25} \Omega^{1/3} \left(\frac{M}{10^9 g}\right)^{5/2} \quad (10^{11} g < M < 10^{13} g) \quad (48)$$

The most interesting aspect of this result is that it means only a tiny fraction of the Universe would need to go into PBHs in the mass range $10^{10} g < M < 10^{13} g$ in order to generate the observed deuterium abundance. This means that the usual conclusion that Ω must be less than 0.1 in order for the cosmological production of deuterium to be correct[109] can be circumvented. However, the value of β required is consistent with the γ-ray limit, $\beta(M_*) < 10^{-26}$, only if the spectrum falls off faster than $M^{-3.5}$ or has an upper cut-off below M_*.

3.11 CONCLUSION

We have seen that there are a wide variety of ways in which PBH evaporations could have affected the history of the Universe and that there are several cosmological problems which they could resolve. However, most of these problems have other possible resolutions, so it would be premature to infer that evaporating PBHs <u>must</u> have existed. Indeed most of the preceding discussion only serves to indicate that the fraction of the Universe going into PBHs must have been tiny on all scales above 10^5 g, at least if the early Universe had a hard equation of state. The limits on $\beta(M)$ in this context are summarized in Fig. 6, which is reproduced from Novikov <u>et al.</u>[143]. The limits in the case of a universe which is dust-like at early times are somewhat weaker but they still require that $\beta(M)$ be tiny for $M > 10^{10}$ g. We have seen that one would <u>expect</u> β to be tiny unless the early Universe was

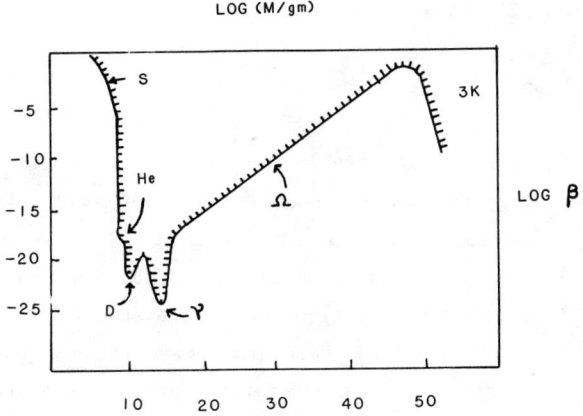

Figure (6): Constraints on fraction of Universe contained in PBHs

chaotic. Indeed the main point of the limits discussed above is that they reinforce the conclusion that the Universe was quiescent[217], containing horizon-epoch fluctuations no larger than 1% in amplitude, at all times after 10^{-33}s.

Of course the striking point is that PBH evaporations could have interesting cosmological effects even if β is tiny. However, we have seen that it is <u>a priori</u> unlikely that β would have the value required for Ω_{PBH} to be significant without being negligible. If PBHs did form abundantly, it is more likely that they did so at a phase transition. In this case, the holes would not be of the evaporating kind if the phase transition occurred after 10^{-23}s, but even large PBHs could have interesting cosmological consequences. In particular, PBHs in the mass range 10^{16}g - 0.1 $M_\odot$ could still be viable candidates for the dark matter in galactic halos.

4. POPULATION III STARS

4.1 Introduction

Population III stars may be defined as the stars that formed before the Universe had any appreciable heavy element content. The existence of a few such stars is inevitable since heavy elements can only be generated through stellar nucleosynthesis. In this section we will focus on the much more exciting and controversial possibility that most of the Universe may have been processed through them, perhaps before galaxies formed. We have seen that the prime motive for this suggestion is the possibility that the dark matter in galactic halos may consist of Population III remnants. We will examine why such a large number of stars might form in Section 4.2. However, the cosmological interest in Population III stars is not confined to the dark matter issue. They would also be expected to produce radiation,

Table (4): Cosmological consequences of Population III stars

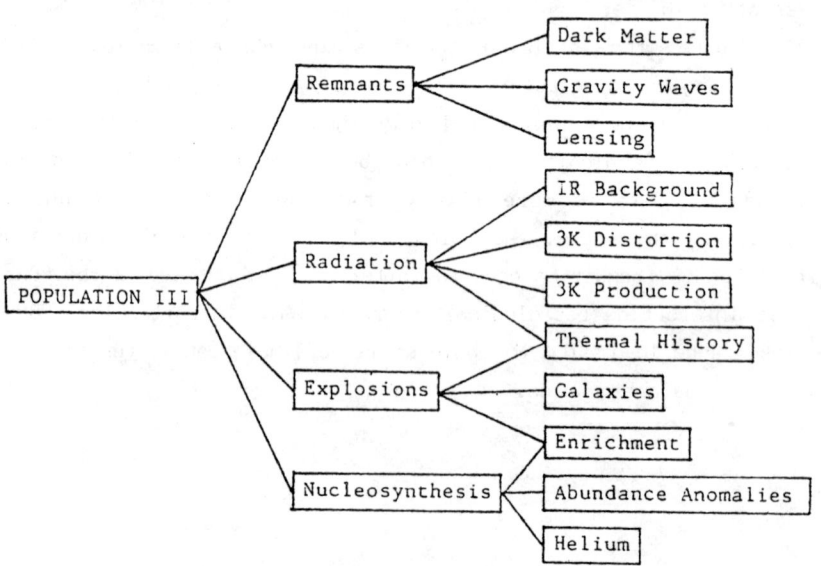

explosions, and nucleosynthesis products. As illustrated in Table (4), each of these could have important cosmological consequences. It must be stressed at the outset, however, that the e are no observations which definitely require their existence. The most conservative approach therefore is to use some of the effects indicated in Table (4) to constrain the Population III hypothesis; this we do in Section 4.3. The more positive aspect of what Population III stars could explain will be explored in Section 4.4.

4.2 Scenarios For Population III Formation

The existence of galaxies and clusters of galaxies implies that density fluctuations must have been present in the early Universe. The likelihood that these fluctuations could give rise to a population of pregalactic stars depends on what model one adopts for the nature of the fluctuations.

4.2.1 Isothermal fluctuations.

In this case, the fluctuations are entirely in the baryons and one expects the smallest surviving scale after decoupling to be

$$M_{Jb} \simeq 10^6 \, \Omega_b^{-1/2} \, M_\odot \tag{49}$$

corresponding to the baryon Jeans mass then.[42] (Here Ω_b is the baryon density in units of the critical density.) Larger structure would then build up through a process of hierarchical clustering [81]. The form of the galaxy correlation function[181,182] suggests that the fluctuations at decoupling must have had the form

$$\delta_{dec} = \left(\frac{M}{M_1}\right)^{-\beta}, \quad M_1 = 10^6 - 10^8 M_\odot, \quad \beta = \frac{1}{3} - \frac{1}{2}. \tag{50}$$

The precise values of M_1 and β required depend on the value of Ω_b. This implies that one would expect the first pregalactic clouds (of mass M_{Jb}) to bind shortly after decoupling ($z \simeq 10^3$). These clouds would presumably fragment into stars, though the typical fragment mass and the efficiency of star formation is very uncertain. The lack of heavy elements[183,184] and substructure,[185] as well as the effects of the microwave background,[186] would tend to make the final fragments larger than at the present epoch and perhaps in the VMO range. Possibly there would be no fragmentation at all, in which case one could end up with a single SMO. On the other hand, enhanced molecular hyudrogen formation[187] and associated thermal instabilities[188] could reduce the fragment mass to $0.01\ M_\odot$. The mass of the first stars is thus uncertain by 10^8 orders of magnitude. A hybrid scheme has been suggested in which each cloud forms a disc due to rotational effects; one then gets stars of mass $0.01\ M_\odot$ in the disc, together with a VMO which forms through relaxation at the center.[186]

4.2.2 <u>Adiabatic fluctuations in a "cold" universe</u>. If the fluctuations are contained in the total density and the Universe's mass is presently dominated by collisionless "cold" particles, like axions or photinos or primordial black holes, one expects bound clumps of these particles to form down to very small scales. Baryons would then fall into the potential wells, forming bound clouds, on baryon scales above

$$M_{Ja} = 10^6\ \Omega_b\ \Omega_a^{-3/2}\ M_\odot \qquad (51)$$

where Ω_a is the cold particle density.[42] These clouds could then form pregalactic stars just as in the previous case. In fact, the formation of the pregalactic clouds is even easier in this case because the cold particle fluctuations grow by an exta factor of $10\Omega_a$ between the time when the cold particles dominate the density and decoupling.

4.2.3 <u>Adiabatic fluctuations in a "hot" universe</u>. If the Universe's mass is dominated by baryons or collisionless "hot" particles, like neutrinos with non-zero rest mass, then adiabatic fluctuations are erased on subgalactic scales by photon diffusion for[189]

$$M < M_s \simeq 10^{13}\ \Omega_b^{-5/4}\ M_\odot \qquad (52)$$

or neutrino free-streaming for[90]

$$M < M_\nu \simeq 10^{15}\ \Omega_\nu^{-2}\ M_\odot \qquad (53)$$

Thus, the first objects to form are "pancakes" of cluster scale.[190] However, one still expects these pancakes to fragment into clumps of mass $10^8 M_\odot$ and these clumps might in turn fragment into stars before clustering into galaxies.[191] Even in this case, therefore, one might expect pregalactic stars to form, albeit at a relatively low redshift (z < 10). The mass of the resulting stars is still very uncertain but at least two of the previous arguments for high mass fragments (lack of metallicity and substructure) would still pertain.

This discussion emphasizes that pregalactic stars could form in all of the most likely scenarios for the density fluctuations. However, it is now clear how much of the Universe would be expected to go into the stars. If they formed very efficiently, cosmological nucleosynthesis constraints on the baryon density ($\Omega_b < 0.1$) would permit their remnants to provide the dark matter in halos but not a critical density.[109] On the other hand, in the present epoch (e.g. in giant molecular clouds) star formation seems to be rather inefficient, so - if one wants to argue that most of the Universe has been processsed through Population III stars - one needs to suppose that the conditions for star formation are very different at early times. One possibility is that the stars may be massive enough to generate a lot of explosive energy; in this case, the explosions may have amplified the fraction of the Universe going into stars,[192] as discussed in Section 4.4.10.

4.3 Cosmological Constraints

4.3.1 Remnants.

Only stars larger than $M_c \simeq 200 M_\odot$ can generate black holes with high efficiency (i.e., with the fraction of the initial star mass which collapses ϕ_B being close to 1). Providing the Population III spectrum is such that most of the mass is in stars this large, the fraction of the Universe ending up in black hole remnants should be $f_B = f_* \phi_B$, where f_* is the fraction of the Universe going into the stars. Observationally, we require $0.9 \leq f_B \leq 0.99$ (the lower limit being required to explain the dark mass in halos [51] and the upper limit to ensure that enough visible material is left over). It might seem unlikely that f_B would be this high since we could not expect both ϕ_B and f_* to be so close to 1. In particular, ϕ_B must be less than 0.5 if envelope-ejection occurs at hydrogen-shell burning.[27] However, one could in principle boost f_B by invoking black hole accretion or many generations of stars. We have seen that dynamical constraints, associated with the puffing up of the disc in our own galaxy by traversing halo holes,[117,192] require that the mass of the hole which dominates the halo not exceed $10^6 M_\odot$. More specifically, we require

$$\Omega_B(M) < 0.1 \min \left[1, \left(\frac{M}{10^6 M_\odot} \right)^{-1} \right] . \tag{54}$$

as indicated in Fig. 7. We therefore conclude that the mass of the halo holes must lie between $10^2 M_\odot$ and $10^6 M_\odot$ (i.e., they must derive from VMOs or low mass SMOs). The other possibility for explaining the dark matter is to suppose that the Population III stars are so small[112] that their mass-to-light ratio exceeds 100; this requires $M < 0.1 M_\odot$. As discussed in Section 2.3.3, infrared observations of other spiral galaxies may exclude any hydrogen-burning stars.

4.3.2 Background Light.
Since stars turn $\varepsilon_R \approx 0.01$ of their rest mass into radiation over their nuclear-burning time, the background light density they generate should be[192-194]

$$\Omega_R \approx 10^{-2} \Omega_* f_R (1+z_*)^{-1} \left(\frac{\varepsilon_R}{0.01} \right) \tag{55}$$

where the densities are in units of the critical density, z_* is the redshift at which they burn most of their fuel, and f_R - the fraction of the radiation which goes into background light rather than heating effects - should be close to 1. Since the observed background density over all wavebands cannot exceed about 10^{-4} (with the possible exception of the far IR band, which is presently unobserved), this implies

$$\Omega_* < 0.03 \max \left[1, \left(\frac{M}{10^2 M_\odot} \right)^{-1/2} \right] (1+z_*) \tag{56}$$

Thus the stars must form before a redshift of 10 if $\Omega_* > 0.1$. This also requires that the stars be larger than about $10 M_\odot$ in order to burn out by then; the precise value depends on rather uncertain cosmological parameters. The value of Ω_* permitted for lower values of z_* is indicated in Fig. 7. Of course, these limits do not apply for stars with $M < 1\,M_\odot$ since these are still burning. For these, the background light constraint requires [192]

$$\Omega_* < 0.07 \left(\frac{M}{M_\odot}\right)^{-3} \tag{57}$$

which is somewhat weaker than the constraint associated with infrared observations of individual galaxies. This limit is also shown in Fig. 7. The background light limits would be more interesting if one knew that the starlight should presently be in a waveband where the density is less than 10^{-4}. However, as discussed in Section 4.4.2, the effects of dust make this assumption questionable.

4.3.3 <u>Enrichment</u>. One of the strongest constraints on the spectrum of Population III stars comes from the fact that stars in the mass range $4\,M_\odot - M_c$ should eventually explode, producing an appreciable heavy element yield. The dependence of the yield on M can be expressed as

$$Z_{ej} \approx \begin{cases} 0.5 - (M/6M_\odot)^{-1} & (8M_\odot < M < 100M_\odot) \\ 0.1 & (4M_\odot < M < 8M_\odot) \\ 0.5 & (100\,M_\odot < M < M_c) \end{cases} \tag{58}$$

Since Population II stars are observed with metallicity as low as 10^{-5}, this implies that the pregalactic enrichment cannot exceed this amount unless Population II stars are themselves pregalactic.[197] In any case, it cannot exceed the lowest metallicity observed in Population I stars ($Z = 10^{-3}$). The associated constraint on the Population III mass spectrum, assuming $\Omega_g = 0.1$, is shown in Fig. 7. If one wants to produce the dark matter without contravening this constraint, the most straightforward solution is to assume that the spectrum either begins above M_c or ends below $4\,M_\odot$. As mentioned in Section 4.2, it is possible that the first stars were either very large or very small, so this solution is not unreasonable.

The effects discussed above place interesting constraints on the density of stars with mass M which could have formed at a redshift z; these constraints are summarized in Fig. 7. The black hole constraint is most important for high M, the nucleosynthetic constaint for intermediate M, and the background light constraint for low M. Only the last constraint is dependent on z and it is interesting in the intermediate mass range only for z < 10. Fig. indicates that the only mass ranges in which a large fraction of the Universe can be processed through Population III stars are $M < 0.1 \, M_\odot$ and $M_c < M < 10^6 M_\odot$. It also restricts the form of the mass spectrum: for example, if the spectrum encompasses the exploding mass range, then the nucleosynthetic constraints require a stellar mass spectrum which is either very steep if the dark matter is in low mass stars or very shallow if it is in high mass stars.

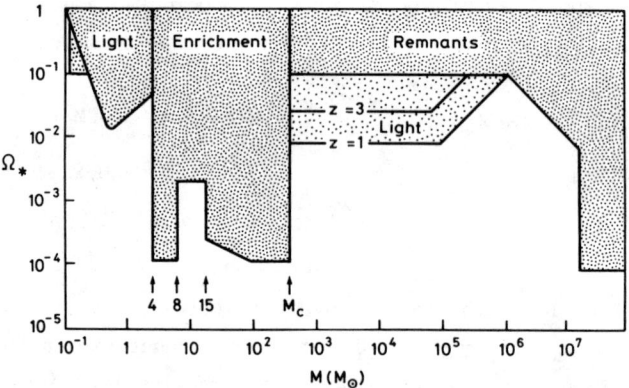

Figure (7): Constraints on Population III mass spectrum

4.4 What Population III Stars Can Achieve

In this section we will discuss various cosmological observations (some of them rather tentative) which Population III stars may be able to explain. It should be stressed, however, that Population III stars are not the only explanation of these observations, so the considerations below do not provide unequivocal evidence for their existence.

4.4.1 Near infrared background.

We have seen that an important constraint on the number of Population III stars is associated with the observational upper limit on the background radiation density. On the other hand, the existence of large background radiation density would be one of the chief hallmarks of the Population III scenario. Recently the detection of an IR background in the waveband $2 - 5\mu$ with a density $\Omega_{IR} \simeq 10^{-4}$ and a black-body spectrum with temperature 1500 K has been reported.[198] Although this claim remains to be confirmed, it is obviously interesting to inquire what sort of stars would be required to explain the data. If we assume that the stars are VMOs with $M > M_c$ (in order to avoid overenrichment) and that their radiation is not modified by grain absorption, then the expected spectrum can be shown to be[67]

$$\Omega_R(\nu) \equiv \frac{4\pi\nu i(\nu)}{\rho_{crit}c^3} \simeq 6 \times 10^{-4} \left(\frac{f_b X_o}{0.6}\right) \left(\frac{\Omega_*}{1+z_*}\right) \left(\frac{x^4}{e^x - 1}\right) \quad (59)$$

where $i(\nu)$ is the intensity, $x = h\nu(1 + z_*)/kT_s$, and $f_b X_o$ is the fraction of the star's mass burnt to helium (normalized to the value appropriate for a VMO with $X_o = 0.75$). Since VMOs have a surface temperature of $T_s \simeq 10^5$ K (independent of M), the quantity $\Omega_R(\nu)$ peaks at a present wavelength

$$\lambda_{max} \simeq 0.04 \, (1+z_*)\mu \quad (60)$$

which lies in the observed waveband for $40 < z_* < 140$. A comparison with the data shows that no single curve passes through all the data points. If one excludes the 5μ point (which is the most dubious since it could be contaminated by interplanetary dust emission), the best fit curve [67] corresponds to $z_* = 75$ and $\Omega_* = 1.6h^{-2}$ where $h \equiv (H_o/50)$. The value of Ω_* is rather high if one believes the cosmological nucleosynthesis constraint.

4.4.2 <u>Microwave distortions</u>. The above discussion fails if absorption by dust occurs. For some range of values of z and the dust fraction χ, the starlight would be absorbed and re-radiated in the far IR. One can show [199] that the dust temperature should have a value

$$T_d = T_c \left[1 + 10 \left(\frac{\Omega_R}{\Omega_c}\right) \left(\frac{r}{0.1}\right)^{-1} \left(\frac{1+z_*}{10^3}\right)^{-1} \right]^{1/5} \quad (61)$$

where T_c and Ω_c are the temperature and density of the 3 K background, Ω_R is the density of the stellar radiation, and r is the grain radius. This means that the re-radiated light should presently reside in the waveband 300 - 900μ, with only a weak dependence on z_* and χ. This is presently unobservable, but could become so with future space experiments (e.g. SIRTF and LDR). For sufficiently high values of z_* and χ, the dust temperature would be held so close to the microwave background temperature that the re-radiated starlight would distort its black-body spectrum.[200-202] A few years ago, such a distortion was, in fact, reported: the best-fit black-body temperature around the peak was found to be 2.9 K (significantly higher than the 2.7 K found in the Rayleigh-Jeans region) and there also appeared to be a trough just shortward of the peak.[203] This could be explained if the 2.9 K component was thermalized pregalactic starlight, with the trough reflecting the reduced absorption efficiency of grains in the 1 - 10μ region; the trough would be redshifted to the observed wavelength for $z_* \simeq 100$. This proposal would require that 25% of the 3K background energy come from stars, which in turn would require $\Omega_* \simeq 0.1$. However, more recent data suggest that the distortion may be going away.[204]

4.4.3 <u>Generation of the 3K background</u>. Even if the 3K background has an undistorted spectrum, it could still have been generated entirely by stars providing one can invoke a suitable thermalizing agent[205]; this may be feasible even at low redshifts if one adopt special types of elongated grains.[206-209] However, these models seem rather contrived: it is very difficult to ensure thermalization at both long and short wavelengths. As discussed in Section 1.3, an alternative scheme is to assume that the black hole remnants of the stars generate the radiation through accretion [48]. Of course, any scheme which envisages the 3K background deriving from Population III stars or their remnants must also require that the early Universe was cold or tepid [45] (with the primordial photon-to-baryon ratio being much less than its present value of 10^9). The advantages originally claimed for the cold and tepid models were: (i) that the photon-to-baryon ratio could be explained naturally rather than being assigned arbitrarily as an initial condition of the Universe; and (ii) that the existence of a prolonged matter-dominated phase before decoupling would permit galactic-scale density fluctuations a longer period of growth. However, both these arguments have become less compelling in recent years. The grand unified theories[211] permit alternative explanations for the value of S by invoking baryon-non-conserving processes at 10^{-35} s; and the currently favoured picture of galaxy formation requires that the early Universe was cold anyway, but in the sense of being dominated by non-relativistic particles for some period before decoupling rather than in the sense of having no radiation.

4.4.4 Thermal history. The light generated by Population III stars could have an important effect on the thermal history of the Universe even in the conventional hot Big Bang scenario. During its main-sequence phase, each star (or cluster of stars) would be surrounded by an HII region. The fraction of the Universe in such regions would progressively grow on account of both the increasing number of stars and the increasing size of the individual HII regions. For $z < 60 \delta^{-1/3}$, where δ is the gas clumpiness, the recombination time within each HII region exceeds the lifetime of the stars, so the whole Universe is ionized once the number of ionizing photons generated exceeds the number of atoms [212,213]; the fraction of the gas in stars is then just

$$f_* \simeq 3 \times 10^{-5} \delta. \qquad (62)$$

For $z > 60 \delta^{-1/3}$, one gets a fully developed HII region (whose structure is determined by the balance of ionizations and recombinations) and the situation is more complicated; one now needs a larger value of f_* to reionize the Universe but it is still small.[192] Such reionization could have important cosmological implications (e.g., in reducing anisotropies in the 3K background). After the HII regions have merged, the Universe would maintain a high degree of ionization,

$$1-x \simeq 10^{-10} f_*^{-1} \left(\frac{T}{10^4 K}\right)^{-0.8} \tag{63}$$

and the gas temperature T would tend to a value of $10^4 - 10^5 K$ until the stars cease burning.[192] If the stars finally explode, or if they leave black holes which heat the Universe through accretion,[56,64] the temperature may be boosted even higher.

4.4.5. Helium production by VMOs. Because the pulsational instability leads to mass-shedding of material convected from its core, a VMO is expected to return helium to the background medium during core-hydrogen burning.[214] The net yield depends sensitively on the mass loss fraction ϕ_L. If this is very high, the yield will be low because most of the mass will be lost before significant core burning occurs. However, for $\phi_L < (1-Y_i)/(2-Y_i)$, the mass loss is always slower than the shrinkage of the convective core and one can show that the fraction of mass returned as new helium is[215]

$$\Delta Y = (1-\tfrac{1}{2}Y_i)\phi_L^2 \lesssim 0.25 \frac{(1-Y_i)^2}{(1-Y_i/2)} \tag{64}$$

Here Y_i is the initial (primordial) helium abundance and the equality sign on the right applies only if ϕ_L has the critical value. This does not necessarily impose a constraint on the number of VMOs since ΔY is very small if ϕ_L is well below the critical value. However, there is some indication from numerical calculations that hydrogen-shell burning may produce a super-Eddington luminosity which completely ejects the stellar envelope.[26,27] If true, this means that the maximal helium production permitted by eqn (64) is guaranteed. This would have profound cosmological implications. If $Y_i = 0.22$, corresponding to the conventional primordial value, $\Delta Y = 0.17$, so one would substantially overproduce helium if much of the Universe went into VMOs. In this case, the only remaining candidate for the dark matter would be SMOs in the mass range $10^5 - 10^6 M_\odot$. On the other hand, if $Y_i = 0$, corresponding to no primordial production, $\Delta Y = 0.25$, which is tantalizingly close to the standard primordial value. One can show that, if a fraction F of the Universe is processed through VMOs, then the resulting helium abundance is given by[215]

$$F = \int_0^Y \frac{2(2-Y)}{(1-Y)^2} \exp\left(\frac{2Y}{Y-1}\right) dY . \qquad (65)$$

For small Y, this just gives $F \simeq 4Y$, so one naturally gets the sort of value required if F is close to 1. More precisely, Y lies between 0.20 and 0.25 for $0.8 < F < 0.9$. This raises the question of whether the Population III VMOs invoked to produce the dark matter might not also generate the helium which is usually attributed to cosmological nucleosynthesis. Unlike stars smaller than M_c, which could not do so without overproducing heavy elements, stars larger than M_c may produce no appreciable yield since the heavy elements they generate collapse with the core to form a black hole. One still has to find a way to suppress primordial helium production. One way to do this would be to suppose that the early Universe was cold. In this case, the amount of helium production is determined entirely by the neutrino degeneracy factor[216,217] and one could avoid any helium production at all for a lepton-to-baryon ratio exceeding 1.5.

4.4.6 Heavy element production. We saw in Section 4.3.3 that one of the strongest constraints on the Population III hypothesis derives from the requirement that the stars not produce too many metals. Nevertheless, for some purposes it would be advantageous to have a slight pregalactic enrichment (e.g., to explain the G-dwarf problem[218] or the small grain abundance required to produce alleged distortions in the 3K spectrum[200] or the possible lack of a metallicity gradient in globular clusters[219]), so the question arises of whether it is likely that Population III stars could generate just a small metallicity (e.g., $Z \simeq 10^{-5}$). One natural way in which this would occur might be to suppose that the first stars do indeed explode but that their formation is self-limited due to reionization suspending star formation once f_* reaches the value given by eqn (62). This would naturally generate an enrichment of about 10^{-5}. Another possibility is that the heating effect of the first stars might shift the typical stellar mass into the non-exploding range once f_* gets sufficiently large. In this case, most of the Population III density could end up in dark remnants, even though[220] some enrichment is produced. If the first stars are unusual, one might expect their metal yield to dispay unusual abundance ratios. It is therefore of interest that various chemical anomalies have been observed for metal-poor stars. For example, the high oxygen-to-iron ratio at low Z[221] suggests that the first stars were either MOs or VMOs (the latter producing a lot of oxygen since they explode in their oxygen burning phase)[222]; and the possible evidence for primary nitrogen[223,224] might also be explained by VMOs if the carbon and oxygen produced during helium-core burning could be convected through the hydrogen-burning shell and there CNO-processed to nitrogen.[215,225]

4.4.7 **Dynamical effects of halo holes.** It was mentioned in Section 4.3.1 that the most interesting constraint on the mass of any holes in our own halo is associated with their tendency to puff up the disc. A more detailed calculation of this effect suggests that halo holes could actually be responsible for the amount of disc-puffing which is observed.[117] The velocity dispersion of the disc stars in the radial, transverse, and vertical directions should evolve according to

$$\sigma_u(t) = (\sigma_{uo}^2 + D_e t)^{1/2}, \quad \sigma_v(t) = \beta^{-1}\sigma_u(t), \quad \sigma_w(t) = (\sigma_{wo}^2 + D_z t)^{1/2} \quad (66)$$

respectively, where

$$D_{e(z)} = 2\pi G^2 n M^2 \ln\left(\frac{V^3 T_{orb}}{GM}\right) F_{e(z)}(V/\sigma, \beta). \quad (67)$$

Here σ_o is the initial velocity dispersion of the stars, σ is the velocity dispersion of the holes, V is the relative velocity of the holes and the stars (close to the circular velocity in the disc if the halo is assumed to have little rotation), n and M are the mass and number density of the holes, β is the ratio of the circular to epicyclic frequency, and F_e and F_z are determined in terms of the error function. The important features are: (i) that all three components of σ grow asymptotically as $t^{1/2}$, thus explaining the empirical relationship between the observed scale heights of stars and their age; and (ii) that the predicted ratio of the vertical and radial components σ_w/σ_u agrees with observation.[226] Attempts to explain the puffing up of the disc through heating by giant molecular clouds, for example, would seem to be less successful on both accounts: one then expects $\sigma \propto t^{1/4}$ and σ_w/σ_u is too small.[230] In order to normalize the $\sigma(t)$ relationship correctly, one needs

$$n M^2 \simeq 3 \times 10^4 M_\odot^2 \text{ pc}^{-3}. \tag{68}$$

Combining this with the inferred halo density at our own galactocentric distance, implies a hole mass $M \simeq 2 \times 10^6 M_\odot$. Note that this argument does not strictly require that the halo object be a single hole; even a cluster of smaller holes - or indeed a cluster of any other sort of object - would suffice. In this context, it may be relevant that M is close to the size of the first clouds one would expect to form in the "isothermal" or "cold" scenarios discussed in Section 4.2.

4.4.8 <u>Gravity waves from single holes</u>. The formation of black holes would be expected to generate bursts of gravitational radiation. The characteristic period and density of the waves would be

$$P_o \simeq 10 \frac{GM}{c^3}(1+z_B) \simeq 10^{-2} \left(\frac{M}{10^2 M_\odot}\right)(1+z_B) \text{ s}, \quad \Omega_g = \varepsilon_g \Omega_B (1+z_B)^{-1} \tag{69}$$

where z_B is the redshift at which the holes form and ε_g is the efficiency with which the collapsing matter generates gravity waves. One can show that the expected time between bursts (as seen today) would be less than their characteristic duration providing $\Omega_B > 10^{-2} \Omega^{-2}$. If the holes make up galactic halos, one would therefore expect the bursts to form a background of waves.[227] This background can be detectable by laser interferometry if M is below $10^2 M_\odot$ or by the Doppler tracking of interplanetary spacecraft if M is in the range $10^5 - 10^{10} M_\odot$.

4.4.9 <u>Gravity waves from binary holes</u>. The prospects of detecting the gravitational radiation would be even better if the holes formed in binaries.[228] This is because two sorts of radiation would then be generated: (i) continuous waves as the binaries gradually spiral inwards due to quadrupole emission; and (ii) a final burst of waves when the components finally merge. The burst would have the same characteristics as that associated with isolated hole formation except it would be postponed to a lower redshift and ε_g would be larger ($\varepsilon_g \approx 0.08$) because of the larger asymmetry; from eqn (69) both factors would increase Ω_g. The continuous waves would also be interesting since they would extend the spectrum to considerably longer periods, thus making the waves detectable by a wider variety of techniques. Over most wavebands, the spectrum of the waves would be dominated by the holes whose initial separation was such that they are coalescing at the present epoch. This corresponds to a separation

$$a_{crit} \simeq 10^2 \left(\frac{M}{10^2 M_\odot}\right)^{3/4} R_\odot \qquad (70)$$

If the fraction of holes which form in the binaries with this separation is f_{crit}, the spectrum should have the form:[228]

$$P\frac{d\Omega_g}{dP} \cong 0.08\ \Omega_B f_{crit} \left(\frac{P}{P_0}\right)^{-2/3} \left[10^{-2}\left(\frac{M}{10^2 M_\odot}\right) \text{ s} < M < 10^5 \left(\frac{M}{10^2 M_\odot}\right)^{5/8} \text{s}\right] \quad (71)$$

Providing $\Omega_B f_{crit}$ is not too small, this background should be detectable by laser interferometry for $M < 400\ M_\odot$ and by Doppler tracking of interplanetary spacecraft for $4 \times 10^4 M_\odot < M < 6 \times 10^{10} M_\odot$. One could also hope to observe the coalescences which are occurring in our own galactic halo at the present epoch. The average time between halo bursts and their expected amplitude would be

$$t_{burst} \cong 10\left(\frac{M}{10^2 M_\odot}\right)^{-1} f_{crit}^{-1} h^{-1} y, \quad h_{burst} \cong 7 \times 10^{-17} \left(\frac{M}{10^2 M_\odot}\right), \qquad (72)$$

respectively. The crucial issue, of course, is whether one can expect black holes to form in binaries. Since at least 50% of O stars (the largest stars forming in the present epoch) appear to be in binaries,[229] it does not seem implausible that VMO binaries should also be abundant. While O stars appear to have a separation spectrum[231] which peaks at $a \simeq 30 R_\odot$, this could be raised by about a factor of 4 for VMOs due to mass loss effects. Therefore, a reasonably large fraction of the resulting black holes binaries could have the separation a_{crit}.

4.4.10 Pregalactic explosions

Stars in the mass range $4 - 200 M_\odot$ should produce explosive energy with an efficiency $\varepsilon = 10^{-5} - 10^{-4}$; those in the range $200 - 10^5 M_\odot$ may explode with comparable efficiency providing the shell ejection mechanism discussed in § 4.4.5 works.[232] This explosive release could have an important effect on the large-scale structure of the Universe.[233-237] One would expect the shock-wave generated by each exploding star (or cluster of stars) to sweep up a shell of gas. Under suitable circumstances, this shell could eventually fragment into more stars. If the new stars themselves explode, one could then initiate a bootstrap process in which the shells grow successively larger until they overlap. This mechanism has been proposed in three contexts: (i) as a means to boost the fraction of the Universe being processed through pregalactic stars[192]; (ii) as a means to generate many galaxies from a few seed galaxies[234]; and (iii) as a way of producing the giant voids and filaments, whose existence is indicated by observational data[238,239], from much smaller scale initial structure. During the Compton cooling era ($z > 10$), one can show that the maximum amplification which can be achieved in a single step is

$$\xi_1 \equiv \frac{M_{shell}}{M_1} \approx 4 \times 10^5 (1+z)^{-1.7} \left(\frac{\varepsilon}{10^{-4}}\right)^{0.6} \left(\frac{M_1}{10^6 M_\odot}\right)^{-0.4} \quad (73)$$

where M_1 is the seed mass and z is the redshift at which the shell fragments. The shell mass at the nth step is[237]

$$M_n \approx \xi_1^{2.5} \{1-(0.6)^{n-1}\} M_1 \quad (74)$$

and this tends asymptotically to

$$M_\infty \approx \xi_1^{2.5} M_1 \approx 1 \times 10^{20} (1+z)^{-4.3} \left(\frac{\varepsilon}{10^{-4}}\right)^{1.5} M_\odot \quad (75)$$

providing the shells do not overlap first. Equation (75) specifies the maximum scale of structure that can be attained at a redshift z. By the end of the Compton era, M_∞ is already of order $10^{15} M_\odot$, independent of the original seed mass. For $z < 10$, radiative cooling dominates and the situation becomes more complicated. An upper limit to the final shell size in all circumstances is $\sqrt{\varepsilon} ct$. This is 30 Mpc for $\varepsilon \approx 10^{-4}$, which is just about large enough to explain the largest voids.

4.5 Conclusion

The shaded regions in Table (5) indicate what sorts of Population III stars are required to explain the various cosmological effects alluded to in this section. We have ordered these effects according to the extent to which they require a deviation from the standard Big Bang picture; only the last two require that one gives up the conventional hot scenario. The "IR background" refers specifically to the Japanese measurement and the "GW background" is required to be detectable. The stars which solve most of the problems would seem to be the VMOs: indeed, the combination of exploding and collapsing VMOs would appear to explain everything. Therefore, if one is prepare to forego some of the attractions of the standard picture, VMOs would seem to be the Population III candidate with most explanatory power. On the other hand, if one wants to preserve the standard picture as much as possible, the most attractive Population III candidate would be SMOs of $10^6 M_\odot$: such objects could avoid making helium, light, enrichment, and explosions, while at the same time producing dark halos, disc heating, and detectable gravitational waves.

Table (5): Explanatory power of different Population III candidates

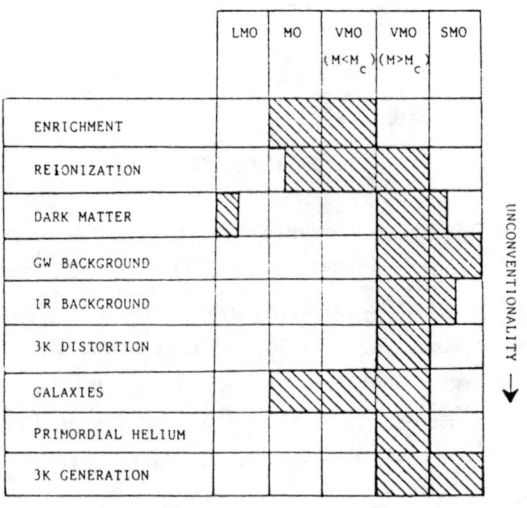

<u>Acknowledgments</u>. I would like to thank Dr. J.L. Sanz and Dr.L.J. Goicoechea for inviting me to give these lectures and organizing such a pleasant conference. I am especially grateful to Dr. Sanz for convincing me that "monopoles" do exist, at least in Spain!

REFERENCES

1. T. Gold, Nature, 218, 731 (1968).
2. J.R. Oppenheimer & H. Snyder, Phys.Rev. 56, 455 (1939).
3. S.W. Hawking & R. Penrose, Proc.R.Soc.London A. 314, 529 (1969).
4. R. Penrose, Nuovo Cimento 1, 252 (1969).
5. W. Israel, Phys.Rev. 164, 1776 (1967).
6. B. Carter, Phys.Rev.Lett. 26, 331 (1970).
7. D.C. Robinson, Phys.Rev.D. 10, 458 (1974).
8. R.H. Price, Phys.Rev.D.5, 2439 (1972).
9. R.P. Kerr, Phys.Rev.Lett. 11, 237 (1963).
10. S.W. Hawking, in Black Holes, eds. B.S. DeWitt & C. DeWitt (Gordon & Breach, New York, 1973).
11. S.W. Hawking, Nature 248, 30 (1974).
12. S. Chandrasekhar, Astrophys.J. 74, 81.
13. W.D. Arnett, Astrophys.Space Sci. 5, 180 (1969).
14. T.A. Weaver, G.B. Zimmerman & S.E. Woosley, Astrophys.J. 225, 1021 (1978).
15. B.L. Webster & P. Murdin, Nature 235, 37 (1972).
16. A. Toor, Astrophys.J.Lett. 215, L57 (1977).
17. W. Forman, et al. Astrophys.J. 208, 849 (1976).
18. C. Motch, S.A. Ilovaisky & C. Chevalier, Astron. Astrophys., 109, L1 (1983).
19. M. Schwarzschild & R. Harm, Astrophys.J. 243, 1 (1959).
20. I. Appenzeller, Astron.Astrophys. 9, 216 (1970).
21. J. Papaloizou, Mon.Not.R.Astron.Soc. 162, 169 (1973).
22. J.P. Cassinelli, J.S. Mathis & B.D. Savage, Science 212, 1497 (1981).
23. C.D. Andriesse, B.D. Donn & R. Viotti, Mon.Not.R.Astron.Soc. 185, 771 (1978).
24. R. Chevalier, Fund Cosmic Phys. 7, 1 (1981).
25. W.A. Fowler & F. Hoyle, Astrophys.J.Supp. 9, 201 (1964).
26. S.E. Woosley & T.A. Woosley, in Supernovae: A Survey of Current Research, eds. M.J. Rees & R.J. Stoneham (Reidel, Dordrecht, (1982).

27. J.R. Bond, W.D. Arnett & B.J. Carr, Astrophys.J. 280, 825 (1984).
28. W.W.Ober, M.F.El Eid & K.L. Fricke, Astron.Astrophys. 119, 61 (1983).
29. W. Glatzel, M.F. El Eid & K.J. Fricke, preprint (1985).
30. W.A. Fowler, Astrophys.J. 144, 180 (1966).
31. K.J. Fricke, Astrophys.J. 183, 941 (1973).
32. M.C. Begelman & M.J. Rees, Mon.Not.R.Astron.Soc. 185, 847 (1978).
33. M.J. Rees, Phys. Script 17, 193 (1978).
34. M.J. Duncan & S.L. Shapiro, Astrophys.J. 268, 565 (1983).
35. J.G. Hills, Nature 254, 294 (1975).
36. D. Lynden-Bell, Nature 223, 690 (1969).
37. W.L. Sargent, P.L. Young, A. Boksenberg, K. Shortridge, C.R. Lynds & F.D.A. Hartwick, Astrophys.J. 221, 731 (1978).
38. J.H. Lacy, C.H. Townes & D.J. Hollenbach, Astrophys.J. 262, 120, (1983).
39. A.C. Fabian, D. Maccagni, M.J. Rees & W.R. Stoeger, Nature 260, 683 (1976).
40. S.W. Hawking, Mon.Not.R.Astron.Soc. 152, 75 (1971).
41. P.J.E. Peebles & R.H. Dicke, Astrophys.J. 154, 891 (1968).
42. B.J. Carr & M.J. Rees, Mon.Not.R.Astron.Soc. 206, 315 (1984).
43. B.J. Carr, Astrophys.J. 201, 1 (1975).
44. B.J. Carr, Mon.Not.R.Astron.Soc 181, 293 (1977).
45. B.J. Carr & M.J. Rees, Astron.Astrophys. 61, 705 (1977).
46. M.J. Rees, Nature 275, 35 (1978).
47. E.L. Wright, Astrophys.J. 250, 1 (1981).
48. B.J. Carr, Mon.Not.R.Astron.Soc. 195, 669 (1981).
49. R.V. Wagoner, Astrophys.J. 179, 343 (1973).
50. G.E. Miller & J.M. Scalo, Astrophys.J.Supp. 41, 513 (1979).
51. S.M. Faber & J.S. Gallagher, Ann.Rev.Astron.Astrophys. 17, 135 (1979).
52. B.J. Carr, Comm.Astrophys. 7, 161 (1978).
53. A.A. Jackson & M.P. Ryan, Nature 245, 88 (1973).
54. K.S. Thorne, in Theoretical Principles in Astrophysics and Relativity, eds. N.R. Lebovitz, W.H. Reid & P.O. Vandervoort (University of Chicago Press, 1978).

55. J.M. Bardeen & R.V. Wagoner, Astrophys.J. 167, 359 (1971).
56. P. Meszaros, Astron.Astrophys. 38, 5 (1975).
57. B.J. Carr, Astron.Astrophys. 56, 377 (1977).
58. B.J. Carr & M.J. Rees, Mon.Not.R.Astron.Soc. 206, 801 (1984).
59. J.D. Beckenstein, Astrophys.J. 183, 657 (1973).
60. V. Moncrieff, Astrophys.J. 238, 333 (1980).
61. M.J. Fitchett, Mon.Not.R.Astron.Soc. 203, 1049 (1983).
62. I.D. Novikov & K.S. Thorne, in Black Holes, eds. B.S. DeWitt & C. DeWitt (Gordon & Breach, New York, 1973).
63. H. Bondi, Mon.Not.R.Astron.Soc. 112, 195 (1952).
64. B.J. Carr, Mon.Not.R.Astron.Soc. 194, 639 (1983).
65. B.J. Carr, Nature 284, 326 (1983).
66. E. Boldt & D. Leiter, Nature 290, 483 (1981).
67. B.J. Carr, J. McDowell & H. Sato, Nature 306, 666 (1983).
68. J.N. Bahcall, Astrophys.J. 276, 169 (1984).
69. V.C. Rubin, W.K. Ford & N. Thonnard, Astrophys.J. 238, 471 (1980).
70. N. Krumm & E.E. Salpeter, Astrophys.J.Lett. 208, L7 (1976).
71. L. Blitz, Astrophys.J.Lett. 231, L115 (1979).
72. C.S. Frenk & S.D.M. White, Mon.Not.R.Astron.Soc. 198, 173 (1982).
73. D.N.C. Lin & D. Lynden-Bell, Mon.Not.R.Astron.Soc. 198, 707 (1982).
74. M. Petrou, Mon.Not.R.Astron.Soc. 191, 767 (1980).
75. J.P. Ostriker & P.J.E. Peebles, Astrophys.J. 186, 467 (1973).
76. A. Tubbs & A.H. Sanders, Astrophys.J. 230, 736 (1979).
77. R. Davies, Mon.Not.R.Astron.Soc. 194, 876 (1981).
78. D.N.C. Lin & S.M. Faber, Astrophys.J.Lett. 266, L21 (1983).
79. F. Zwicky, Helv.Phys.Acta 6, 110 (1933).
80. W. Forman & C. Jones, Ann.Rev.Astron.Astrophys. 20, 547 (1982).
81. S.D.M. White & M.J. Rees, Mon.Not.R.Astron.Soc. 183, 341 (1978).
82. A.H. Guth, Phys.Rev.D. 23, 347 (1981).
83. C.B. Collins & S.W. Hawking, Astrophys.J. 180, 317 (1973).
84. J.A. Wheeler, in Gravitation and Relativity, eds. H.Y. Chiu & W.F. Hoffman (Benjamin, New York, 1964).

85. M. Davis & P.J.E. Peebles, Astrophys.J. 267, 465 (1983).
86. S. Weinberg, in Gravitation and Cosmology (Wiley, New York, 1972).
87. B.W. Lee & S. Weinberg, Phys.Rev.Lett. 39, 165 (1977).
88. A.S. Szalay & G. Marx, Astron.Astrophys. 49, 437 (1976).
89. R. Cowsik & J. McClelland, Phys.Rev.Lett. 29, 669 (1972).
90. J.R. Bond, G. Efstathiou & J. Silk, Phys.Rev.Lett. 45, 1980 (1980).
91. A.G. Doroshkevich, Ya.B. Zeldovich, R.A. Sunyaev & M. Yu. Khlopov, Sov.Astron.Lett. 6, 252 (1981).
92. V.A. Lyubimov, E.G. Novikov, V.Z. Nozik, E.F. Tretyakov & V.S. Kosik, Phys.Lett.B. 94, 266 (1980).
93. F. Boehm, in Proc. Fourth Worksho on Grand Unification
94. J. Bahcall et al, Rev.Mod.Phys. 54, 767 (1982).
95. J.R. Bond, A.S. Szalay & M.S. Turner, Phys.Rev.Lett. 48, 1636 (1982).
96. G.R. Blumenthal, H. Pagels & J.R. Primack, Nature 299, 37 (1982).
97. K.A. Olive & M.S. Turner, Phys.Rev.D. 25, 213 (1982).
98. S. Weinberg, Phys.Rev.Lett. 50, 387 (1983).
99. H. Goldberg, Phys.Rev.Lett. 50, 1419 (1983).
100. L.A. Ibanez, Phys.Lett.B. 137, 160 (1984).
101. J. Preskill, M. Wise & F. Wilczek, Phys.Lett.B. 120, 127 (1983).
102. L. Abbott & P. Sikivie, Phys.Lett.B. 120, 133 (1983).
103. M.S. Turner, F. Wilczek & A. Zee, Phys.Lett.B. 125, 35 (1983).
104. M. Dine & W. Fischler, Phys.Lett.B. 120, 137 (1983).
105. E.W. Kolb, D. Seckel & M.S. Turner, Nature 314, 415 (1985).
106. E.N. Parker, Astrophys.J. 160, 383 (1970).
107. D.J. Hegyi, K.A. Olive, Phys.Lett.B. 126, 28 (1983).
108. J. Yang, D.N. Schramm, G. Steigman & R.T. Rood, Astrophys.J. 277, 697 (1979).
109. D.N. Schramm & G. Steigman, Astrophys.J. 243, 1 (1981).
110. R.F.A. Staller & D. de Jong, Astron.Astrophys. 98, 140 (1981).
111. G. Gilmore & P. Hewett, Nature 306, 669 (1983).

112. S.P. Boughn, P.R. Saulson & M. Seldner, Astrophys.J.Lett. 250, L15 (1981).
113. W.H. Press & J.E. Gunn, Astrophys.J. 185, 397 (1973).
114. J.R. Gott, Astrophys.J. 243, 140 (1981).
115. C.R. Canizares, Astrophys.J. 263, 508 (1982).
116. J.N. Bahcall, P. Hut & S. Tremaine, Astrophys.J. 290, 15 (1985).
117. C.G. Lacey, in *Formation and Evolution of Galaxies and Large Structures in the Universe*, eds. J. Audouze & J. Tran Thanh Van (Reidel, Dordrecht, 1984).
118. S. Van der Bergh, Nature 224, 891 (1969).
119. S. Tremaine & J.E. Gunn, Phys.Rev.Lett. 42, 403 (1979).
120. D.K. Nadejin, I.D. Novikov & A.G. Polnarev, Sov.Astron. 22, 129 (1978).
121. Ya.B. Zeldovich & I.D. Novikov, Sov.Astron.A.J. 10, 602 (1967).
122. B.J. Carr & S.W. Hawking, Mon.Not.R.Astron.Soc. 168, 399 (1974).
123. D.N.C. Lin, B.J. Carr & S.M. Fall, Mon.Not.R. Astron.Soc. 177, 51 (1976).
124. G.V. Bicknell & R.N. Henriksen, Astrophys.J. 219, 1043 (1978).
125. S. Hacyan, Astrophys.J. 229, 42 (1979).
126. D. Lindley, Mon.Not.R.Astron.Soc. 196, 317 (1981).
127. L.S. Marochnik, P.O. Nasel'skii & N.A. Zabotin, SRI Preprint 564 (1980).
128. J.D. Barrow & B.J. Carr, Mon.Not.R.Astron.Soc. 182, 537 (1978).
129. V. Canuto, Mon.Not.R.Astron.Soc. 184, 721 (1978).
130. M.Yu. Khlopov & A.G. Polnarev, SRI Preprint 578 (1980). ???
131. A.F. Grillo, Phys.Lett.B. 94, 364 (1980).
132. M.Yu. Khlopov & A.G. Polnarev, SRI Preprint 646 (1981). ???
133. S.W. Hawking, I.G. Moss & J.M. Stewart, Phys.Rev.D. 26, 2681, (1982).
134. H. Kodama, M. Sasaki, K. Sato, & K. Maeda, Prog.Theor.Phys. 66, 2052 (1981).
135. M. Crawford & D.N. Schramm, Nature 298, 538 (1982).
136. B.J. Carr & J.D. Barrow, Gen.Rel.Grav. 11, 383 (1979).
137. S.W. Hawking, Commun.Math.Phys. 43, 199 (1975).
138. G.W. Gibbons, Commun.Math.Phys. 44, 245 (1975).

139. D.N. Page, Phys.Rev.D. **13**, 198 (1976).
140. D.N. Page, Phys.Rev.D. **14**, 3260 (1976).
141. D.N. Page, Phys.Rev.D. **16**, 2402 (1977).
142. R. Hagedorn, Astron.Astrophys. **5**, 184 (1970).
143. I.D. Novikov, A.G. Polnarev, A.A. Starobinskii & Ya.B. Zeldovich, Astron.Astrophys. **80**, 104 (1979).
144. B.J. Carr, Astrophys.J. **206**, 8 (1976).
145. D.N. Page & S.W. Hawking, Astrophys.J. **206**, 1 (1976).
146. C.E. Fichtel, R.C. Hartman, D.A. Kniffen, D.J. Thompson, G.F. Bignami, H. Ogelman, M.F. Ozel & T. Tumer, Astrophys.J. **198**, 163 (1975).
147. G.F. Chapline, Nature **253**, 251 (1975).
148. Ya.B. Zeldovich & R.A. Sunyaev, Astrophys.Space Sci. **4**, 301 (1969).
149. Ya.B. Zeldovich & A.A. Starobinskii, JETP Lett. **24**, 571 (1976).
150. P.D. Nasel'skii, Pisma Astron.Zh. **4**, 387 (1978).
151. B. Carter, G.W. Gibbons, D.N.C. Lin & M.J. Perry, Astron. Astrophys. **52**, 427 (1976).
152. A.I. Belyaevskii, V.L. Bokov, V.K. Bocharkin, I.F. Bugakov, G.M. Gorodinskii, Yu.G. Derevitskii, E.M. Kruglov, G.A. Pyatigorskii & E.I. Chuikin, JETP Lett. **21**, 345 (1975).
153. N.A. Porter & T.C. Weekes, Mon.Not.R.Astron.Soc. **183**, 205 (1978).
154. N.A. Porter & T.C. Weekes, Nature **277**, 199 (1979).
155. D.J. Fegan, B. McBreen, D. O'Brien & C. O'Sullivan, Nature **271**, 731 (1978).
156. M.J. Rees, Nature **266**, 333 (1977).
157. R.D. Blandford, Mon.Not.R.Astron.Soc. **181**, 489 (1977).
158. S. Phinney & J.H. Taylor, Nature **277**, 117 (1979).
159. W.P.S. Meikle, Nature **269**, 41 (1977).
160. N.A. Porter & T.C. Weekes, **267**, 501 (1977).
161. J.V. Jelley, G.A. Baird & E. O'Mongain, Nature **267**, 499 (1977).
162. C.L. Bhat, H. Razdan & M.L. Sapru, Astron.Space Sci. **73**, 513 (1980).

163. A.C. Cummings, E.C. Stone & R.E. Vogt, 13th International Cosmic Ray Conference, Denver (1973).
164. M. Leventhal, C.J. MacCallum & P.D. Stang, Astrophys.Lett. 225, L11 (1978).
165. P.N. Okeke & M.J. Rees, Astron.Astrophys. 81, 263 (1980).
166. R.R. Daniel & S.A. Stephens, Space Sci.Rev. 17, 45 (1975).
167. P.D. Nasel'skii & N.V. Pelikov, Sov.Astron. 23, 402 (1979).
168. K. Greisen, Texas Symposium on Relativistic Astrophysics (1967).
169. A. Buffington, S.M. Schindler & C.R. Pennypacker, Astrophys.J. 248, 1179 (1981).
170. R.L. Golden et al. Phys.Rev.Lett. 43, 1196 (1979).   ???
171. P. Kiraly, J. Szabelski, J. Wdowczyk & A.W. Wolfendale, Nature, 293, 120 (1981).
172. M.S. Turner, Nature, 297, 379 (1982).
173. J. MacGibbon, preprint (1985).
174. S. Miyama & K. Sato, Prog.Theor.Phys. 59, 1012 (1978).
175. B.V. Vainer & P.D. Nassel'skii, Sov.Astron. 22, 138 (1978).
176. D. Lindley, Mon.Not.R.Astron.Soc. 193, 593 (1980).
177. B.V. Vainer & P.D. Nasel'skii, Sov.Astron.Lett. 3, 76 (1977).
178. Ya.B. Zeldovich, A.A. Starobinskii, M.Yu. Khlopov & V.M. Chechetkin, Sov.Astron.Lett. 3, 110 (1977).
179. B.V. Vainer, O.V. Dryzhakova & P.D. Nasel'skii, Sov.Astron.Lett. 4, 185 (1978).
180. T. Rothman & R. Matzner, Astrophys.Space Sci. 75, 229 (1981).
181. P.J.E. Peebles, Astrophys.J.Lett 189, L51 (1974).
182. S.M. Fall, Rev.Mod.Phys. 51, 21 (1979).
183. J. Silk, Astrophys.J. 211, 638 (1977).
184. R. Terlevich, Ph.D. thesis (Cambridge University, 1983).
185. J.E. Tohline, Astrophys.J. 239, 417 (1980).
186. A. Kashlinsky & M.J. Rees, Mon.Not.R.Astron.Soc. 205, 955 (1983).
187. F. Palla, E.E. Salpeter & S.W. Stahler, Astrophys.J. 271, 632 (1983).
188. J. Silk, Mon.Not.R.Astron.Soc. 205, 705 (1983).
189. J. Silk, Astrophys.J. 151, 459 (1968).

190. Ya.B. Zeldovich, Astron.Astrophys. 5, 84 (1970).
191. J.R. Bond & A. Szalay, Astrophys.J. 274, 443 (1983).
192. B.J. Carr, J.R. Bond & W.D. Arnett, Astrophys.J. 277, 445 (1984).
193. P.J.E. Peebles & R.B. Partridge, Astrophys.J. 200, 527 (1975).
194. J.R. Thorstensen & R.B. Partridge, Astrophys.J. 200, 527 (1975).
195. P.W. Tarbet & M. Rowan-Robinson, Nature 298, 711 (1982).
196. T.B. Weaver & S.E. Woosley, Ann.New York Acad.Sci. 336, 335 (1980).
197. H.E. Bond, Astrophys.J. 248, 606 (1981).
198. T. Matsumoto, M. Akiba & M. Murakami, in COSPAR Journal: Advances in Space Research, eds. G.F. Bignami & R.A. Sunyaev (Pergamon Press, Oxford, 1984).
199. J.R. Bond, B.J. Carr & C. Hogan, preprint (1985).
200. M. Rowan-Robinson, J. Negroponte & J. Silk, Nature 281, 635 (1979).
201. J. Negroponte, M. Rowan-Robinson & J. Silk, Astrophys.J. 248, 38 (1981).
202. J.L. Puget and J. Heyvaerts, Astron.Astrophys. 83, L10 (1980).
203. D.P. Woody & P.L. Richards, Astrophys.J. 248, 18 (1981).
204. P.L. Richards, in Proc.Inner Space/Outer Space Conference (University of Chicago Press, 1984).
205. D. Layzer & R.M. Hively, Astrophys.J. 179, 361 (1973).
206. E.L. Wright, Astrophys.J. 255, 401 (1982).
207. N.C. Rana, Mon.Not.R.Astron.Soc. 197, 1125 (1981).
208. N.C. Wickramasinghe, M.G. Edmunds, S.M. Chitre, J.V. Narlikar & S. Ramadurai, Astrophys.Space Sci. 35, L9 (1975).
209. H. Alfven & A. Mendis, Nature 266, 699 (1977).
210. S.M. Chandler & J.M. Scalo, preprint (1984).
211. E.W. Kolb & M.S. Turner, Ann.Rev.Nuc.Part.Sci. 33, 645 (1983).
212. T.W. Hartquist & A.G.W. Cameron, Astrophys.Space Sci. 48, 145 (1977).
213. C. Hogan, Mon.Not.R.Astron.Soc. 188, 781 (1979).
214. R.J. Talbot & W.D. Arnett, Nature 229, 250 (1971).
215. J.R. Bond, B.J. Carr & W.D. Arnett, Nature 304, 514 (1983).

216. M. Kaufman, Astrophys.J. 160, 459 (1970).
217. B.J. Carr, Astron.Astrophys. 56, 377 (1977).
218. J.W. Truran & A.G.W. Cameron, Astrophys.Space Sci. 14, 179 (1971).
219. C.A. Pilachowski, G.D. Bothun, E.W. Olszewski & A. Odell, Astrophys.J. 273, 187 (1983).
220. J.W. Truran, in Formation and Evolution of Galaxies and Large Structures in the Universe, eds. J. Audouze & J. Tran Thanh Van (Reidel, Dordrecht, 1984).
221. C. Sneden, D. Lambert & R.W. Whitaker, Astrophys.J. 234, 964 (1979).
222. C. Chiosi & F. Matteucci, in Formation and Evolution of Galaxies and Large Structures in the Universe, eds. J. Audouze & J. Tran Thanh Van (Reidel, Dordrecht, 1984).
223. B.E.J. Pagel & M.G. Edmunds, Ann.Rev.Astron.Astrophys. 19, 77 (1981).
224. B. Barbuy, Astron.Astrophys. 123, 1 (1983).
225. J. Klapp, Astrophys.Space Sci. 93, 313 (1983).
226. R. Wielen, Astron.Astrophys. 60, 263 (1977).
227. B. Bertotti & B.J. Carr, Astrophys.J. 236, 1000 (1980).
228. J.R. Bond & B.J. Carr, Mon.Not.R.Astron.Soc. 207, 585 (1984).
229. C.D. Garmany, P.S. Conti & P. Massey, Astrophys.J. 242, 1063 (1980).
230. C.D. Lacey, Mon.Not.R.Astron.Soc. 208, 687 (1984).
231. V. Trimble & C. Cheung, in IAU Symposium No. 73, eds. P. Eggleton, S. Mitton & J. Whelan (1976).
232. J. Bookbinder, L.L. Cowie, J.H. Krolik, J.P. Ostriker & M.J. Rees Astrophys.J. 237, 647 (1980).
233. A.G. Doroskevich, Ya.B. Zeldovich & I.D. Novikov, Sov.Astron AJ. 11, 231 (1967).
234. J.P. Ostriker & L.L. Cowie, Astrophys.J.Lett. 243, L127 (1981).
235. S. Ikeuchi, Pub.Astron.Soc. Japan 33, 211 (1981).
236. E. Bertschinger, Astrophys.J. 268, 17 (1983).
237. B.J. Carr & S. Ikeuchi, Mon.Not.R.Astron.Soc., 207, 585 (1985).
238. M. Davis, J. Huchra, D.W.Latham & J. Tonry, Astrophys.J. 253, 423 (1982).
239. J. Einasto, M. Joeveer & E. Saar, Mon.Not.R.Astron.Soc. 193, 353 (1983).

PRIMORDIAL NUCLEOSYNTHESIS AND THE GRAVITINO PROBLEM

Rosa Domínguez Tenreiro
Departamento de Física Teórica
Universidad Autónoma de Madrid
Cantoblanco, 28049-Madrid
SPAIN

ABSTRACT

A short review of the standard primordial nucleosynthesis (PN) process and its comparison with light element observed abundances is first made. Then it is explained how PN can be used to constrain some astrophysical and particle physics parameters. In particular, some results concerning the gravitino problem are presented.

1. INTRODUCTION

There exist three sets of observations which in order to be interpreted require a model of the early stages of the universe. These observations are: i) the recession of galaxies with velocity proportional to their distance (Hubble law), ii) the microwave background of thermal radiation at 2.7°K, and iii) the observed abundances of the light elements D, ^{3}He, ^{4}He and ^{7}Li.

The standard Friedmann-Robertson-Walker (FRW) cosmologies explain in rather a natural way these three sets of observations.

The recession of galaxies and the background radiation provide us with information on the universe evolution up to the recombination era. Light element abundances allow us to go back in time until epochs as remote as about 10^2 sec. This is the reason why the study of primordial nucleosynthesis (PN) is of major importance in order to decipher the physical conditions in the early universe and to set up important limits on several astrophysical and particle physics parameters[*]. It must be emphasized that the microphysics involved in the PN process (Fermi theory of weak interactions and analytical fittings to measured nuclear cross sections) is reliably well know.

In the frame of FRW models, the theoretical PN calculation is a closed problem. There remain some uncertainties in the nuclear cross

---
[*] See, however, Bond, Carr and Arnett[1] and Audouze and Silk[2] for an alternative picture of light element origins.

sections, but the resulting uncertainties in the light element yields have been evaluated by Beaudet and Reeves[3] and Yang et al[4], so that the theoretical aspects of the problem are under control.

The open question is the determination of the observed abundances of the light elements, to be compared with the computed ones. An enormous amount of work has been devoted in recent years to the measurement of these abundances in different astrophysical sites, and to the construction of chemical evolution models for the galaxy which allow us to extrapolate back in time from the present observations up to the big-bang.

On the other hand, light element production in a coherent way is a condition to be fulfilled by every newly proposed early universe model or microphysics theory. Supersymmetric theories of particle physics have been proposed in the last years, and even though some qualitative works have been devoted to the study of PN in such a frame, more detailed computations were not available.

In this talk a short review of the standard PN process and its comparison with observations will first be made. Then it will be explained how PN can be used to constrain some astrophysical and particle physics parameters. In particular, some results concerning the PN products evolution in the frame of supersymmetric theories will be presented.

2. PRIMORDIAL NUCLEOSYNTHESIS IN STANDARD FRW COSMOLOGIES

2.1 Assumptions And Dynamics.

The simplest models of PN, such us those investigated by Peebles[5] and Wagoner, Fowler and Hoyle[6], have been constructed in the framework of what are often called standard big-bang models. They are defined by the following assumptions:

i) General relativity describes correctly the evolution of the universe.

ii) The universe was isotropic and homogeneous at PN time. From assumption i), this implies that the evolution of the cosmic scale factor R is controlled at early times by the equation:

$$\frac{1}{R}\frac{dR}{dt} = \left[\frac{8\pi G}{3}\rho\right]^{1/2} \qquad (2.1)$$

where G is the gravitational constant and ρ is the total energy density of the universe.

Concerning the universe content, it is assumed that:

iii) The baryon number asymmetry was present at PN time, and antibaryon annihilation had been completed at this time.

iv) The lepton number is much lower than the photon number density.

The early universe dynamics is dominated by relativistic particles. Any particle of mass m will be dynamically significant as long as the temperature of the particles in thermal equilibrium, T, is higher than m. During the radiation dominated period, the energy density may be written in terms of the photon energy density, $\rho_\gamma = aT^4$ (a is the black-body constant), as:

$$\rho(T) = \frac{1}{2} g(T) \rho_\gamma(T) \qquad (2.2)$$

where

$$g(T) = \sum_B g_B \left(\frac{T_B}{T}\right)^4 + \frac{7}{8} \sum_F g_F \left(\frac{T_F}{T}\right)^4 \qquad (2.3)$$

$g_{B,F}$ is the effective number of degrees of freedom of the bosons(B) and fermions (F) present at radiation temperature T, that is, the number of spin (and color) states for each boson or fermion type. $T_{B,F}$ is the temperature of particles B or F, eventually decoupled from the thermal component.

Equations (2.1) and (2.2) give us the dynamics of the universe. The age-temperature relation can be deduced from them, and reads:

$$t_{sec} T^2_{MeV} = 2.42 \left[g(T)\right]^{-1/2} \qquad (2.4)$$

2.2 The Nucleosynthesis Process.

Let us now qualitatively describe the nucleosynthesis process.

First, we will examine the neutron-proton abundance ratio as a function of time. This is controlled by the weak interaction reactions:

$$n + \nu_e \leftrightarrow p + e^- \quad , \quad n + e^+ \leftrightarrow p + \bar{\nu}_e \quad , \quad n \rightarrow p + e^- + \bar{\nu}_e \qquad (2.5)$$

whose rates per nucleon λ (i+j → k+l) are given, for example, by Weinberg[7]). The evolution of the ratio X_n of neutrons to all nucleons results from the equation:

$$\frac{dX_n}{dt} = -\lambda(n \rightarrow p) X_n + \lambda(p \rightarrow n)(1 - X_n) \qquad (2.6)$$

where $\lambda(n \rightarrow p)$ and $\lambda(p \rightarrow n)$ are the total n → p and p → n transition rates, respectively. The main qualitative features of $X_n(t)$ may be outlined without numerically integrating (2.5); they are the following:

i) Let us assume that $KT \simeq Q = m_n - m_p = 1.293$ MeV. The rates are to be compared with the universe age at these temperatures. From (2.3) and (2.4), and taking three low mass neutrino pairs ($N_\nu = 3$), the age is

$$t \simeq 1.0 \left(\frac{T}{10^{10} \, ^\circ K}\right)^{-2} sec \qquad (2.7)$$

If $\lambda(T) t(T) \gg 1$, the neutron fraction X_n should be given by the equilibrium solution of (2.6):

$$X_n \simeq \frac{\lambda(p \to n)}{\lambda(p \to n) + \lambda(n \to p)} \qquad (2.8)$$

This is correct in standard big-bang models with $N_\nu = 3$ up to a temperature $T_* \simeq 3 \; 10^{10} \, °K$ (freeze-out temperature).

ii) As long as the neutrinos are in equilibrium with the thermal component (that is for $T > 10^{10} \, °K$), the rates verify:

$$\frac{\lambda(p \to n)}{\lambda(n \to p)} = \exp[-Q/kT] \qquad (2.9)$$

so that, if at the same time (2.8) holds ($T > T_*$), we will have

$$X_n = [1 + \exp(Q/kT)]^{-1} \qquad (2.10)$$

which implies that $X_n(t) \to 1/2$ for $kT \gg Q$, and then decreases slowly as the universe cools. That gives the initial condition for (2.6), which follows from the microphysical theory of weak interactions and does not depend on the particular cosmological model describing the evolution of the early universe.

A faster expansion rate means freeze-out at a higher temperature, resulting (see eq.2.10) in a higher neutron fraction at T_*, and as we will see, a higher ^{4}He production.

iii) For $T \lesssim 10^9 \, °K$, the weak processes (2.5) are switched off, and the only remaining reaction up to the onset of nucleosynthesis is the neutron decay:

$$n \longrightarrow p + e^- + \bar{\nu}_e \qquad (2.11)$$

Let us now consider the nuclear reactions leading to complex nuclei formation.

Baryon number density is too low to allow many-body collisions, so that nuclei have to be synthetized by sequences of two-body processes. The first step is to build up deuterium:

$$p + n \longrightarrow D + \gamma \qquad (2.12)$$

and only if D is abundant enough, the following steps can proceed. At high energies, the different nuclei are in thermal equilibrium, and their abundance is negligible until the temperature drops to a value T_i given by[7]:

$$T_i = \frac{B_i}{k(A_i - 1)|\ln \varepsilon|} \qquad (2.13)$$

where B_i is the binding energy of nucleus i, A_i its mass number and ε a quantity depending on the baryon density and cosmological model, and typically of order $O(10^{-12})$. For D this temperature is $T_D \simeq .8 \; 10^9 \, °K$, at which time the nucleosynthesis starts.

The outcome of the nuclear reactions is the incorporation of almost all available neutrons into ^{4}He, which have the highest binding energy among light elements nuclei. The reason is that the Coulomb barrier in the reactions: $^4\text{He} + {}^3\text{He} \to {}^7\text{Be} + \gamma$ and $^4\text{He} + {}^3\text{H} \to {}^7\text{Li} + \gamma$, and the absence of stable nuclei with A = 5 and A = 8, prevent ^{4}He destruction.

To have a quantitative picture of the synthesis process in the big-bang, one must carry out a numerical integration of the equations of nuclear reaction kinetics. This has been first done by Peebles (1966)[5] Wagoner, Fowler and Hoyle (1967)[6], and Wagoner (1969)[8], who first considered a nuclear reaction network including all elements with mass number $A \leq 32$. This work is the classical PN paper.

2.3 Results.

In the standard big-bang models the light element final yields depend on the neutron half-life $\tau_{1/2}$, the number of lepton families N_ν and the present baryon density, which is usually parametrized by:

$$\rho_B = 7.15 \cdot 10^{-27} h_o^2 \, g\,cm^{-3} = 6.64 \cdot 10^{-22} \eta \, g\,cm^{-3} \quad (2.14)$$

The main qualitative traits of the PN computations outcome in standard big-bang models have not been altered since Wagoner's 1973 work[9]. They are the following:

i) Only ^{4}He, D, ^{3}He and ^{7}Li are significantly produced in the big-bang.

ii) The ^{4}He abundance does not appreciably depend on the present baryon density. On the contrary, D, ^{3}He and ^{7}Li final yields are remarkably sensitive to the parameter h_o.

iii) The ^{4}He production is an increasing function of $\tau_{1/2}$ and N_ν, while the other nuclei production is almost insensitive to those parameters.

Figure 1 shows theoretical light element abundances (by number relative to H) as a function of $\eta = n_B/n_\gamma$. Boxes are observed abundances with their errors (see section 3). For ^{4}He we show the results for $\tau_{1/2} = 10.8$ min (———), $\tau_{1/2} = 10.6$ min (-----) and $\tau_{1/2} = 10.4$ min (-.-.) and two boxes corresponding to the upper bounds $Y_p \leq 0.25$ and $Y_p \leq 0.24$ and the lower limit $Y_p \leq 0.22$. The η interval allowed from D and D + ^{3}He abundance measurements is shown at the bottom of the figure. PN yields have been taken from Schramm[10] and Yang et al[4].

3. OBSERVED ABUNDANCES

The determination of light element abundances to be compared with the predicted PN production is a difficult open question, which deserves much attention because of the information on the early universe physics and microphysics carried by these abundances. In evaluating primordial abundances, there exist two sources of difficulties: the measurement itself and the extrapolation of what is observed at present up to the PN time, extrapolation that necessarily has to be done in the frame of

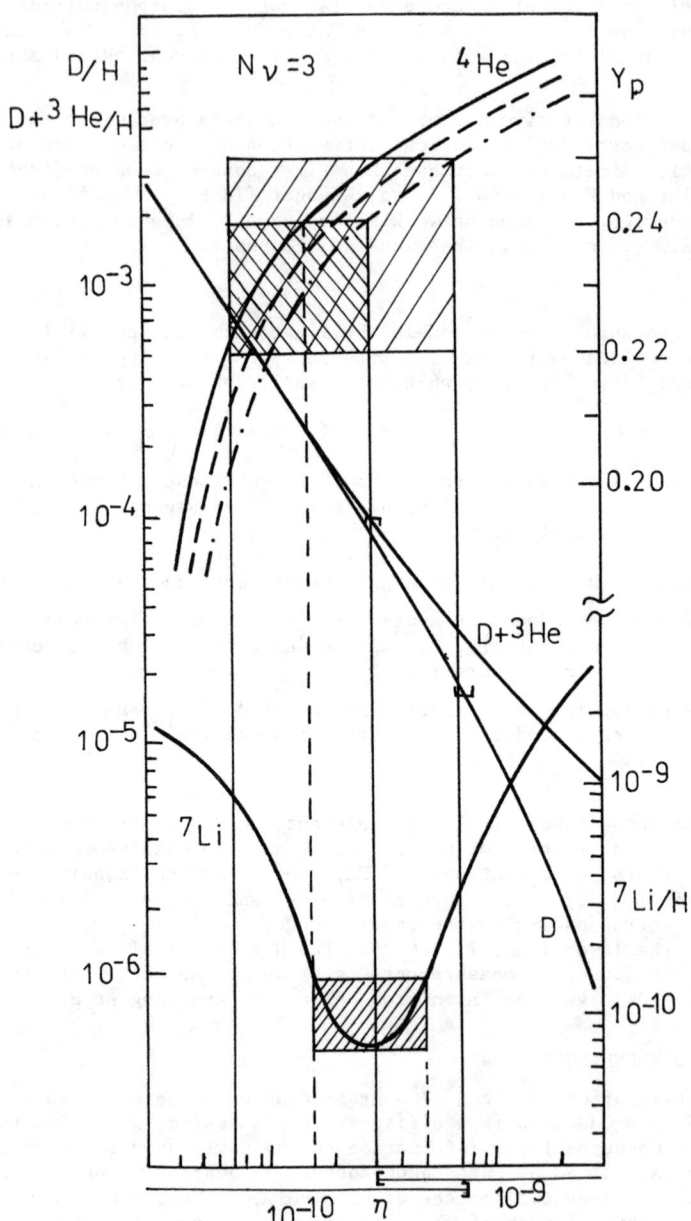

Figure 1.- Theoretical light element abundances as a function of η. Boxes are observations with their errors (see text).

a galactic evolution model.

We present a short review of the observed abundances of the light elements produced by PN. A recent extensive and critical review can be found in Yang et al[4]).

3.1 Deuterium

($I/H \equiv N(I)/N(H)$; $N(I) \equiv$ atoms number density of element I).

Deuterium is observed in two main sites: inside the solar system and in interstellar space.

3.1.1 Solar system (abundances in the protosolar nebula).

i) Assuming that D was burned to ^3He in the sun, solar wind ^3He is the sum of pregalactic D and ^3He. Quoting Geiss and Reeves[11]):

$$\left(\frac{D + {}^3He}{{}^4He}\right)_{Pre\odot} = (4.3 \pm 0.3) \, 10^{-4} \tag{3.1}$$

In carbonaceous chondrites, D is unprocessed. They have been used by Black[12] to estimate:

$$\left(\frac{{}^3He}{{}^4He}\right)_{Pre\odot} = (1.5 \pm 1.0) \, 10^{-4} \tag{3.2}$$

Yang et al[4]) conclude that:

$$\left(\frac{D}{{}^4He}\right)_{Pre\odot} = (2.8 \pm 0.8) \, 10^{-4} \tag{3.3}$$

Taking $0.08 \lesssim ({}^4He/H)_\odot \lesssim 0.1$, this gives

$$\left(\frac{D}{H}\right)_{Pre\odot} \gtrsim 1.6 \, 10^{-5} \tag{3.4}$$

ii) Giant planets atmosphere:

Evaluation of D/H have been recently carried out from Voyager data. We quote the following results:

Jupiter $(D/H)_{Pre\odot} = 3.6 \, {}^{+1.0}_{-1.4} \, 10^{-5}$ Kunde et al.[13] (3.5)

$(D/H)_{Pre\odot} = 1.2 - 3.1 \, 10^{-5}$ Encrenaz and Combes[14] (3.6)

Saturn $(D/H)_{Pre\odot} = 2.4 \, {}^{+2.4}_{-1.3} \, 10^{-5}$ Courtin[15] (3.7)

They are all consistent with the bound (3.4). As D is destroyed during galactic evolution, these are lower limits on the pre-galactic D/H.

3.1.2 Interstellar space (abundances found now).

i) Observation of deuterated molecules in molecular clouds show that the ratio deuterated molecules/hydrogenic molecules decreases when one goes from external to internal regions of the galaxy. This would be an argument favouring a pregalactic D origin[16].

ii) Observation of the D line in the UV at 918 Å with Copernicus satellite.
Data show a big scatter in the D/H determinations in the lines of sight of different stars: differences of as much as a factor of more than ten are obtained[17].
Vidal-Madjar et al[18)19)] attemppted to explain this scatter by a segregation mechanism, which would produce a spatial separation of D and H. On the basis of this mechanism, Brunston et al[20] conclude that the unperturbed interstellar D/H value should be rather high

$$(D/H)_{IS} \simeq 2.25 \cdot 10^{-5} \tag{3.8}$$

Nevertheless, observations towards ϵ Persei showed a drastic variation in less than four hours[21]. This variation has been interpreted as the effect of a high velocity transient H component, ejected by the star. The same situation might have occurred in other determinations, and as it blends D lines inducing an apparent increase in the D/H values, the unperturbed value should be rather low. Vidal-Madjar et al[21] give:

$$(D/H)_{IS} = (5. \pm 3.) \cdot 10^{-6} \tag{3.9}$$

significantly lower than the protosolar value. This difference has to be explained by a galactic evolution model.

York[22] maintains a different point of view. He argues that for the cleanest lines of sight there is no scatter, and he suggests taking

$$(D/H)_{IS} \simeq 2. \cdot 10^{-5} \tag{3.10}$$

as the average interstellar abundance.

3.2 ^{4}He.

The ^{4}He abundance can be determined with much higher precision than the other light element abundances, because of its higher cosmic concentration. This is the reason why ^{4}He is a key constraint in physics theories, by comparing in detail PN predictions and data. Many observations have been made in different astrophysical sites (sun and solar system, old stars, ionized regions of the interstellar medium, other galaxies, etc.). Only the most significative will be mentioned in this talk.

3.2.1 <u>Jovian atmosphere</u> ($Y = X_{^4He}$ = mass fraction of ^{4}He). Gautier et al[23] have determined the ^{4}He in Jupiter from Voyager data. They found:

$$Y_J = 0.19 \pm 0.05 \tag{3.11}$$

As the jovian ^{4}He content has a stellar component, they conclude that the pregalactic value is upper bounded by:

$$Y_p \leq 0.24 \tag{3.12}$$

This limit is reliable as long as a differentiation of helium from hydrogen has not occurred in the jovian atmosphere (see ref.24 for a discussion).

3.2.2 <u>H II regions</u>. Y can be determined through optical and ratio recombination lines in galactic or extragalacti H II regions, with different metal content (that is, in different stages of chemical evolution). A correlation between Y and the metallicity Z would be expected on the basis of current evolution models, so that an extrapolation towards Z = 0 would give the pregalacti ^{4}He abundance. Nevertheless, presently the observational situation is ambiguous. While some authors find this correlation (Peimbert and Torres-Peimbert[25][26]; Lequeux et al.[27]; French[28]; Peimbert et al[29])), no evidence for an un ambiguous correlation is found by other observers (Dufour[30]; Talent and Dufour[31]; Kinman and Davidson[32], Kunth[33]; Kunth and Sargent [34]; Shaver et al[35])). Among them, Kunth's data are particularly significant, because of their accuracy; only an upper limit, as an average of different galaxies content, can be derived. He gives

$$Y_p \leq 0.245 \pm 0.003 \tag{3.13}$$

A review of the measurements of different authors has been made by Pagel[36]. As an average of all results, he concludes that

$$Y_p = 0.23 \pm 0.01 \tag{3.14}$$

The low boundary $Y_p \geq 0.22$ is consistent with all current data.

3.3 ^{7}Li.

^{7}Li has been measured at the surface of very old halo stars (Spite and Spite[37]; Spite et al.[38]). On the basis of statistical arguments (their statistics have been improved by their recent observations), these authors maintain that ^{7}Li has not been depleted by nuclear reactions in the inner layers of these stars, and in consequence, their determination corresponds to the ^{7}Li abundance at the begining of the galaxy life. They give:

$$^7Li/H = (1.12 \pm 0.38) \, 10^{-10} \tag{3.15}$$

The solar system and interstellar value is 10-20 times higher[39].

3.4 ^{3}He.

As has been said, presolar ^{3}He abundance is determined from carbonaceous chondrites (see (3.2)).

Recent observations in H II regions[40] show a higher ^{3}He abundance, as would be expected after stellar production (see ref. 4 for an outline of ^{3}He abundance evolution in the galaxy).

3.5. Pregalactic Abundances.

In order to make a guess at the pregalactic abundance of light elements, we need a model for the chemical evolution of the galaxy, which allows us to extrapolate observed abundances up to pregalactic times.

In the frame of Vigroux's models[41], ^{7}Li and D observations can be fitted (Gry et al.[42]; Andouze et al.[43]) by assuming ^{7}Li production by novae and/or red giants, and D free material infall. In this situation, in order to explain the difference between the protosolar and interstellar D abundances (see (3.5) and (3.9)), Gry et al.[42] find a primordial D concentration of:

$$(D/H)_p = (1.33 \pm 0.67) \, 10^{-4} \tag{3.16}$$

We have to keep in mind that this value, as a model dependent quantity, is uncertain. Table 1 gives a summary of limits and guesses for primordial D, ^{4}He, ^{7}Li and ^{3}He abundances.

Table 1.- Pregalactic abundances of big-bang elements.

Element	Pregalactic abundance	References
D	$(D/H)_p = (1.33 \pm 0.67) 10^{-4}$ $(D/H)_p \geq 1.6 \; 10^{-5}$	42 11, 12
^{4}He	$Y_p \leq 0.24$ $Y_p = 0.245 \pm 0.03$	23, 36 33
^{7}Li	$(^7Li/H)_p = (1.12 \pm 0.38) 10^{-10}$	37, 38
^{3}He	$(^3He/H)_p \leq 8.4 \; 10^{-5}$	4

4. COMPARISON BETWEEN THEORETICAL AND OBSERVED ABUNDANCES

In fig. 1 we show the theoretical yields for D, ^{7}Li and ^{4}He production as a function of $\eta = n_B/n_\gamma$, and the boxes for the observed values. As the light elements have been synthesized at the same time, a single interval of η values should be able to account for the observed abundances. We see that this is the case (Yang et al.[4]; see however, Rana[44], Gautier and Owen[24], Vidal-Madjar and Gry[17]) for the opposite point of view). Nevertheless, the ^{7}Li and ^{4}He abundances are only marginally coherent if $\tau_{1/2}$ = 10.8 min and $Y_p \leq 0.24$, and if future ^{4}He determinations favoured a lower Y_p value, this compatibility would be destroyed.

5. SOME IMPLICATIONS OF BIG-BANG NUCLEOSYNTHESIS

5.1 The Baryon Density Of The Universe

Thermonuclear reaction rates involved in PN calculations depend on the baryon density, which implies a dependence on η of the light element production. The D abundance is particularly sensitive to this parameter, as fig. 1 shows. The primordial D yield decreases steeply as the baryon density grows, so that a lower limit on X_{Dp} results in an upper limit on η. Taking the protosolar value:

$$(D/H)_p \gtrsim (D/H)_{Pre\odot} \gtrsim 1.6 \; 10^{-5} \tag{5.1}$$

(suggested by solar system data) as a lower limit, we have

$$\eta \leq 8 \cdot 10^{-10} \tag{5.2}$$

This value of the present baryon density is lower than the critical density needed to close the universe:

$$\rho_c = 1.9 \; 10^{-29} \; h^2 \; gr \; cm^{-3} \; , \; \frac{1}{2} \leq h \leq 1 \tag{5.3}$$

Big-bang nucleosynthesis then tells us that the universe cannot be closed by baryons.

Recently Yang et al[4] have derived a lower bound on η from ^{3}He and D abundance determinations taken together. They find:

$$\eta \gtrsim 3 \cdot 10^{-10} \tag{5.4}$$

This limit is important because, together with ^{4}He measurements, it allows limiting of the number of different lepton familes (Olive et al.[45]; Yang et al.[4]). Let us now turn to this problem.

5.2 The Number Of Two Component Neutrino Species.

The neutron to proton abundance ratio freezes out at a value

$$\left(X_n/X_p\right)_f = e^{-Q/kT_*} \tag{5.5}$$

where T_* is the freeze-out temperature (see section 2), determined by a competition between the expansion rate $t^{-1}(T)$ and the reaction rate $\lambda(T)$.

The presence of new neutrino species implies a faster expansion rate and hence a freeze out at higher temperatures, which leaves a higher abundance of surviving neutrons. Since most neutrons are incorporated into ^{4}He nuclei in the synthesis process, more ^{4}He will be formed, and comparison with observed ^{4}He abundance results in limits on the number of different two component neutrino families.

Table 2.- *Maximum number of light neutrino species allowed for different values of* $\tau_{1/2}$, $\eta_{10} \equiv 10^{10}\eta$ *and* Y_p.

$\tau_{1/2}$ =10.4min		$N_\nu \leq$	$\tau_{1/2}$ = 10.8min		$N_\nu \leq$
η_{10}	$Y_p \leq$		η_{10}	$Y_p \leq$	
3	0.25	4	3	0.25	3
3	0.24	3	3	0.24	2
6	0.25	3	6	0.25	2
6	0.24	2	6	0.24	2
10	0.25	2	10	0.25	2
10	0.24	2	10	0.24	1

Table 2 displays these limits for different values of the neutron half life $\tau_{1/2}$, the nucleon-to-photon ratio η (allowed interval) and the upper bounds on Y_p depicted in fig. 1. Only one more light neutrino type is allowed beyond the τ neutrino, and even this fourth family is excluded for $\tau_{1/2} > 10.4$ min.[46] and/or $\eta_{10} > 3$.

6. HELIUM ABUNDANCE AND THE GRAVITINO PROBLEM.

Among the particle physics theories newly proposed, the supersymmetric extensions of the standard model engage a great deal of interest at present. (For recent reviews, see Haber and Kane[47]; Ibáñez[48] and Nilles[49]). One of the most interesting and appealing such theories at this time is N = 1 supergravity coupled to matter, with the supersymmetry broken via gravitational couplings. In these frames, the gravitino (supersymmetric partner of the graviton) has a mass of order $m_G \sim 10^2$ GeV. However, and in order to ensure coherence with PN yields and the average mass density of the universe, in a standard (without inflation) cosmological scenario, gravitino mass ranges from about 10^4 GeV to 1KeV are excluded. The reason for the first limit is that gravitino-gravitino annihilation is negligible as long as $m_G < 0.3$ Mp (Mp is the Planck mass)[50], so that the gravitino number density n_G may only be depleted by decays into lighter supersymmetric particles. The decay rate has been calculated by Krauss[51] and is given by:

$$\tau^{-1} \simeq \frac{N_D \, m_G^3}{2\pi \, M_P^2} \tag{6.1}$$

(N_D is the number of possible decay modes). We see that for $m_G \lesssim 10^4$ GeV, gravitino decays after PN, disturbing its out-puts, which have been shown to be in reasonable agreement with observations. We need a n_G suppression by many orders of magnitude to account for observed light element abundances. (That is what is known in literature as "the gravitino problem").

Primordial inflation is a natural way to solve this question, which is similar to the monopole problem[52]. However, and unlike the monopole case, gravitinos may be produced after inflation, mainly by processes of the form $X + \tilde{\gamma} \to X + G$, where $\tilde{\gamma}$ is a gauge fermion and X are scalar particles. We get in this way a gravitino abundance just before decay[53]:

$$Y_G = 2\alpha \, T_R / \sqrt{N(T_R)} \, M_P \tag{6.2}$$

where $Y_G = n_G/n_\gamma$, T_R is the reheating temperature, $\alpha \sim 10^{-2}$ is the fine structure constant and N(T) the effective number of degrees of freedom at temperature T.

In order to create a baryon asymmetry we need $T_R \gtrsim O(10^{10})$ GeV[53], so that cosmological inflation might have depressed Y_G as much as

$$Y_G \gtrsim 10^{-12} \tag{6.3}$$

without conflicting with baryon number production requirements. But we must ask ourselves if this lower limit on Y_G is coherent with light element abundance observations, which result in upper limits on Y_G. The evaluation of these upper bounds is set out below.

PN yields are affected by the presence and decay of a gravitino population because:

a) They increase the total density of the universe.

b) They modify the baryon density at nucleosynthesis time. In fact, an effective baryon density parameter h_{eff} appears given by:

$$h_{eff} = \left[\frac{(TR)_a}{(TR)_b}\right]^3 h_o \quad (6.4)$$

which plays the same role as h_o in the adiabatic case (Here "a" and "b" mean after and before the decay, respectively). As $(TR)_a \geq (TR)_b$, due to entropy production, more ^{4}He and less D will be synthesized.

c) The decay products act on the preexisting nuclei causing photodissociation[54)55)] and/or ^{4}He annihilation[56)57)]. Let us examine the possible decay modes into gauge bosons and their superpartners, which are the dominant modes:

(i) $G \to \tilde{\gamma}\gamma$ (photino-photon)

The photino is generally assumed to be absolutely stable, as the lightest supersymmetric particle[58)].

(ii) $G \to \tilde{g}g$ (gluino-gluon)

This channel is the prototype for the situation where the superpartner is unstable, and since $\tilde{g} \to q\bar{q}\tilde{\gamma}$ is likely to be the dominant $\tilde{g}$ decay mode[47)], this gives rise to a significant amount of proton-antiproton (p-$\bar{p}$) pairs. The analogy between the processes:

$$G \to p\bar{p} \text{ anything}$$

and

$$e^-e^+ \to p\bar{p} \text{ anything}$$

suggests taking the relative yield of $\bar{p}$ per gravitino decay:

$$r_{\bar{p}} = \frac{\Gamma(G \to p\bar{p} \text{ anything})}{\Gamma(G \to \text{anything})} \quad (6.5)$$

to be of the same order as[56)]:

$$r'_{\bar{p}} = \frac{\sigma(e^-e^+ \to p\bar{p} \text{ anything})}{\sigma(e^-e^+ \to \text{anything})} \quad (6.6)$$

which has been measured to be 0.3 - 0.4[59)], and 0.6[60)].

Annihilation of these $\bar{p}$ with H and ^{4}He (as the most abundant nuclei), will result in a ^{4}He destruction and a post-big bang creation of ^{3}He and D.

p$\bar{\text{p}}$ pairs may also be found in decays of SU(2) gauge bosons superpartners[61)62)].

A fraction f of the gravitino energy will end up at a distribution of high energy photons, in any of the previous situations. They lose their energy through e^-e^+ pair production and Compton scattering and then they photodissociate nuclei.

The consequences of the gravitino decay on preexisting element abundance differ according to whether it results in p$\bar{\text{p}}$ production or not. They need a separate treatment and we consider them in turn:

A.- No p$\bar{\text{p}}$ production ($G \to \tilde{\gamma}\gamma$).
In this case the dynamics of the universe is modified (effects a) and b)) and nuclei are photodisintegrated.

This last effect has been studied by Lindley[54)] and Ellis et al.[55)]. The most stringent limit comes from D destruction, and it conflicts with baryon number production requirements, $Y_G \lesssim 10^{-12}$, only for low present baryon densities and/or high m_G values.

Let us now properly evaluate the Y_G suppression needed in order that effects a) and b) are not operative.

As the decay proceeds, n_G decreases. Its evolution is described by the equation:

$$\frac{dn_G}{dt} = -3\frac{\dot{R}}{R} n_G - \frac{1}{\tau} n_G \qquad (6.7)$$

The decay photons thermalize instantaneously on a cosmological scale. As a consequence, an energy transfer $\mathcal{E}_\gamma = m_G [1 - (m_{\tilde{\gamma}}/m_G)^2]/2$ from the population of decoupling gravitinos to the photon sea, follows each decay, which leads to an entropy production. The temperature evolution follows from the second principle of Thermodynamics, and is given by

$$\frac{dT}{dt} = -T\frac{\dot{R}}{R} + \frac{1}{4\rho_\gamma} T \frac{n_G}{\tau} \mathcal{E}_\gamma \qquad (6.8)$$

As for the decay photinos, the universe is too cold for them to annihilate or get thermalized[63)], so that they will follow the expansion as a decoupled component. Their number density, $n_{\tilde{\gamma}}$, is controlled by the equation:

$$\frac{dn_{\tilde{\gamma}}}{dt} = -3\frac{\dot{R}}{R} n_{\tilde{\gamma}} + \frac{1}{\tau} n_G \qquad (6.9)$$

The dynamical description of the universe is completed by the Einstein equation (2.1), with

$$\rho = \rho_\gamma + \rho_\nu + n_G m_G + n_{\tilde{G}} m_{\tilde{G}} + \rho_B \qquad (6.10)$$

The relevant quantity here with regard to light element production is h_{eff} (see eq.(6.4)). It has been computed by numerically integrating eqs. (6.7), (6,8), (6.9) and (2.1) and the results of this calculation show that if $Y_G \sim 10^{-10}$, no entropy will be significantly produced by gravitino decay; in this case, the dynamics of the universe (effect a)) will not be disturbed by the G population at PN time.

B.- Let us now analyze the situation where $p\bar{p}$ pairs are produced, which will result in ^{4}He destruction and D and ^{3}He formation.

The abundances of ^{4}He, p and $\bar{p}$ as the decay occurs are given by the equations of nuclear reaction kinetics, which can be written:

$$\frac{dY_{^4He}}{dt} = - Y_{^4He} Y_{\bar{p}} \rho_B N_A \langle \sigma_{^4He,\bar{p}} v \rangle \qquad (6.11)$$

$$\frac{dY_{\bar{p}}}{dt} = - Y_{^4He} Y_{\bar{p}} \rho_B N_A \langle \sigma_{^4He,\bar{p}} v \rangle - Y_p Y_{\bar{p}} \rho_B N_A \langle \sigma_{p,\bar{p}} v \rangle$$
$$+ \frac{\sigma_{\bar{p}} n_G^b}{N_A \rho_B^b} \frac{e^{-t/\tau}}{\tau} \qquad (6.12)$$

$$\frac{dY_p}{dt} = - Y_p Y_{\bar{p}} \rho_B N_A \langle \sigma_{p,\bar{p}} v \rangle + \alpha Y_{^4He} Y_{\bar{p}} \rho_B N_A \langle \sigma_{^4He,\bar{p}} v \rangle$$
$$+ \frac{\sigma_{\bar{p}} n_G^b}{N_A \rho_B^b} \frac{e^{-t/\tau}}{\tau} \qquad (6.13)$$

where $Y_i = n_i/\rho_B N_A$, N_A is Avogrado's number, n_i is the number density of nucleus i, $<\sigma_{i,\bar{p}} \cdot v>$ is the product of cross section for the annihilation reaction of nucleus i by $\bar{p}$ and the relative velocity averaged over the velocity distribution, α represents the fraction of $\bar{p}$ resulting from the process $\bar{p} + {}^4He \to$ anything, and a 'b' superindex means before G decay.

The annihilation cross-section for $p\bar{p}$ has been taken to be [64]:

$$\beta \sigma_{p,\bar{p}} = 32.83 + 0.096 E \text{ millibarns} \tag{6.14}$$

with $\beta = v/c$ and E in MeV. This fit is valid for E lower than 10 MeV.

As for $\sigma_{{}^4He,\bar{p}}$, it can be evaluated by means of the Born approximation to be:

$$\sigma_{{}^4He,\bar{p}} \simeq 2.5 \, \sigma_{p,\bar{p}} \tag{6.15}$$

The system of eqs. (6.11), (6.12) and (6.13) has been numerically integrated. We have taken as initial conditions (prior to decay) the PN theoretical yields relative to a baryon density determined by h_{eff} (see eq. (6.4)); the $\bar{p}$ initial abundance has been taken to be equal to zero. Fig. 2 depicts the results of the calculations: the amount of ${}^4He$ destroyed by antiprotons as a function of Y_G, for $r_{\bar{p}} = .6$ and different values of the present baryon density (this amount is not appreciably sensitive to the parameters α and m_G).

Table 3.- Maximum Y_G allowed for a pregalactic ${}^4He$ mass concentration of $X_{{}^4He} \geq .22$ for different values of the present baryon density parameter h_o and $r_{\bar{p}} = .6$. For other $r_{\bar{p}}$ values, Y_G^{max} can be deduced taking into account that $Y_G^{max} r_{\bar{p}} = cte.$

log h_o	-3.5	-4.0	-4.5	-5.0
Y_G^{max}	5.10^{-10}	$1.3 \, 10^{-10}$	$2.5 \, 10^{-11}$	$2.5 \, 10^{-12}$

Table 3 gives the maximum Y_G allowed in order to have a pregalactic ${}^4He$ mass concentration of $(X_{{}^4He})_p \geq .22$, for a neutron half life of 10.4 sec. and $N_\nu = 3$. These Y_G^{max} values are compatible with baryon asymmetry production requirements, except for the case of low baryon density, where Y_G^{max} is close to the limit $Y_G \gtrsim 10^{-12}$. We conclude that the inflationnary scenario might solve the gravitino problem insofar as the ${}^4He$ abundance observations are concerned.

Now, if we intend to limit Y_G by requiring that not too much D or ${}^3He$ would be produced by ${}^4He$ destruction, two uncertainties are found: the pregalactic D or ${}^3He$ abundance and the fraction of D or ${}^3He$ formed

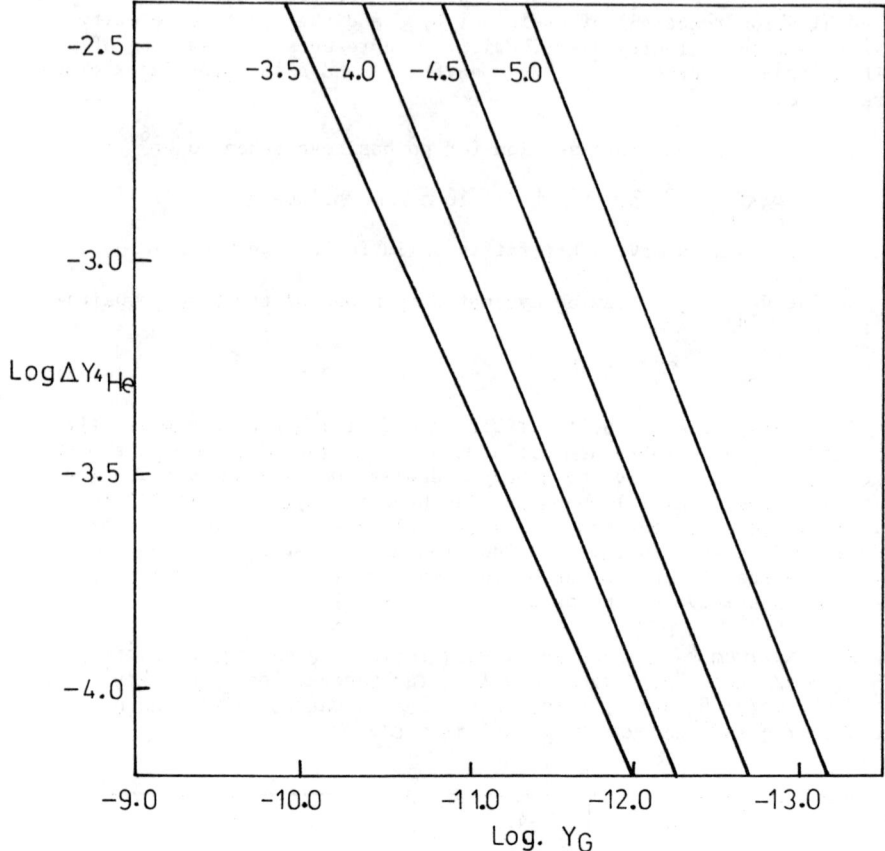

Figure 2.- The number fraction of ^4He destroyed by $\bar{p}$ against the relative gravitino number density. $\Delta Y_{4He} = Y_{4He}(T_b) - Y_{4He}(T_a)$, b = before and a = after gravitino decay. The results are depicted for log h_o (present baryon density parameter) = -3.5, -4.0, -4.5 and -5.0, and $r_{\bar{p}}$ = .6.

by ^{4}He annihilation (f_D or f_{3He}), now being measured in CERN[56].

As for the pregalactic value of X_D or X_{3He}, only upper and lower bounds are available. Following the analysis by Yang et al.[4], taking the fraction of ^{3}He which survives processing through stars to be $\geq 1/2$ and $0.75 \leq X_H \leq 0.78$, we find:

$$(X_D)_p \gtrsim 2.4 \cdot 10^{-5} \tag{6.16}$$

$$(X_{3He})_p \lesssim 1.1 \cdot 10^{-4} \tag{6.17}$$

As an example, let us suppose that

$$(X_D)_p = 4 \cdot 10^{-5} \tag{6.18}$$

$$f_D = .1 \tag{6.19}$$

The amount of D formed by ^{4}He destruction is given by:

$$\Delta Y_D = f_D \Delta Y_{^4He} \tag{6.20}$$

Fig. 2 then tells us that $Y_G^{max} \lesssim 10^{-12} - 10^{-13}$, which conflicts with baryosynthesis requirements.

7. CONCLUSIONS

We have examined the physics involved in the PN process and made a short review of the present status of the abundance determination problem. Comparison of predicted and observed abundances allows us to constrain astrophysical and particle physics parameters. In the frame of the standard big-bang model, these are, among others, the present baryon density and the number of two component light neutrino families.

In the frame of supersymmetric theories, gravitino decay disturbs PN outcomes if the gravitino density is not suppressed. Cosmological inflation is a natural way to solve this problem, but as gravitinos are produced after inflation, we have to look for compatibility between baryogenesis and gravitino number deplection. The results presented in this talk lead to the following conclusions:

i) If only the $G \to \tilde{\gamma}\gamma$ channel is open, the most restrictive upper limit on Y_G known at present comes from D photofission. In this case a gravitino of $m_G \sim O(100 \text{ GeV})$ can live in an inflationary scenario except for low present baryon density and/or high m_G values, in which case the compatibility with baryon number generation is only marginal.

ii) If $\bar{p}p$ pairs are produced (which is certainly the case if $m_G > m_{\tilde{g}}$), some difficulties arise, that imply constraints on the class of theories where the gravitino problem has a solution.

NOTE AND ADDED IN PROOF

While this manuscript was being edited, results for the $f_{3_{He}}$ measurement at CERN became available[65]. Defining $f_{3_{He}}$ as:

$$f_{3_{He}} = \frac{\sigma(\bar{p} + {}^4He \rightarrow {}^3He + anything)}{\sigma_r(\bar{p} + {}^4He \rightarrow anything)}$$

where σ_r is the reaction cross section, they give:

$$f_{3_{He}} = 0.355 \pm 0.044$$

Taking the limit (6.17) for $X_{3_{He}}$, expression (6.20) and fig. 2 result in the bounds on Y_G displayed in Table 4 for different values of h_o. The general conclusions previously stated are not changed.

Table 4.- Y_G^{max} compatible with ${}^3He$ (and D) abundance observations, for various baryon densities and $r_{\bar{p}/p} = .6$.

log h_o	-3.5	-4.0	-4.5	-5.0
Y_G^{max}	2.1 10^{-12}	1. 10^{-12}	3.5 10^{-13}	1.2 10^{-13}

REFERENCES

1.- J.R. Bond, B.J. Carr and W.D. Arnett, Nature 304, 514 (1983).

2.- J. Audouze and J. Silk, in *Proceedings of the ESO Workshop on Primordial Helium*, ed. P.A. Shaver, D, Kunth and K. Kjär (Garching, European Southern Observatory, 1983).

3.- G. Beaudet and H. Reeves, in *Proceedings of the ESO Workshop on Primordial Helium*, ed. P.A. Shaver, D. Kunth and K. Kjär (Garching, European Southern Observatory, 1983).

4.- J. Yang, M.S. Turner, G. Steigman, D.N. Schramm and K.A. Olive Ap. J. 281, 493 (1984).

5.- P.J.E. Peebles, Ap. J. 146, 542 (1966).

6.- R.V. Wagoner, W.A. Fowler and F. Hoyle, Ap. J. 148, 3 (1967).

7.- S. Weinberg, in *Gravitation and Cosmology*, Wiley, New York (1972).

8.- R.V. Wagoner, Ap. J. Suppl. 18, 247 (1969).

9.- R.V. Wagoner, Ap. J. 179, 343 (1973).

10.- D.N. Schramm, Phil. Trans. Roy. Soc. A. 307, 43 (1982).

11.- J. Geiss and H. Reeves, Astron. Astrophys. 18, 126 (1972).

12.- D.C. Black, Geochim. Cosmochim. Acta 36, 347 (1972).

13.- V. Kunde et al., Ap. J. 263, 443 (1982).

14.- T. Encrenaz and M. Combes, Icarus 52, 54 (1982).

15.- R. Courtin, thesis, Univ. Paris VII (1982).

16.- J.P. Ostriker and B.M. Tinsley, Ap. J. Letters 201, L51 (1975).

17.- A. Vidal-Madjar and C. Gry, Lab. de Physique Stellaire et Planétaire (Verrieres-le-Buisson) preprint (1983).

18.- A. Vidal-Madjar, J. Audouze, P. Bruston and C. Laurent, La Recherche 8, 617 (1977).

19.- A. Vidal-Madjar, C. Laurent, P. Bruston and J. Audouze, Ap. J. 223, 589 (1978).

20.- P. Bruston, J. Audouze, A. Vidal-Madjar and C. Laurent, Ap. J. 243, 161 (1981).

21.- A. Vidal-Madjar, C. Laurent, C. Gry, P. Bruston, R. Ferlet and D.G.

York, Astron. Astrophys. 120, 58 (1983).

22.- D.G. York, private communication to J. Yang et al. 1984 (1983).

23.- D. Gautier et al., J. Geophys. Res. 86, 8713 (1981).

24.- D. Gautier and T. Owen, Nature 302, 215 (1983).

25.- M. Peimbert and S. Torres-Peimbert, Ap. J. 193, 327 (1974).

26.- M. Peimbert and S. Torres-Peimbert, Ap. J. 203, 581 (1976).

27.- J. Lequeux, M. Peimbert, J.F. Rayo, A. Serrano and S. Torres-Peimbert, Astron. Astrophys. 80, 155 (1979).

28.- H.B. French, Ap. J. 240, 41 (1980).

29.- M. Peimbert, S. Torres-Peimbert and J.F. Rayo, Ap. J. 220, 516 (1978).

30.- R.J. Dufour, Ap. J. 195, 315 (1975).

31.- D.L. Talent and R.J. Dufour, Ap. J. 233, 888 (1979).

32.- T.D. Kinman and K. Davidson, Ap. J. 243, 127 (1981).

33.- D. Kunth, Ph. D. thesis, Université Paris VII (1982).

34.- D, Kunth and W.L.W. Sargent, Ap. J. 273, 81 (1983).

35.- P.A. Shaver, M.X. Mc Gee, L.M. Newton, A.C. Danks and S.R. Pottash, M.N.R.A.S. 204, 53 (1983).

36.- B.E.J. Pagel, Phil. Trans. Roy. Soc. London A 307, 19 (1982).

37.- F. Spite and M. Spite, Astron, Astrophys. 115, 357 (1982).

38.- M. Spite, J.P. Maillard and F. Spite, Obs. Meudon preprint (1984).

39.- J. Audouze and H. Reeves in *Essays in Nuclear Astrophysics*, eds. C.A. Barnes et al., Cambridge University Press (1982).

40.- T.L. Wilson, R.T. Rood and T.M. Bania, in *Proceedings of the ESO Workshop on Primordial Helium*, ed. P.A. Shaver, D. Kunth and K. Kjär (Garching, European Southern Observatory, 1983).

41.- L. Vigroux, Ph.D. thesis, Université de Paris XI (1979).

42.- C. Gry, G. Malinie, J. Audouze and A. Vidal-Madjar, in *Proceedings of the ESO Workshop on Primordial Helium*, ed. P.A. Shaver, D. Kunth and K. Kjär (Garching, European Southern Observatory, 1983).

43.- J. Audouze, O. Boulade, G. Malinie and Y. Poilane, Astrophys. 127, 164 (1983).

44.- N. Rana, Phys. Rev. Lett. 48, 209 (1982).

45.- K.A. Olive, D.N. Schramm, G. Steigman, M.S. Turner and J. Yang, Ap. J. 246, 557 (1981).

46.- D.H. Wilkinson, Progress in Particle and Nuclear Physics, 6, 328 (1979).

47.- H.E. Haber and G.L. Kane, Michigan preprint UM HE TH 83-17(1984).

48.- L.E. Ibañez, Madrid preprint FTUAM 84/17 (1984).

49.- H.P. Nilles, University of Geneva preprint UGVA-DPT 1984/412 (1984).

50.- S.Weinberg, Phys. Rev. Lett. 48, 1303 (1982).

51.- L.M. Krauss, Nucl. Phys. B227, 556 (1983).

52.- J. Ellis, A.D. Linde and D.V. Nanopoulos, Phys. Lett. 118B, 59 (1983).

53.- D.V.Nanopoulos, K.A. Olive and M. Srednicki, Phys. Lett. 127B, 30 (1983).

54.- D. Lindley, M.N.R.A.S. 193, 593 (1980).

55.- J. Ellis, J.E. Kim and D.V. Nanopoulos, CERN preprint TH.3839 (1984).

56.- I.V. Falomkin, G.B. Pontecorbo, M.G. Sapozhnikov, M.Yu Khlopov, F. Balestra and G. Piragino, Nuovo Cimento 79, 193 (1984).

57.- M. Yu Khlopov and A.D. Linde, Phys. Lett 138B, 265 (1983).

58.- J. Ellis, J.S. Hagelin, D.V. Nanopoulos, K. Olive and M. Srednicki, Nucl. Phys. B238, 453 (1984).

59.- S.L. Wu, in *Proceedings of the Conference on Particles and Fields 1981*, Santa Cruz, 1982.

60.- G. Goldhaber and J.M. Weiss, in *Proceedings of the Conference on Particles and Fields 1981*, Santa Cruz, 1982.

61.- J. Ellis, J.S. Hagelin, D.V. Nanopoulos and M. Srednicki, Phys. Lett. 127B, 233 (1983).

62.- W. Bartel et al., JADE Collaboration, DESY preprint 84-038 (1984).

63.- H. Goldberg, Phys. Rev. Lett. <u>50</u>, 1419 (1984).

64.- C.B. Dover, 10 International Conference on Few Body Problems in Physics, Karlsruhe, Germany, 1983.

65.- F. Balestra et al., CERN preprint EP/84-108 (1984).

The X-ray Background

A.C. Fabian

Institute of Astronomy

Madingley Road

Cambridge CB3 OHA

England

INTRODUCTION

The major source of the cosmic X-ray energy density above the earth's atmosphere is the X-ray background. It dominates the intensity of the quiet Sun above ~ 5 keV. The brightest extragalactic X-ray sources are about 1000 times fainter but as their surface brightness is much higher, they are made detectable by the use of collimators and telescopes. Unfortunately for the X-ray background specialist, the recent drive for high sensitivity on small angular scales has meant little progress on studies of the overall properties of the background. Significant advances have been made, however, on the contribution of point source populations to the background intensity. The origin of the X-ray background is still an open question.

The best data on the 3-300 keV X-ray background are from the HEAO-1 satellite (Marshall et al. 1980, Boldt 1981, Gruber 1984). This energy range is our prime concern here. The spectrum of the X-ray background over that band is known with much greater precision than that of the microwave background (De Zotti et al. 1982) and fits a thermal bremsstrahlung model remarkably well (Marshall et al. 1980). This has raised the possibililty that most of the background is truly diffuse and originates in a hot intergalactic plasma (Cowsik & Kobetich 1972, Field & Perrenod 1977, Sherman 1979, Fabian 1981, Fabian & Kembhavi 1982).

The detection of many extragalactic sources, and in particular quasars, with the Einstein Observatory reopened the case for an origin entirely in resolvable sources (Setti & Woltjer 1979, Giacconi et al. 1979, Tananbaum et al. 1979). As the data have been consolidated, however, the situation has again become less certain. There is some contribution by known classes of point source but it need not exceed 20 percent above 3 keV.

Below 1 keV, the X-ray background is dominated by a Galactic contribution which is due, at least in part, to a combination of hot gas and stars (see Fried et al. 1982, Marshall & Clark 1984). The spectrum in the range 1-3 keV is uncertain (Garmire & Nousek 1981). A galactic contribution amounting to a few percent is also observed in the 2-10 keV range (Warwick, Pye & Fabian 1979, Schwartz 1980, Protheroe & Wolfendale 1980). This may be due to inverse Compton scattering of starlight by low energy cosmic-ray electrons. The remaining background is highly isotropic, with any fluctuations remaining after accounting for observed classes of sources being less than 2 percent on scales of 25 square degrees (Shafer 1983, Shafer & Fabian 1983).

An additional component becomes apparent in the spectrum above 100 keV, leading to the so-called MeV bump. This may well be due to the high energy end of the contribution by active galactic nuclei.

A HOT INTERGALACTIC GAS

The spectrum of the X-ray background remains in good agreement with that from thermal bremsstrahlung, even after an active galaxy contribution and adiabatic cooling of the gas have been accounted for (Fabian & Kembhavi 1982). Field & Perrenod (1977) and Sherman (1979) considered a hot intergalactic medium (IGM) heated by quasars at a redshift of $\sim$ 3 which has subsequently cooled adiabatically with the expansion of the Universe. The particle density n varies as $(1+z)^3$ and the temperature T as $(1+z)^2$. The characteristic temperature of the background spectrum is $\sim$ 40 keV. Field & Perrenod (1977) required a maximum temperature at $z_m = 2$ of $\sim$ 120 keV and a particle density now of $\sim 10^{-6}$ cm^{-3}, assuming $q_o = 1/2$ and $H_o = 50$ km s^{-1} Mpc^{-1}. The cooling time of the gas is always much greater than the age of the Universe, H^{-1} and the process is thus very inefficient. It is therefore also very extravagant in energy and, on these grounds, several workers find that a hot IGM is implausible (e.g. McKee 1980).

Guilbert and I (1985) have reconsidered this hypothesis including the enhanced emissivity at the implied mildly-relativistic temperatures of kT approaching $m_e c^2$. Electron-electron bremsstrahlung is then important and electron-ion bremsstrahlung is enhanced. The particle densities at the current epoch are up to a factor of 2 smaller than those required by Field & Perrenod (1977).

We find that $0.2 < \Omega_g < 0.3$ where Ω_g is the fraction of the closure density due to this hot gas; the product of $H^3 \Omega_g$ is constant. The hot IGM could be the dominant baryonic matter in the Universe. Its present temperature and density may be lower than 10^8 K and 7×10^{-7} cm^{-3} respectively. Compton scattering of the microwave background is a severe coolant of the hot gas at $z_m > 5$. The gas has to be heated at lower redshifts in order that the resultant Compton cooling does not steepen the X-ray background spectrum to an unacceptable degree. The microwave background develops a short wavelength tail.

There are major problems. The cooling time is still $\gg H^{-1}$ and $\sim 10^{64}$ erg/large spiral galaxy is required. This far exceeds the binding energy of such galaxies, and constitutes $\sim 10^{-2}$ of their rest mass and $< 10^{-3}$ of the rest mass in the IGM. This is extravagant, but so are most active galaxies. The good agreement of thermal bremsstrahlung models relies to a certain extent on the underlying assumption of a Maxwellian electron distibution. This is reasonable as the 2-body equilibration timescales for electron-electron energy exchange $< H^{-1}$. (The electron-ion timescale $> H^{-1}$, however.) Perhaps the worst problem is the requirement that $\Omega_g > 0.2$ which conflicts with current deuterium abundance estimates of $\Omega_b < 0.19$ (Yang etal. 1984). The hot gas hypothesis for the X-ray background is ruled out unless the energy and matter density problems can be overcome. (Some, otherwise unobserved, epoch of enormous energy and deuterium production is required.)

Clumping can make the gas radiate much more efficiently but generates observable spatial fluctuations in the X-ray background (McKee 1980, Fabian 1981) and, through the Sunyaev-Zeldovich effect, also in the microwave background (Guilbert & Fabian 1985).

UNEVOLVED EXTRAGALACTIC X-RAY SOURCES

Although we are primarily concerned here with their contribution to the X-ray background, I shall briefly outline some of the observational characteristics and consequences of each class of source that are of particular relevance to the rest of this School.

a) Normal Galaxies

Many nearby normal galaxies have been detected as X-ray sources by the Einstein Observatory. The X-ray emission from the normal stellar component is presumably negligible but populations of X-ray binaries and supernova remnants as well as any hot interstellar medium were observable. The X-ray luminosity of a normal galaxy is typically about 0.01 percent of its visual luminosity. The emission from late-type galaxies, is mostly due to X-ray binaries. Early-type galaxies contain substantial amounts of X-ray emitting gas at $10^7 K$.

The X-ray appearance of our own Galaxy is probably typical of many spiral galaxies. Two major populations of luminous X-ray binaries are evident. Those in the disc of the galaxy are mainly associated with OB stars and are due to accreting neutron star or black hole binary companions (e.g. Cen X-3 and Cyg X-1). The Population II sources concentrated in the 'bulge' and in the globular clusters are in orbit about lower mass stars. The most luminous sources are not much more powerful than 10^{38} erg s^{-1}, which is the Eddington limit for 1 $M_\odot$ objects. Galactic supernova remnants appear to be at least a factor of 10 - 100 times less luminous.

The Magellanic Clouds are much richer, per unit mass, in luminous Population I X-ray binaries than our galaxy (Clark et al. 1978). M31 is much richer in X-ray luminous globular cluster and central bulge sources (van Speybroeck et al. 1979). There are no clear answers to these differencs, although abundance variation may be an important factor. A number of other nearby spiral galaxies have been resolved in X-rays (Long & Van Speybroeck 1983 and references cited therein).

Some irregular galaxies in which star formation is particularly active have ratios of X-ray to visual luminosity of between 10^{-1} and 10^{-2} (Fabbiano et al. 1982, Stewart et al. 1982). It is presumed that many massive X-ray binaries are formed in these regions, although there is no firm evidence.

Extended X-ray emission from a number of early-type galaxies in the Virgo cluster was discovered soon after the Einstein Observatory was launched (Forman et al. 1979). This was in addition to the well-known diffuse X-ray source centred on M87. The X-ray 'plume' to M86 (Forman et al. 1979) suggested that much of the emission was due to hot gas, which in that case was being pushed out of M86 by the ram pressure of the intergalactic (or circum-M87) gas. Feigelson et al. (1981) argue convincingly that the diffuse X-ray emission found in Centaurus-A (NGC

5128) is most unlikely to be due to stars or to X-ray binaries.

Further observations by Bechtold et al. (1983), Biermann & Kronberg (1983), Nulsen et al. (1984) and, in particular, Forman, Jones & Tucker 1985, have produced a list of 30 or more X-ray detected elliptical and S0 galaxies. Although many of these are in the Virgo cluster, there appears to be no strong environmental influence provided they are not in a rich cluster. Reasonably isolated elliptical galaxies have similar X-ray to optical luminosity ratios (L_X/L_V) to those in groups and loose clusters. The dependence of L_X on L_V is steeper ($L_X \propto L_V^\alpha$, where $3/2 \leq \alpha \leq 2$) than for spiral galaxies, where $\alpha \sim 1$. This argues further against a binary source origin (Forman, Jones & Tucker 1985).

The X-ray spectra of early-type galaxies is best fit by radiation from hot gas with a temperature of about 10^7 K. The emission extends for at least 10 kpc in most cases and may extend much further in some well-studied cases (Forman, Jones & Tucker 1984). The mass of X-ray emitting interstellar gas in elliptical galaxies is at least $10^8 M_\odot$ within about 10 kpc, and may be considerably more at larger radii. Most of the gas must be close to hydrostatic equilibrium. If its motion approaches the sound speed, then the mass flow rate must exceed $\sim 10\ M_\odot\ yr^{-1}$ - far too high for either an outflow or an inflow. Galactic winds do not occur in most galaxies.

Consequently, the equation of hydrostatic support can be used to obtain accurate mass profiles for elliptical galaxies from X-ray observations. Basically

$$\frac{dP_{gas}}{dr} = -\rho_{gas}\ g(r)$$

where $g(r)$ is the acceleration due to gravity ($GM(r)/r^2$). The X-ray surface brightness profile and gas temperature distribution give $\rho_{gas}(r)$

and $P_{gas}(r)$ and thus $M(R)$.

$$M = 10^{12} \left(\frac{T}{10^7 K}\right) \left(\frac{R}{20 \text{ kpc}}\right) M_\odot .$$

The relatively high density of the interstellar gas means that its radiative cooling time t_{cool} is less than the Hubble time within $R \sim 20 - 30$ kpc. $t_{cool} \lesssim 10^8$ yr within $\cong 10$ kpc (Nulsen et al. 1984). It is likely that a cooling inflow of gas occurs (for a review see Fabian et al. 1984) involving about $0.1 - 1$ $M_\odot$ yr^{-1}. A rough steady-state may be reached within the galaxy where stellar mass-loss into the interstellar medium is compensated for by cooling (see also White & Chevalier 1984).

The cooled gas presumably forms stars. The conditions are however different from spiral galaxies and star formation may there proceed with a different initial-mass-function (IMF) than is inferred for our Galaxy (Jura 1977; Fabian et al. 1982; Sarazin & O'Connell 1983). In particular, the pressure is 100 - 1000 times higher than in the local interstellar medium so that the Jeans mass (at a given low temperature of say 10 K or less) is 10 times smaller. The bulk of the star formation may therefore involve low-mass stars which contribute little optical light. This is particularly necessary for the large-sale cooling flows observed in clusters of galaxies and involving hundreds of solar masses per year as discussed in the next subsection. The implication of this is that the star formation rate in elliptical galaxies is only slightly less than it is in spiral galaxies. The major difference lies in the stellar IMF. As inflow is involved - individual gas elements probably do not move in more than a factor of two or so in radius - the structure of early-type galaxies must be continuing to evolve. The implied steep IMF for cooling flows provides a source of baryonic matter with a high mass-to-light ratio. The cooling flow may be the 'tail-end' of galaxy formation. There are, of course, further consequences of such inflows for nuclear activity in elliptical galaxies, for the propagation of radio jets, and for the signs of star formation observed at ultraviolet wavelengths (e.g. Bertola et al. 1980).

The contribution of normal galaxies to the X-ray background is expected to be very small (< few percent).

b) Clusters of Galaxies

The X-ray luminosities of rich clusters of galaxies generally lie in the range of $10^{43} \lesssim L \lesssim 10^{45.5}$ erg s^{-1}, with only loose clusters and groups appearing fainter. The dominant source of X-rays is thermal bremsstrahlung in the intracluster gas which has a temperature 2 keV $\lesssim$ kT $\lesssim$ 15 keV. The sound speed of this gas is then comparable with the velocity dispersion of the galaxies moving in the potential well of the cluster ($\sigma_c \simeq 10^3$ km s^{-1}) and the intracluster medium is close to hydrostatic equilibrium. Any net radial flow must be highly subsonic otherwise prohibitively high mass flow rates are required (exceeding $\sim 10^5$ M$_\odot$ yr^{-1}).

Once again, X-ray studies of intracluster gas will eventually enable us to map the gravitational potential - and thus the profile of the so-called 'missing mass' - with some accuracy and to carry out abundance studies. Such work requires imaging X-ray spectroscopy, which is unfortunately not yet available. Most of the cluster maps have so far been made in one wavelength band whereas the available spectra apply to the whole source. Some success has, however, been achieved with M87 in the Virgo cluster (Fabricant & Gorenstein 1983; Stewart et al. 1984a).

Cooling flows are common in both rich and poor clusters (Stewart et al. 1984b; Canizares et al. 1983) with mass flow rates of hundreds of solar masses per year. The mean mass-to-light ratio of the stars (or objects) formed in NGC 1275 in the Perseus clusters must be at least 10. Very substantial masses are accumulated,

$$\dot{M} t_H \simeq 5 \times 10^{12} \left(\frac{H_o}{50}\right)^{-3} \left(\frac{\dot{M}}{250 \, M_\odot \, yr^{-1}}\right) M_\odot \; .$$

and thus most, or all, of the central galaxy in a cooling flow cluster may be due to the flow.

The presence of cooling flows suggests that the formation out of gas of the largest galaxies is continuing today. It is possible that they were more common in the past and have since been disrupted by subcluster collisions (Stewart et al. 1984b). Optical filaments and star formation found around galaxies at redshifts of 1 and more (Spinrad & Djorgovski 1984; Lilley et al. 1983; Butcher & Oemler 1984) may be evidence for this. It is unimportant that these galaxies are bright radio sources, for even Cygnus-A is surrounded by a massive intracluster atmosphere and a cooling flow of 100 $M_\odot$ yr (Arnaud et al. 1984).

Taking the process further back, we note (Fabian et al. 1985) that the formation of all galaxies from gas only requires enough high mass stars to form metals. The rest of the gas can participate in high-pressure (low stellar mass) star formation. This is one way of producing baryonic matter with a high M/L.

c) Active Galaxies

Most Seyfert galaxies and BL objects are detectable X-ray sources. Varibaililty on timescales of 100 sec (Tennant et al. 1981) to days shows that the X-ray emission region is compact. Seyfert 1 galaxies generally have power-law X-ray spectra with an energy index of 0.6 extending from ≤ 0.7 to ≥ 100 keV (Mushotzky 1982; Petre et al. 1984; Rothschild et al. 1984). This is significantly steeper than the spectrum of the X-ray background which has an index of 0.4 over the energy range 3 - 30 keV.

d) Luminosity Functions

The work of Piccinotti et al. (1981) McKee et al. (1980) and Abramopoulos & Ku (1983) using the HEAO-1 data-base and of Maccacaro et al. (1983), Gioia et al. (1983) and Soltan & Henry (1983) is yielding detailed luminosity functions for low redshift sources. Those for clusters and active galaxies integrate to give a contribution of $\sim$ 5 percent and $\sim$ 15 percent respectively, at $\sim$ 3 keV and assuming no evolution, which could increase or decrease these estimates. The active galaxy contribution, assumed to have a mean energy index of 0.6, extrapolates to higher energies to give the 'MeV bump'. This implies either that the Seyfert galaxies do not evolve (and thus are distinct from QSOs) or that a simple extrapolation is inappropriate (Boldt 1981). Low-luminosity active galaxies may contribute a further 5 percent or so (Elvis et al. 1983).

QSOs

QSOs are generally considered to be prime candidates for the origin of the X-ray background. Most of the work centres on observations made with the Einstein Observatory and has involved many groups (including Tananbaum et al. 1979, Zamorani et al. 1981, Ku et al. 1980, Avni & Tananbaum 1982, Kembhavi & Fabian 1982, Cheney & Rowan-Robinson 1982, Setti & Woltjer 1979, 1982, Reichert et al. 1983, Chanan 1983, Anderson & Margon 1983, Stocke et al. 1983, Gioia et al. 1983, Kriss & Canizares 1982, Marshall et al. 1983, 1984, and Cheng et al. 1984). The observations are made in the range 0.5 - 3 keV and are generally referred to 2 keV. This makes comparisons with the 3-50 keV X-ray background intensity dependent upon the assumed source spectrum. Many of the earlier workers assume that QSO spectra are similar to that of the background in the 3-10 keV range (energy index 0.4), although there is no evidence to support it and even some evidence against it. Where spectra are available, the energy index is generally steeper than 0.4 (e.g. Halpern & Grindlay 1983, Worrall & Marshall 1984; Petre et al. 1984; but see 3C273, Worrall et al. 1979). Elvis, Wilkes & Tananbaum (1985) hav started to obtain spectra from the Einstein Observatory data on the brightest QSOs. The spectra are significantly softer than the X-ray background.

Radio-loud quasars are generally more X-ray luminous than radio-quiet QSOs and most observed radio-loud quasars have been detected. Some confidence can then be placed in a mean radio/X-ray luminosity ratio (or spectral index α_{RO}) which can be convolved with radio source counts to predict the X-ray counts from radio-loud quasars. As these counts converge, the contribution of such quasars is fairly well determined (within the asumptions) at $\sim$ 15 percent (Kembhavi & Fabian 1982). Note, however, that this is still an X-ray-spectrum-dependent quantity.

The radio-quiet QSO contribution is <u>very</u> uncertain. Less than 20 percent of the radio-quiet QSOs with $z > 1$ (i.e. the majority of QSOs) observed with the Einstein Observatory (<u>including</u> the Marshall <u>et al.</u> 1984 survey) were actually detected. This is hardly a secure base from which to extrapolate or combine observational upper limits. At the present time, the QSO contribution may safely be assumed to lie anywhere betweeen 10-90 percent, with other classes of source contributing the remainder. Elvis, Wilkes & Tananbaum (1985) estimate that $\leq$ 15 percent of the X-ray background need originate in QSOs wth B < 19.8 and z < 2.2.

Most 'serendipitous' X-ray active galaxies have redshifts less than 1 and, at a density of $\sim$ 1 per square degree, contribute little to the background intensity. The deep surveys (Giacconi et al. 1979, Griffiths et al. 1983) provide the best evidence for point sources making a significant fraction of the background (20 10 percent, if their spectra have an index 0.4). Further X-ray surveys that especially can provide spectra are crucial. The ROSAT survey in 1988 followed by studies with the broader-band AXAF and XMM observatories should supply the answer.

LIMITS ON ACTIVE GALAXY EVOLUTION FROM THE X-RAY BACKGROUND

Whether a hot IGM exists or not, the spectrum and intensity of the X-ray background limit the X-ray luminosity evolution of active galaxies. Cavaliere and De Zotti et al. (1982) have shown that simple 2 component power-law spectra (with an index and spectral break) cannot fit the overall spectrum after evolution and redshift. The sources must be flatter and have a lower mean energy at high redshifts. No known class of extragalactic point X-ray source has spectra similar to that of the X-ray background. Spectral evolution is thus required (Leiter & Boldt 1982).

If much of the X-ray background is due to active galaxies then the high luminosity ones at large redshifts may have $\ell > 1$, where ℓ is the compactness parameter $L/4\pi R^2 \sigma_r c$ (Guilbert, Fabian & Rees 1983). Electron-positron pairs may be present if the spectra extend to $h\nu > m_e c^2$ and these pairs can cause most of the observed luminosity to emerge at lower energies. I (1984) have explored the non-thermal case outlined by Bonometto & Rees (1972), Guilbert, Fabian & Rees (1983) and Kazanas (1984). A simple computational approach has been taken in which monoenergetic relativistic electrons at γ_{max} and monochromatic photons at $\epsilon_o \ll m_e c^2$ are continuously injected into a volume of radius R. In each time step, $\Delta t \ll R/c$, the following processes are computed: relativistic

Compton scattering which initially gives an energy index for the photons of 0.5, photon-photon collision with the resultant pairs added to the injection spectrum, $e^{\pm}$ annihilation with a broadened line added to the photon spectrum, photon leakage in a time R/c (1 + τ_{es}) and non-relativistic Compton scattering. The pairs help to steepen the photon spectrum to an index of around 0.8. It is remarkable that such a simple situation yields X-ray spectra similar to those observed. Of particular importance to spectral evolution is the dramatic change which occurs as exceeds unity. The source becomes optically thick to electron scattering in the pairs, which then cool by Comptonizing the low energy photons. Much further work is needed but it is evident that the most luminous sources can be the softest. Spectral evolution is expected. Whether it is able to explain the X-ray background, or just means that QSOs may be strong sources below 1 keV remains to be seen. Note that a large contribution to the 2 keV X-ray background from sources with spectral index of 0.6 extending to 30 keV or more would require a most unusual source for the remaining flux. It would have to increase with energy. QSOs, for example, must therefore have spectra which are predominantly flat (0.4) or they must be mostly very steep above 2 keV!

OTHER POSSIBLE SOURCES

Young galaxies may contribute to the X-ray background (Bookbinder et al. 1980). As discussed earlier, nearby galaxies with evident bursts of star formation are particularly X-ray luminous (Fabbiano et al. 1982). NGC 5408 emits $\sim 10^{40}$ erg s^{-1} in the 0.5-3 keV range despite being 2 magnitudes intrinsically fainter than the SMC (Stewart et al. 1982). Massive X-ray binaries are presumably responsible and these do often have hard spectra, although not generally extending out to 40 keV or more. If the metals in galaxies are associated with OB binaries (and not just OB stars), then an admittedly uncertain extrapolation shows that the X-ray production rate can make up the X-ray background.

Intergalactic cosmic ray electrons were once a strong contender for the origin of the X-ray background (Felten & Morrison 1966). Unfortunately, there has been little subsequent observational evidence of any such electron population.

FLUCTUATIONS IN THE X-RAY BACKGROUND

The results on large scale fluctuations have been reviewed by Shafer & Fabian (1983). The dipole component due to the Earth's motion is detected at the 95 percent confidence level ($\ell = 282°$, $b = +30°$, $v = 475 \pm 165$ km s^{-1}). No other large scale variations have been confirmed. Potentially, they can provide the best limit on lumpiness in the universe on scales of a few hundred Mpc (Rees 1981, Fabian 1981, Kaiser 1982). Fluctuations do not at present constrain the origin of the X-ray background, although the observed dipole does show that it is mostly extragalactic.

REFERENCES

Abramapoulos, F. & Ku, W.H.M., 1983, Astrophys. J., 271, 446.
Anderson, S.F. & Margon, B., 1983, 24th Liege Intl. Astrophys. Symp., 68.
Arnaud, K.A., Fabian, A.C., Eales, S.A., Jones, C. & Forman, W., 1984,
    Mon. Not. R. astr. Soc., 211, 981.
Avni, Y. & Tananbaum, H., 1982. Astrophys. J., 262, L17.
Bechtold, J., Forman, W., Giacconi, R., Jones, C., Schwarz, J., Tucker, W.
    & Van Speybroeck, L., 1983, Astrophys. J., 265, 26.
Biermann, P. & Kronberg, P., 1983, Astrophys. J. Lett., 268, L69.
Boldt, E.A., 1981, Comments on Astrophys., 9, 97.
Bonometto, S. & Rees, M.J., 1971, Mon. Not. R. astr. Soc., 152, 21.

Bookbinder, J., Cowie, L.L., Krolik, J.H., Ostriker, J.P. & Rees, M.J., 1980, Astrophys. J., 237, 647.

Canizares, C.R., Stewart, G.C. & Fabian, A.C., 1983, Astrophys. J., 272, 449.

Chanan, G.A., 1983, Astrophys. J., 275, 45.

Cheney, J. & Rowan-Robinson, M., 1981, Mon. Not. R. astr. Soc., 197, 313.

Cheng, F.Z., Danese, L., De Zotti, G. & Lucchini, F., 1983, Mon. Not. R. astr. Soc., 208, 799.

Clark, G., Doxsey, R., Li, F., Jernigan, F.G. & Van Paradijs, J., 1978, Astrophys. J.Lett., 221, L37.

Cowsik, R. & Kobetich, E.J., 1972, Astrophys. J., 177, 585.

De Zotti, G., Boldt, E.A., Cavaliere, A., Danese, L., Marshall, F.E., Swank, J.H. & Symkowiak, A.E., 1982, Astrophys. J., 253, 47.

Elvis, M., Soltan, A. & Keel, W.C., 1984, Astrophys. J., 283, 479.

Elvis, M., Wilkes, B. & Tananbaum, H., 1985, Astrophys. J., 292, 357.

Fabbiano, G., Feigelson, E. & Zamorani, G., 1982, Astrophys. J., 256, 397.

Fabian, A.C., 1981, Ann. .Y. Acad. Sci., 375, 235.

Fabian, A.C., Nulsen, P.E.J. & Canizares, C.R., 1982, Mon. Not. R. astr. Soc., 201, 933.

Fabian, A.C., Nulsen, P.E.J. & Canizares, C.R., 1984, Nature, 310, 733.

Fabian, A.C., Arnaud, K.A., Nulsen, P.E.J. & Mushotzky, .F., 1985, preprint.

Fabian, A.C. & Kembhavi, A.K., 1982, in Extragalactic Radio Sources (eds. D.S. Heeschen & C.M. Wade), IAU 97, 453.

Fabian, A.C., 1984, in Proceedings of Manchester Conference on Active Galaxies, M.U.P. in press.

Fabricant, D. & Gorenstein, P., 1983, Astrophys. J., 267, 535.

Feigelson, E.D., Schreier, E.J., Delvaille, J.P., Giacconi, R.,
    Grindlay, J.E. & Lightman, A.P., 1981, Astrophys. J., 251, 31.
Felten, J. & Morrison, P., 1966, Astrophys. J., 146, 686.
Field, G.B. & Perrenod, S.C., 1977, Astrophys. J., 215, 717.
Forman, W., Schwarz, J., Jones, C., Liller, W. & Fabian, A., 1979,
    Astrophys. J. Lett., 234, L29.
Forman, W. & Jones, C. & Tucker, W., 1985, Astrophys. J., in press.
Fried, P.M., Nousek, J.A., Saunders, W.T. & Kraushaar, W.L., 1980,
    Astrophys. J., 242+, 987.
Garmire, G. & Nousek, J., 1981, Bull. A.A.S., 12, 853.
Giacconi, R., Bechtold, J., Branduardi, G., Forman, J., Henry, J.P.,
    Jones, C., Kellogg, E., Van der Laan, H., Liller, W., Marshall, H.,
    Murray, S.S., Pye, J., Schreier, E., Sargent, W.L.W., Seward, F.
    & Tananbaum, H., 1979, Astrophys. J., 234, L1.
Gioia, I.M., Feigelson, E.D., Maccacaro, T., Schild, R., & Zamorani, G.,
    1983, Astrophys. J., 271, 524.
Griffiths, R.E., Murray, S.S., Giacconi, R., Bechtold, J., Murdin, P.,
    Smith, M., McGillivray, H., Ward, M., Danziger, J., Lub, J.,
    Peterson, B.A., Wright, A.E., Batty, M.J., Jauncey, D.L. &
    Malin, D.F., 1983, Astrophys. J., 269, 375.
Gruber, D. in 'UV and X-ray emission from Active Galaxies' MPI, 129.
Guilbert, P.W., Fabian, A.C. & Rees, M.J., 1983, Mon. Not. R. astr. Soc.,
Guilbert, P.W. & Fabian, A.C., 1985, preprint.
Halpern, J.P. & Grindlay, J.E., 1982, Bull. Am. Astr. Soc., 13, 849.
Jura, M., 1977, Astrophys. J., 212, 634.
Kaiser, N., 1982, Mon. Not. R. astr. Soc., 198, 1033.
Kazanas, D., 1984, Astrophys. J., 287, 112.
Kembhavi, A.K. & Fabian, A.C., 1982, Mon. Not. R. astr. Soc., 198, 921.
Kriss, G.A. & Canizares, C.R., 1982, Astrophys. J., 261, 51.

Ku, W.H-M., Helfand, D.J. & Lucy, L.B., 1980, Nature, 288, 323.

Leiter, D. & Boldt, E.A., 1982. Astrophys. J., 260, 1.

Lilly, S., Longair, M.S. & McLean, I.S., 1983, Nature, 301, 488.

Long, K. & Van Speybroeck, L., 1983, in Accretion-Driven Stellar X-ray Sources, (ed. Lewin & van den Heuvel), CUP.

Maccacaro, T., Avni, Y., Gioia, I.M., Giommi, P., Griffiths, R.E., Liebert, J., Stocke, J. & Danziger, J., 1983, Astrophys. J., 266, L73.

McKee, J.D., Mushotzky, R.F., Boldt, E.A., Holt, S.S., Marshall, F.E., Pravdo, S.H. & Serlemitsos, P.J., 1980, Astrophys. J., 242, 843.

Marshall, F.E., Boldt, E.A., Holt, S.S., Miller, R., Mushotzky, R.F., Rose, L.A., Rothschild, R. & Serlemitsos, P.J., 1980, Astrophys. J., 235, 4.

Marshall, H.L., Tananbaum, H., Zamorani, G., Huchra, J.P., Braccesi, A., & Zitelli, V., 1983, Astrophys. J., 269, 42.

Marshall, H.L., Avni, Y., Braccesi, A., Huchra, J.P., Tananbaum, H., Zamorani, G. & Zitelli, V., 1984, preprint.

Marshall, F.J. & Clark, G.W., 1984, Astrophys. J., 287, 633.

McKee, C.F., 1980, Phys. Scripta, 21, 738.

Mushotzky, R.F., 1982, Astrophys. J., 256, 92.

Nulsen, P.E.J., Stewart, G.C. & Fabian, A.C., 1984, Mon. Not. R. astr. Soc., 208, 185.

Petre, R., Mushotzky, R.F., Krolik, J.H. & Holt, S.S., 1984, Astrophys. J., 280, 499.

Piccinotti, G., Mushotzky, R.F., Boldt, E.A., Holt, S.S., Marshall, F.E., Serlemitsos, P.J. & Shafer, R.A., 1982, Astrophys. J., 263, 495.

Protheroe, R.J., Wolfendale, A.W. & Wdowczyk, J., 1980, Mon. Not. R. astr. Soc., 192, 445.

Rees, M.J., 1980, in Objects at High Redshifts (eds. G. Abell & J. Peebles), IAU 92, 209.

Reichert, G.A., Mason, K.O., Thorstenson, J.R. & Bowyer, S., 1982, Astrophys. J., 260, 437.

Rothschild, R.E., Mushotzky, R.F, Baity, W.A., Gruber,D.E., Matteson, J.L. & Peterson, L.E., 1983, Astrophys. J., 269, 423.

Sarazin, C.L. & O'Connell, R.W., 1983, Astrophys. J., 268, 552.

Schwartz, D.A., 1980, Phys. Scripta, 21, 644.

Setti, G. & Woltjer,L., 1979, Astr. Astrophys., 76, L1.

Shafer, R.A., 1983, Ph.D. Dissertation, Univ. of Maryland

Shafer, R.A. & Fabian, A.C., 1983, in Early Evolution of the Universe and its Present Structure (eds. G. Abell & G. Chincarini), IAU 104, 333.

Sherman, R.D., 1979. Astrophys. J., 232, 1.

Soltan, A. & Henry, J.P., 1983, Astrophys. J., 271, 442.

Spinrad, H. & Djorgovski, G., 1984, Astrophys. J., 280, L9.

Stewart, G.C., Fabian, A.C., Terlevich, R.J. & Hazard, C., 1982, Mon. Not. R. astr. Soc., 200, 61P.

Stewart, G.C., Canizares, C.R., Fabian, A.C & Nulsen, P.E.J., 1984a. Astrophys. J., 278, 47.

Stewart, G.C., Fabian, A.C., Jones, C. & Forman, W., 1984b. Astrophys. J., 285, 1.

Stocke, J.T., Liebert, J., Gioia, I.M., Griffiths, R.E., Maccacaro, T., Danziger, I.J., Kunth, D. & Lub, J., 1983, Astrophys. J., 273, 458.

Tananbaum, H., Avni, Y., Branduardi, G., Elvis, M., Fabbiano, G., Feigelson, E., Giacconi, R., Henry, J.P., Pye, J.P., Soltan, A. & Zamorani, G., 1979, Astrophys. J., 234, L9.

Tennant, A.F., Mushotzky, R.F.,Boldt, E.A. & Swank, J.H., 1981, Astrophys.

J., 251, 15.

Tennant, A.F. & Mushotzky, R.F., 1983, Astrophys. J., 264, 92.

Van Speybroeck, L., Epstein, A., Forman, W., Giacconi, R., Jones, C., Liller, W. & Smarr, L., 1979, Astrophys. J., 234, L45.

Warwick, R.S., Pye, J.P. & Fabian, A.C., 1980, Mon. Not. R. astr. Soc., 190, 243.

White, R.E. & Chevalier, R.A., 1983. Astrohys. J., 280, 561.

Worrall, D.M., Mushotzky, R.F., Boldt, E.A., Holt, S.S. & Serlemitsos, P.J., 1979, Astrophys. J., 232, 683.

Worrall, D.M., & Marshall, F.E., 1984, Astrophys. J., 276, 434.

Yang, J., Turner, M.S., Steigman, G., Schramm, D.N. & Olive, K.A., 1984, Astrophys. J., 281, 493.

Zamorani, G., Henry, J.P., Maccacaro, T. & Tananbaum, H., Soltan, A., Liebert, G., Stocke, J., Strittmatter, P.A., Weymann, R.J., Smith, M.G. & Condon, J.J., 1981, Astrophys. J., 245, 357.

COSMOLOGY AND GALAXY FORMATION
An Introduction to some Recent Ideas

Bernard J.T. Jones

and

Enrique Martinez Gonzalez

NORDITA
Blegdamsvej, 17
2100 Copenhagen
Denmark

INTRODUCTION

The aim of the present series of lectures is to be unashamedly pedagogical and present, in simple terms, an overview of our current thinking about our universe and the way in which we believe galaxies have formed. There have in the past been a number of fine "summer schools", workshops and conferences on various aspects of cosmology and there is little point in re-iterating what has been said so well there. There have, however, been a number of recent developments in our thinking about cosmology which are worth putting in perspective.

The most recent impetus to the subject has come from high energy physics. Understanding the nature of the elementary particle fields that pervade the universe at the earliest times allows us the possibility of solving a number of cosmological conundrums such as "why is the universe so isotropic?", or "why is the universe not made up of equal amounts of matter and anti-matter?". High energy physics has also introduced us to a remarkable zoo of exotic elementary particles, some of which may be of importance in understanding how the structure of our universe has evolved. These particles go by names such as "axions", "paraphotons", "massive neutrinos", "gravitinos", and "photinos". With the possible exception of the massive neutrino, we have no direct evidence for the existence of any of these particles and it may even be that cosmology provides the only laboratory in which the consequences of their existence may be observed. If nothing else, this is a sobering reflection of our possible level of ignorance about the origin and evolution of the universe in which we live.

THE UNIVERSE IN WHICH WE LIVE

(The subject of this section is covered at various levels in many of the books cited in the reference list at the end of this chapter)

Let us, for the sake of simplicity, restrict ourselves to discussing models of the universe which are overall homogeneous and isotropic, and seek to describe the evolution of the radius $l(t)$ of a spherical piece of the universe that expands with the universe. Einstein's theory of general relativity then provides a remarkably simple set of equations for $l(t)$:

$$\frac{\ddot{l}}{l} = -\frac{4\pi G}{3}\left(\rho + 3\frac{p}{c^2}\right) + \frac{1}{3}\Lambda \qquad (1a)$$

$$\dot{\rho} + 3\frac{\dot{l}}{l}\left(\rho + \frac{p}{c^2}\right) = 0 \qquad (1b)$$

Here, $\rho(t)$ and $p(t)$ are respectively the density and pressure of the cosmic medium at a given time, and these are themselves related via an equation of state. The equation of state comes, in principle, from a theory for the matter content of the universe and in general that will come from high energy physics which tells us what the constituent particles may be and how they interact. These equations should be referred to as the Friedmann – Lemaitre equations after the physicists who first derived them and worked with them.

There are a number of notable features in these equations. Firstly, had we attempted to derive them from Newton's law of gravitation, we would have obtained exactly the same equations with the exception of the way in which the pressure terms enter. Secondly, we have included the infamous "cosmological constant", Λ. This is a term which Einstein added to his theory of gravitation as an afterthought (he wanted to produced non-expanding cosmological models!). Observational constraints on the value of Λ are not particularly strong – it certainly cannot be large enough to play a role in solar system dynamics, but it could be important for cosmology. Optimistically, we might say that a proper theory of quantum gravity will eventually tell us whether or not Λ should be zero.

A third point is that these equations rest on the assumption that Einsteins theory of gravity is correct at the highest densities. We have good reasons to doubt the theory at epochs so early that quantum mechanics effects become as important as gravitational effects. This is the so-called Planck time,

$$t_{pl} = \left(\frac{G\hbar}{c^5}\right)^{\frac{1}{2}} \simeq 5.4 \times 10^{-44} \text{ sec.} \qquad (2).$$

This time is so early in the history of the universe that it is of little consequence for the phenomena based in high energy physics that we shall discuss here, and moreover there is no generally accepted theory of what the laws of physics may be at this time. The Planck era is thus a mysterious regime on which many of the solutions to the puzzles about our universe may depend.

Relativity may break down in a different way. Einstein's theory is most succinctly summarised in terms of the action principle from which the equations may be derived:

$$\delta \int L \sqrt{-g}\, d^4x = 0 \quad ; \quad L = R$$

Here, the Lagrangian L is a simple linear function of the curvature scalar R. (R is the trace of the Riemann tensor and has dimensions of $[\text{Length}]^{-1}$; g is the determinant of the metric). There is no obvious reason why there should not be terms in R^2 appearing in the Lagrangian L:

$$L = R + \alpha R^2 + \ldots$$

The effect of such terms would be undetectable by laboratory experiments if α were small enough, but it could nevertheless dominate even the first 10^{-10} seconds of cosmic evolution. This could prevent the universe from ever being dense enough for operation of the high energy physics phenomena we are going to discuss here. (It might of course be argued that the fact that we see an asymmetry between baryons and antibaryons demands that the universe had in fact acheived such temperatures and densities, and that this in turn imposes an upper limit on α.) Again, the only hope of resolving this question may lie in a proper theory of quantum gravity.

Let us now apply the Friedmann – Lemaitre equations to the universe today. The present universe appears to be dominated by matter whose pressure is negligible and so the "equation of state" is well approximated by putting

$$p = 0$$

$$\rho l^3 = \text{constant},$$

(the so-called "dust" approximation). The second of these equations expresses the conservation of the number of particles in a volume that expands with the universe (see equation (1b)). These assumptions are not valid for the first million or so years of cosmic history when there is a substantial contribution to the pressure from zero mass particles. We shall also set $\Lambda = 0$ for simplicity. With these assumptions, the first integral of the equations is easily done and we have

$$\left(\frac{\dot{l}}{l}\right)^2 = \frac{8\pi G \rho(t)}{3} - \frac{k}{l^2} \quad (3)$$

where k is a constant of integration that plays a very important role in determining the nature of the cosmic expansion.

This constant is exactly analogous to the energy constant that appears in the first integral of the Newtonian equations for the motion of a body about a central point mass (the "Kepler problem"). In that problem, the energy constant determines whether the orbit is a closed ellipse or a hyperbola, the zero-energy case being the parabolic orbit. If the constant k in the above equations is positive, the universe has a maximum radius of expansion and its initial and final states are both singular. If the constant is negative, the universe evolves out of a singular state and expands forever to a state that is infinitely dispersed. The special case when the constant is zero corresponds to the so-called "Einstein - de Sitter" solution which expands forever and is analogous to the parabolic solution of the two-body problem.

Before going on to ask which type of universe we live in it is useful to recast the free parameters of the equation (3) directly in terms of observable quantities. To do that we note that we should be able observationally to determine the present rate of expansion of the universe (the Hubble "constant"):

$$H_o = \frac{\dot{l}_o}{l_o} \qquad (4)$$

and the present matter density ρ_o. Measuring the Hubble constant requires that we know the absolute distances and recession velocities for a large number of galaxies. The point of debate is always how to estimate the former and so it is conventional to denote the present value of the Hubble Constant by

$$H_o = 100 h \text{ km.s}^{-1}.\text{Mpc}^{-1}. \qquad (5)$$

where h reflects our ignorance. It is generally believed that h lies in the range 0.5 to 1.0.

Measuring the present density of the universe poses an

enormous problem. It is useful to measure that density in terms of the density ρ_c the universe would have were it an Einstein – de Sitter universe. We can define the "critical density" at a time t, when the Hubble parameter was $H = \dot{\ell}/\ell$ by

$$\rho_c = \frac{3H^2}{8\pi G} \qquad (6)$$

This is the relationship between density ρ and time in the "flat" $k = 0$ solution of the Freidmann – Lemaitre equations (see equation (3)). We can then define the cosmic density parameter Ω as

$$\Omega = \frac{\rho}{\rho_c} \qquad (7)$$

The present value, Ω_o is probably not less than 0.1 and probably no greater than 1.0. This appears to represent considerable uncertainty. On the other hand, unless $\Omega_o = 1$ precisely, Ω changes with time and we may wonder why we happen to be living at an epoch when this value is so close to one. This is the essense of the "flatness problem" that we will return to later.

With these definitions, the equation for the cosmic expansion can be written as

$$\left(\frac{\dot{\ell}}{\ell}\right)^2 = \frac{8\pi G}{3} \rho(t) - H_o^2 (\Omega_o - 1) \left(\frac{\ell_o}{\ell}\right)^2 \qquad (8)$$

In this form we see that $\Omega_o = 1$ corresponding to the Einstein – de Sitter model has zero curvature constant k.

The universe today contains a radiation component: it is the cosmic microwave background radiation that provides us with the main evidence that the universe evolved from a hot singular state. The present temperature of the radiation field is

$$T_r = 2.7 \text{ K} \tag{9}$$

and it appears to have a Planckian spectrum. The pressure of the radiation field is

$$P_r = \tfrac{1}{3} \rho_r c^2 = \tfrac{1}{3} a T_r^4. \tag{10}$$

The radiation temperature falls as ℓ^{-1} and so the radiation energy density evolves as

$$\rho_r \propto \ell^{-4} \tag{11}$$

(see equation (1b)). We can include this component in the above dynamical equation:

$$\left(\frac{\dot{\ell}}{\ell}\right)^2 = \frac{8\pi G}{3}\left[\rho_{mo}\left(\frac{\ell_o}{\ell}\right)^3 + \rho_{ro}\left(\frac{\ell_o}{\ell}\right)^4\right] - H_o^2(\Omega_o - 1)\left(\frac{\ell_o}{\ell}\right)^2, \tag{12}$$

where we have explicitly written out the individual contributions from the matter and radiation in terms of the scale factor ℓ. (If ℓ_o is the present value of ℓ, then ρ_{mo} and ρ_{ro} are the present matter and radiation energy densities). The radiation content of the universe today is of little dynamical importance. However, we can see from this equation that in the distant past the universe was fully radiation dominated. The big bang was hot. To take these equations far back into the past we have to modify the simple baryons + radiation model since at ultrahigh energies there are no such things as photons and baryons. We shall find that at extremely high energies the pressure is effectively negative.

As a rough guide we have the formula for the temperature when the universe was t seconds old:

$$T \sim \frac{10^{10} \text{ K}}{\sqrt{t_{sec}}}$$

To within a factor of 10 or so, this works pretty well, and we can estimate the temperature at the Planck time to have been around 10^{32} K ! Even if we use particle physics units and measure temperatures in units of GeV, where 1GeV is about 10^{14} K, this is still a remarkable number.

INFLATION

(Excellent low-brow reviews of this subject are in the Audouze and Tran book "The Birth of the Universe". The high energy physics aspects are discussed in the books by Close and by Dodd. A more technical overview appears in the review articles by Brandenburger and Linde. See reference list at end of chapter)

We are now in a position to explain what has become known as the "flatness puzzle" - why is the present universe so close to being an Einstein - de Sitter universe? Why is the present value of the density parameter not 10^{-3} or 10^{+3}, or some more extreme value? Let us take the universe as described by equation (12) and trace its evolution back in time to the Planck time on the assumption that this equation holds all that way back. Doing that we see that the ratio of the curvature term to the density term was incredibly close to 1 then:

$$1 - \Omega_{pl} \simeq 10^{-59}.$$

This means that the "initial" conditions set at the Planck time has to be finely tuned with a precision of one part in 10^{59} in order that the present universe should look the way it does. Any departure from that level of fine tuning would have led to a universe that was now either very open ($\Omega_o <<< 1$) or very closed ($\Omega_o >>> 1$). The way out of this dilemma (if one accepts that it is indeed a dilemma) is to look for some physical process that would set up this finely tuned state shortly after the Planck time, regardless of what the actual conditions emerging from the Planck era were.

High energy physics has provided us with the possibility of such a solution, and indeed the solution offered is so effective that for general initial value of Ω it demands that we live in a universe today where Ω_o is extremely close to 1. The idea is remarkable in that it is potentially falsifiable (a necessary aspect of any decent scientific theory) simply by demonstrating observationally that Ω_o is not in fact 1.000... to an arbitrary number of decimal places. The measurement of Ω_o is thus one of the principle goals of modern observational cosmology.

There are several variants of the "inflationary cosmology", but all exploit the fact that at extremely high temperatures the equation of state of matter is approximately

$$P_v = - \rho_v c^2 \tag{13}$$

(We use the subscript "v" to indicate that this is not a state of normal matter. Indeed, it is a quantum state described as a "false" vacuum; false because it is not the kind of vacuum we are accustomed to in the laboratory.) The appearance of negative pressures may at first seem peculiar, but it is simply a consequence of the way elementary particles are thought to interact at extremely high energies. There are two remarkable consequences of this equation of state. Firstly, from equation (1b) we have that as the universe expands

$$\rho = \rho_v = \text{constant,} \tag{14}$$

The Value of the constant is determined by high energy physics. Secondly, the expansion takes place exponentially with a time constant determined by ρ_v:

$$\lambda \propto e^{H_v t} \tag{15a}$$

$$H_v^2 = \frac{8\pi G \rho_v}{3} \qquad (15b)$$

Thus, for as long as the universe can remain in this peculiar state it expands dramatically and so the curvature term in equation (12) becomes exceedingly small. (Note that during this era the vacuum is behaving like a cosmological constant). If this "false vacuum" state can survive long enough, we can understand why Ω_o is so "close" to 1.0 (and that it should in fact be very close to 1.0).

Of course there are the problems of how long this "inflation" can last and how we get out of this "inflationary" state, the universe today clearly is not like that. Again that depends in detail on the theory of elementary particles that we adopt for this era. What is important is that when the universe exits from this inflationary state it should find itself at a temperature in excess of 10^{16} GeV, for then, as will be explained below, we can understand the baryon asymmetry of the universe and find an explanation of the present value of the entropy per baryon.

How much inflation do we need? In a given theory for the state of matter in the early universe, the amount of inflation is fixed by the detailed properties of the assumed matter fields. It is in this way that cosmology can impose constraints on theories of ultra-high energy physics. So the question of how much inflation is necessary is an important one. Of course there is the possibility that we need so much inflation that reheating to the baryongenesis temperature is impossible: we then would have to look elsewhere for an explanation of the cosmic baryon asymmetry, or argue that inflation plays no part in the evolution of our universe, or argue that we have got the physics wrong.

The first requirement as regards the amount of inflation is that the density parameter Ω be reduced to a value close to one. That is not a strong constraint since we have no objective basis for knowing what Ω might have been before the onset of the inflationary period. It is not difficult to construct initial conditions such that, despite a large amount of inflation, the universe never looks the way it does today. A more interesting constraint is set by the wish to have the whole of the present observable universe fall into such a small volume near the singularity that opposite sides of the universe could have communicated then. (This is the so-called "horizon problem"). The point is that the inflationary models of the universe become exceedingly small at early times if they are compared with their non-inflationary counterparts. Given large amounts of inflation we can hope to explain such mysteries as the overall homogeneity and isotropy of the present universe. We also have a physical basis for understanding the origin of the perturbations that led to the formation of galaxies. They could arise from the quantum fluctuations of the fields that make up the false vacuum. This in turn provides constraints on the parameters of field theories of matter at ultra-high energies since we know what amplitudes these fluctuation must have.

DO WE LIVE IN AN EINSTEIN – DE SITTER UNIVERSE?

(The galaxy correlation function and the cosmic virial theorem are discussed in detail in Peebles' book "Large Scale Structure of the Universe". Raine's book "The Isotropic Universe" has some good introductory material)

When Hubble first started systematic investigations of the depths of the observable universe, he was struck by its apparent homogeneity and isotropy. Today we have through the isotropy of the cosmic microwave background radiation an even more powerful tool for quantitatively assessing the isotropy of the universe. All attempts have so far failed to reveal evidence for any anisotropy on any angular scale, with the exception of the "dipole anisotropy" which is simply a reflection of our movement relative to the universe. Uson in this volume summarises that situation. By invoking the "Copernican Principle", that we are not in a priviledged place in the universe, a corollary of the observed isotropy is that the universe must be homogeneous.

The large scale homogeneity of the universe is somewhat more difficult to assess observationally since that has to be done by comparing the distribution of galaxies in different places. The simplest quantity describing the nature of the galaxy distribution is the two-point correlation function $\xi(r)$. If we knew the locations of all the galaxies in some sample volume we could calculate the distribution of the separations between every pair of galaxies in the sample, and compare that distribution with what would be expected had the galaxies been distributed randomly (so that the position of any galaxy is independent of where the other galaxies lie).

Expressed mathematically, we can write the probability δP of finding a galaxy in a volume element δV situated at a distance r from a given galaxy in the sample as

$$\delta P = \bar{n}\,\delta V \left[1 + \xi(r)\right] \qquad (16)$$

where $\bar{n}$ is the mean number density of galaxies in the universe. (This definition of $\xi(r)$ assumes that the distribution of pairs is isotropic, and assumes that we have some way of estimating $\bar{n}$. In practise, has to be estimated from the sample itself.) Until the recent redshift surveys the only means of estimating $\xi(r)$ was from the projected distributions of galaxies listed in various catalogues of galaxies. A useful approximation to $\xi(r)$ is

$$\xi(r) \simeq 50\,(hr)^{-1.77}, \quad 0.1\text{ Mpc} \lesssim hr \lesssim 15\text{ Mpc}.$$

Beyond $15h^{-1}$ Mpc. there may be a cut-off where $\xi(r)$ falls to very small or even negative values.

It is possible to test the homogeneity of the universe by measuring the clustering at different distances. If the universe is homogeneous, $\xi(r)$ should be independent of the depth of the galaxy catalogue used to calculate $\xi(r)$. Comparison of the correlation functions for different catalogues of galaxies reveals a remarkable degree of consistency among the estimates of $\xi(r)$. The clustering of galaxies in the Zwicky catalogue that goes down to magnitude 15 or so is the same as the clustering in the galaxy catalogues that look more than 100 times deeper into the universe. This is one of the best pieces of observational evidence for the large scale homogeneity of the universe.

So much for homogeneity and isotropy; the key question at present concerns the value of Ω_o. To assess the value of Ω_o it is useful to consider separately the contribution of the various constituents to Ω_o. The radiation presently makes a negligible contribution, $\Omega_r = 10^{-4}$. The baryonic contribution can only be assessed on the basis of the amount of luminous material we see, and that is mainly in the form of stars in galaxies. This contribution to Ω_o is on the order of $\Omega_L \sim 0.01$, but there is abundant evidence that estimating the mass of a galaxy simply from the light it emits leads to a serious underestimate. If we consider the internal dynamics of galaxies it seems that they have at least three times more mass than we can account for in terms of the stars they obviously contain, and possibly as much as ten times. What the nature this "hidden mass" is we have no idea: it could be anything from very small highly underluminous stars ("Jupiters") or black holes to massive neutrinos or axions.

There is a significant constraint on the amount of baryonic material that can be in the universe from considerations of cosmic nucleosynthesis. It is thought, with good reasons, that most of the Helium and Deuterium we see was made in the first few minutes of the big bang. To get the correct amounts formed requires a baryonic contribution to Ω_o of around 0.1. Too many baryons and we get no deuterium, too few we get no Helium. An interesting observation of the Lithium in very old stars suggests that this too may have been formed in the early universe, and the corrrect amount is formed if is again around 0.1. Thus if Ω_o is close to 1.0, the missing mass has to be made up of non-baryonic material such as massive neutrinos, gravitinos or axions.

We are able to put limits on the amount of clustered dark matter by studying the random motions of galaxies in the universe. The idea is simple: the more matter lying in clusters, the higher will be the random velocities of galaxies. Of course there could be substantial amounts of dark matter distributed homogeneously throughout the universe and that would never be detected by this technique; we would have to resort to the classical cosmological tests like the Hubble diagram and the number magnitude counts which involve almost insuperable difficulties because of the unknown effects of galaxy evolution. (It would however be somewhat strange if the dominant component of the gravitating material did not cluster like the baryons!)

The "cosmic virial theorem" establishes the link between the clustering and the peculiar velocities that clustering produces during the growth of structure in the universe. Mathematically expressed in its simplest form it says

$$\langle v(r)^2 \rangle \simeq \frac{3}{2} \Omega_0 H_0^2 J_2(r) \quad , \quad J_2(r) = \int_0^r s \xi(s) ds \qquad (18)$$

The left hand side is the velocity dispersion of galaxies that are separated by a distance r, and the right hand side tells us the fluctuations in the gravitational potential due to the clustering on that scale. The constant of proportionality has the information about the cosmic mass density and numerical factors that depend on the assumptions as to how precisely the structure grew. The observed correlation function is put into the right hand side and the velocity statistics of complete redshift sample into the left. It is then only a matter of division to get Ω_0. When this is done carefully we have the result that

$$\Omega_0 = 0.15 \pm 0.05 \qquad (r < 4 h^{-1} Mpc) \qquad (19)$$

Thus, unless the dark matter is distributed in a manner that differs substantially from the baryonic component, the universe is open. There is growing evidence that whatever matter cannot be accounted for by stars is nevertheless associated with the galaxies themselves or the clusters in which they lie. We may feel able to conclude that there is no evidence that the universe is closed. The central cosmological question will once again be "why does Ω_o have its present value?"

If indeed we could establish that $\Omega_o \neq 1$, then we could still have the benefits of inflation if we conceded that the cosmological constant were non-zero. The point is that inflation merely produces a geometrically "flat" ($k = 0$) space-time expanding in such a way that

$$3H_o^2 (1 - \Omega_o) - \Lambda = 0 \qquad (20)$$

(this is just the condition that the integration constant $k = 0$ when we leave the Λ-term in (8)). Thus we would have another coincidence to explain: that the present expansion rate happens to be on the order of $\sqrt{(\Lambda/3)}$, a number that is presumably fixed by the laws of physics.

BARYON - ANTIBARYON ASYMMETRY

(The books of Close and Dodd provide excellent introductions to this subject and the background high energy physics. See reference list at end of chapter)

One of the striking observations about our universe is that it appears to be made predominately of matter and not of equal amounts of matter and anti-matter. Some fifteen years ago "symmetric" theories were a hot topic of discussion since, at that time, there was no physical reason why we should expect the universe to be made up of one kind of matter rather than the opposite kind. If the early universe had been made up of equal amounts of matter and antimatter, the radiation we see today could have come from the annihilation of the bulk of the content of the universe, leaving behind disjoint islands of matter and antimatter. Baryon symmetric theories then went on to calculate the expected ratio of photons to baryons (the cosmic entropy per baryon) in the universe. One of the strong arguments in favour of these theories was that there was no other basis for understanding the entropy per baryon of the universe.

The theories foundered on observational grounds: there was no evidence for the existence of substantial amounts of antimatter, and there were also some strong constraints on how much antimatter there could be. The baryon symmetric cosmologies died. In recent years, however, it has become apparent that at very high energies the symmetry between baryons and antibaryons may not be preserved. Thus we can get a situation in the universe where there was a naturally occuring excess of baryons over antibaryons. The annihilation of the antibaryons could then produce the present radiation component of the universe.

We have no direct evidence of the kind of particles that inhabited the universe when its temperature was above 10^{14} GeV. However, there are good theoretical grounds for speculating that among the numerous species of particles present were the so-called X-boson and its anti-particle. The X-particle is one of the particles which generates the force between quarks, and is the agent that may endow the proton with a finite (though nonetheless very long, $> 10^{33}$ yrs.) lifetime. As the universe cooled below 10^{16} GeV, the X and $\overline{X}$ particles would be able to decay into pairs of quarks (q), or a quark (q) plus a lepton (ℓ)

$$X \to \begin{cases} q + q & \text{fraction } r \text{ of decays} \\ q + \ell & \text{fraction } 1-r \text{ of decays} \end{cases}$$

$$\overline{X} \to \begin{cases} \overline{q} + \overline{q} & \text{fraction } \overline{r} \text{ of decays} \\ \overline{q} + \overline{\ell} & \text{fraction } 1-\overline{r} \text{ of decays.} \end{cases}$$

There is in fact no reason why the fraction r of decays of the X into two quarks should equal the fraction of $\overline{r}$ decays of the $\overline{X}$ into antiquarks. Thus we can end up with an excess of quarks over antiquarks. The annihilations of the q and $\overline{q}$ and the ℓ and $\overline{\ell}$ leads to the production of radiation. The slight excess of remaining q and ℓ provide the material from which the present universe is made (baryons, as opposed to antibaryons). The densities of photons and baryons are thus

$$n_B \sim n_q - n_{\bar{q}}$$

$$n_\gamma \sim n_q + n_{\bar{q}} \qquad (21)$$

The computation of the full set of non-equilibrium reactions leads to a "prediction" (actually a "post"-diction) of the cosmic entropy per baryon:

$$S = \frac{n_\gamma}{n_B} \simeq 10^{9 \pm 1} \qquad (22)$$

in terms of the unknown mass of the X-particle. We can in fact say that cosmology appears to demand a mass

$$m_X \sim 10^{14} \text{ GeV}$$

for this particle in order to understand the present cosmic entropy per baryon.

What is especially interesting is that the same X-particle would be involved in the decay of the proton into a π^0 and a e^+. The lifetime of the proton against this decay is related to the X particle mass, and so observations of proton decay would provide a check on the theory of elementary particles, and at the same time say whether this explanation of the present baryon asymmetry of the universe is a plausible one.

INHOMOGENEITIES IN THE EARLY UNIVERSE
(There are no truly introductory works on how to describe random density fluctuations in the early universe. The Peebles book "Large Scale Structure of the Universe" makes extensive use of the concepts presented in this section)

The universe at the earliest times is supposed to have

deviated slightly but significantly from a perfect Friedmann – Lemaitre universe (for otherwise we could not understand the origin of cosmic structure). We do not know what the origin of these inhomogeneities might have been: that is surely one of the most exciting goals of modern cosmology. Current thinking suggests that quantum fluctuations in the fields comprising the material content of the universe shortly after the Planck time led to the deviations from the pure Friedmann state. Precisely how that happened is a subject of much speculation, but despite our ignorance about the precise details we can nevertheless make a few relatively strong statements.

Firstly, we can assert that it is likely that the first perturbations were of the "adiabatic" variety. The fluctuations in the geometry of space-time were such that the entropy per baryon was everywhere the same. This is because the fluctuations of the material fields governing the microphysics conserve the properties of bosons and photons. This is perhaps one of the strong arguments for preferring pancake theories of galaxy formation – we have the possibility of understanding the origin of the initial fluctuations.

Secondly we can say something about the "spectrum" of the inhomogeneities, that is we can say something about what amplitudes of fluctuations we can expect on various mass scales. We need only the assumption that the fluctuations were created in a way that respects the conservation of momentum and energy in small volumes of space-time. It is this point that we wish to pursue in a little more detail here, but to do so we must explain the notion of a power spectrum, a concept that crops up repetedly in discussions of the galaxy formation process. The mathematically noninclined should skip over to the following section.

Suppose that in some coordinate system the density at time t is $\rho(\underline{x},t)$, and that it deviates from some mean value $\bar{\rho}(t)$ by a fractional value δ :

$$\delta = \frac{\rho(\underline{x},t)}{\bar{\rho}} - 1 \qquad (24)$$

Since $\bar{\rho}$ is supposed to be the mean of ρ, we have that the mean of δ is zero:

$$\langle \delta \rangle = 0 \qquad (25)$$

This definition of the mean simply reflects the fact that we average over a finite volume and let the size of the volume tend to infinity. The function $\delta(\underline{x},t)$ is referred to simply as the "relative density fluctuation at a point" and it is the statistical properties of this that we wish to investigate.

The next statistical descriptor of δ is its "correlation function":

$$C(\underline{r}) = \langle \delta(\underline{x})\delta(\underline{x}+\underline{r}) \rangle = \lim_{V \to \infty} \int_V \delta(\underline{x})\delta(\underline{x}+\underline{r}) d^3\underline{x} \qquad (26)$$

This quantifies the extent to which the variations of density at a given point are related with the variations at neighbouring points. A random function $\delta(\underline{x},t)$ has a "patchy" appearance and the scale of the patchiness is often related to the length scale on which $C(\underline{r})$ typically falls to zero. In some sense, C may be regarded as the typical density profile of a patch (in a mean square sense).
The value of the correlation function at the origin is in fact the variance of the random process δ:

$$C(0) = \langle \delta(\underline{x})^2 \rangle \qquad (27)$$

This number quantifies the excursions of the density about the mean.

There are other ways of looking at a random density field which prove useful in cosmology. Of special importance is the Fourier transform of the density field:

$$\delta_{\underline{k}} = \int \delta(\underline{x},t) e^{i\underline{k}\cdot\underline{x}} d^3\underline{x} \qquad (28)$$

(For convenience we drop the explicit reference to the time dependence of $\delta_{\underline{k}}$). This is not the familiar Reimann Integral of the elementary calculus courses since the integrand is a stochastic function. Indeed, we have to be a little careful in how we define this so that the "spectral density" $\delta_{\underline{k}}$ exists.

Generally, it helps to restrict attention to finite spatial volumes, do the desired calculations and then let the volume tend to infinity. We then discretise both the spatial and $\underline{k}$-space variables; we are then talking about discrete Fourier Transforms which in the limit of infinitely fine meshes and infinite volumes tend to the above formulae. Some care is nevertheless needed since we are dealing with stochastic functions. This is the way it is handled in Peebles' book, for example. The $\delta_{\underline{k}}$ are complex numbers defined at all points of the $\underline{k}$-space. It is useful to think in terms of the modulus and argument of these numbers separately, writing

$$\delta_{\underline{k}} = |\delta_{\underline{k}}| e^{i\varepsilon_{\underline{k}}} \tag{29}$$

since we shall make simplifying assumptions about the statistics of the modulus $|\delta_{\underline{k}}|$ in what follows.

In equation (28), the $\delta_{\underline{k}}$ are random functions of $\underline{k}$, simply because the $\delta(\underline{x})$ are random functions of $\underline{x}$, and different realizations of the random process yield different functions. What we are interested in is the statistics of the quantities $\delta_{\underline{k}}$. They clearly have mean value zero:

$$\langle \delta_{\underline{k}} \rangle = 0 \tag{30}$$

where the angular brackets denote a mean value taken over many realizations of the random process. (The fact that we have only one universe means that we can estimate these means by sampling different volumes of space at a given time, thus replacing ensemble averages by spatial averages. It is not clear that this can always be done!) The next descriptor of the random functions $\delta_{\underline{k}}$ is their variance of their amplitudes, $|\delta_k|$, at a given $\underline{k}$:

$$P(\underline{k}) = \langle |\delta_k|^2 \rangle \tag{31}$$

$P(\underline{k})$ is called the "power spectrum" of the random process $\delta(\underline{x},t)$.

There is a very important theorem relating the power spectrum to the correlation function $C(\underline{r})$:

$$P(\underline{k}) = \int C(\underline{r}) e^{i\underline{k}\cdot\underline{r}} d^3\underline{r} \tag{32}$$

One is just the Fourier transform of the other. They both contain the same information, but it is represented differently in each. In situations where the amplitudes are small, the individual Fourier modes are independent and we can view the random process as a superposition of randomly directed plane waves whose amplitudes are fixed by the power spectrum. In the nonlinear situation that develops after recombination it is easier to think of the correlation function since that describes the shape of a typical irregularity.

The inverse Fourier transform of the last equation is

$$C(\underline{r}) = \frac{1}{(2\pi)^3} \int_k P(\underline{k}) e^{-i\underline{k}\cdot\underline{r}} d^3\underline{k} \tag{33}$$

and if we put $\underline{r} = 0$ we have

$$C(0) = \frac{1}{(2\pi)^3} \int P(\underline{k}) d^3\underline{k} \tag{34}$$

The quantity $C(0)$ is simply the variance of the random process $\delta(\underline{x},t)$. So equation (34) says that the power spectrum is the contribution of modes of wavenumber $\underline{k}$ to the total variance, per unit volume of wavenumber space.

It should be remembered that the power spectrum contains only
information about the amplitudes and that specification of the random
process requires some statement about the phases ε_k. The reasonable
assumption is that at the earliest times the phases are random
numbers, uniformly distributed on the interval $[0, \pi)$ (it is hard to
imagine processes that would set up structure in the initial phases,
though of course anything is possible). During the period when the
amplitudes $\delta(\underline{x})$ are small (the linear regime) the phases remain
random, but later on nonlinear interactions produce phase
correlations. The generation of phase correlations during the
collapse is the reason we see discrete well separated objects today
(though some people would argue that this is a perverse way of looking
at the situation!).

The significance of the power spectrum can best be appreciated
by thinking how we might construct, in a computer, a random density
field. We can view the density fluctuations as a superposition of
random sinusoidal waves (this is the inverse Fourier transform):

$$\delta(\underline{x},t) = \frac{1}{(2\pi)^3} \int \delta_{\underline{k}} \, e^{-i\underline{k}\cdot\underline{x}} \, d^3\underline{k} \qquad (35)$$

and if we restrict ourselves to approximating this integral by a sum
over a finite number of such waves (we can do no more in a computer!)
we have

$$\delta(\underline{x},t) = \frac{1}{V_n} \sum_{\underline{k}} \delta_{\underline{k}} \, e^{-i\underline{k}\cdot\underline{x}} \qquad (36)$$

(The normalisation factor V_n depends on the adopted discretisation).
The problem then reduces to specifying the various (complex)
amplitudes $\delta_{\underline{k}}$, for the chosen discretisation of the $\underline{k}$-space. The
amplitude at a given wavenumber is chosen from a Gaussian distribution
of zero mean and with variance equal to the power spectrum at that
wavenumber. The phase of the wave and its direction of propagation

are chosen from uniform random distributions (no preferred phase or direction).

The Fourier representation of the density fluctuations is particularly convenient since we can in linearised theories for the evolution of small amplitude fluctuations treat the individual Fourier components as being independent. This helps at least up until the epoch of recombination, but inevitably fails when large parts of the universe detach themselves from the cosmic background and go into gravitational collapse.

One of the interesting insights that the Fourier representation provides concerns the form of the spectrum of perturbations at the time they were created. If the perturbations are created as a consequence of physical processes that respect the conservation of energy and momentum, we can argue as follows that the primordial spectrum would have a power law of the form $P(k) \propto k^4$. We can formally expand the integral on the right hand side of (28) as

$$\delta_{\underline{k}} = \int \delta(\underline{x}) d^3\underline{x} - i\underline{k} \cdot \int \underline{x}\, \delta(\underline{x}) d^3\underline{x} - \frac{1}{2} k_i k_j \int x_i x_j\, \delta(\underline{x}) d^3\underline{x} + \ldots \tag{37}$$

We then note that the first term must be zero because the mean fluctuation amplitude is zero (definition of the mean!). The second term must also be zero if the process through which the fluctuations are created conserves momentum, since conservation of momentum implies that the mass center of an irregularity must remain fixed. We are then left with the second term being dominant at small wavenumbers, that is at the largest scales. Hence on the largest scales we have a power spectrum

$$P(k) \propto k^4 \tag{38}$$

This is called the "Harrison – Zel'dovich spectrum", and represents our only logical guess for what the primordial power spectrum of density irregularities might have been. An important question is to what extent this spectrum is consistent with the present observed structure in the universe, and that depends largely on our understanding of the galaxy formation process.

There is an obvious connection between wavelength and wavenumber

$$\lambda = \frac{2\pi}{k} \qquad (39)$$

that enables us to translate from the notion of power spectrum in Fourier space to the distribution of mass fluctuations in real space. A density fluctuation in a patch universe of scale λ is made up of contributions from all Fourier components whose frequency exceeds $2\pi/\lambda$. The mass associated with this patch is simply

$$M = \frac{4\pi}{3} \bar{\rho} \lambda^3 \qquad (40)$$

Let us suppose that the power spectrum has the simple form

$$P(k) \propto k^n. \qquad (41)$$

We can then write down the statistics of the fluctuations δM of the mass contained within spheres of a given radius λ :

$$\left\langle \left(\frac{\delta M}{M}\right)^2 \right\rangle \sim \int_{k > \frac{2\pi}{\lambda}} \langle |\delta_k|^2 \rangle k^2 dk$$

$$\sim \int_{k > \frac{2\pi}{\lambda}} k^{n+2} dk \sim k^{n+3} \qquad (42)$$

So since $k \propto \lambda^{-1}$ and $M \propto \lambda^3$ the root mean square mass fluctuation on scale M varies as

$$"\frac{\delta M}{M}" = \left\langle \left(\frac{\delta M}{M}\right)^2 \right\rangle^{1/2} \propto M^{-\frac{1}{2} - \frac{n}{6}} \qquad (43)$$

This is a most important result since it tells us what fluctuations we may expect on a given mass scale at recombination. (This is not directly applicable after substantial nonlinear evolution has taken place). The case $n = 0$ is of particular interest since then $\delta M|_M \propto M^{-1/2}$: the fluctuations in the mass are just the classical "$\sqrt{N}$" statistical fluctuations. We refer to this case as the "white noise" case since all frequencies contribute equally to the power spectrum. (Such a spectrum must of course be cut off at some small scale).

It is not difficult to go one step further and relate the mass fluctuations to the correlation function:

$$\left(\frac{\delta M}{M}\right)^2_r \simeq \frac{3}{r^3} \int_0^r s^2 \xi(s)\, ds \qquad (44)$$

In principle, we could use observations of the correlation function $\xi(s)$ today to evaluate the fluctuations $\delta M/M$ today. (There are a few technical difficulties in correctly interpreting $\delta M/M$ under the present circumstances when the fluctuations in density are manifestly non-Gaussian). We could then extrapolate back to the epoch of recombination to find out what the amplitude of the density fluctuations were at that time. This is an important step in assessing what the amplitude of the initial fluctuations should have been to produce the present cosmic structure, and an important step in assessing the significance of the observed fluctuations of the cosmic microwave background radiation.

BARYONIC PERTURBATIONS: ADIABATIC AND ISOTHERMAL MODES
(The review article by Jones (1976) is a decade old but nevertheless adequate for this section, providing a perspective on the whole field

of galaxy formation. See also the books by Weinberg and Peebles.)

Since the discovery of the evidence for the hot big bang twenty years ago, it has been apparent that we may be able to explain the properties of the large scale cosmic structures as a consequence of physical processes occuring during the "fireball" phase of the expansion when the universe was hot and ionised. This is important since we would otherwise have to say that the properties of the present universe depend entirely on the nature of the initial conditions, and we would not be able to explain anything. There have been a number of outstanding successes in this direction, particularly the realization that the cosmic sound speed and the viscosity of the cosmic medium may have had some role to play in determining the scale of the galaxy formation phenomenon. These physical processes depend in detail on the nature and constituents of the cosmic plasma.

According to the simplest view of our universe, the early universe contains only photons and baryons. More recently, we have reason to believe that other kinds of particle may play an important role. Some of these are the particles associated with recent efforts in high energy physics to unify all the forces of nature in one grand scheme. If neutrinos has a mass in the range 10 - 100eV they too may play a part in determining the scale of present day cosmic structures. Here we shall keep separate what we know and what we simply speculate. Thus in this section we shall confine attention to the "old-fashioned" photon plus baryon plasma picture, keeping the discussion of "exotica" to the following section.

In a two component universe we can consider perturbations in each of the components separately. Thus we could consider fluctuations in the radiation field density while the baryonic density is kept fixed, and conversely fluctuations in the baryonic density while the radiation density is kept uniform. It turns out that this is not the most convenient division of perturbation types, and that if we go through the mathematics a more useful and physically meaningful division emerges. We define two special kinds of mode which are just simple linear combinations of the two just mentioned.

An *adiabatic* perturbation is one in which the radiation and baryons are perturbed together in such a way that the ratio of the number of photons and baryons in a small volume remains everywhere the same. Since this ratio is in fact the entropy per baryon of the cosmic medium, adiabatic fluctuations leave the entropy per baryon constant everywhere, while the density may change from point to point. Under suitable conditions, which we shall come to below, these perturbations behave like sound waves. In the absence of dissipative processes, adiabatic perturbations remain adiabatic, and that is why they are a particularly convenient class of perturbation to consider.

An *isothermal* perturbation is created by changing the baryon density and keeping the photon density constant everywhere. These are isothermal in the sense that the temperature as defined by the photons is everywhere constant. Such perturbations are also "entropy perturbations" because the cosmic entropy per baryon fluctuates. (There is a little variation among authors as to how precisely this isothermal mode is to be defined. Occasionally, authors insist that for a given fluctuation in the baryon density there should be a fluctuation in the photon density of the opposite sign, whose amplitude is adjusted so that there is no net fluctuation in the gravitational field. These are certainly not isothermal in the usual sense of the use of that word, but they are certainly entropy fluctuations.)

The simplest way of describing what happens to perturbations in the early universe is to list the physically important length scales describing the domain of influence of various phenomena. Since length scale at some arbitrary time are of little direct intuitive appeal, it is common to talk in terms of the baryonic mass associated with a particular lengthscale. This is useful since we have some prior notions that mass scales on the order of $10^4 - 10^5 \, M_\odot$ may have something to do with globular clusters, while mass scales of $10^{11} \, M_\odot$ and $10^{13} \, M_\odot$ are associated with galaxies and clusters of galaxies respectively.

The length ct is (to within factors of order unity) the distance over which causal communication is possible when the age of the universe is t. Associated with this radius is a baryonic mass M_H, the "horizon mass":

$$M_H = \frac{4\pi}{3} \rho_B(t) (ct)^3 \qquad (45)$$

Note that if there were an inflationary period, mass scales as large as our present universe would have been causally connected during the period of inflation, since the whole of the present observable universe could have been squashed into a sphere of radius less than M_H at the inflation time. The horizon mass subsequent to inflation is nevertheless a concept that makes sense since that describes the greatest domain of influence that any causal phenomenon may have at a particular time.

Prior to the recombination of the cosmic plasma the universe was ionised and the matter coupled to the radiation via Thomson scattering. The pressure at those epochs is thus given by

$$p = \frac{1}{3} \rho_r c^2 \qquad (46)$$

and an important quantity is the adiabatic sound speed in the cosmic medium

$$c_s = \frac{c}{\sqrt{3}} \left[1 + \frac{3\rho_m}{4\rho_r} \right]^{-1/2} \qquad (47)$$

Associated with this speed is a length λ_J, which is the distance a sound wave can travel in a cosmic expansion time, and a baryonic mass

$$M_J = \frac{4\pi}{3} \rho_B(t) (c_s t)^3 \qquad (48)$$

This is essentially the "Jeans length" in the cosmic plasma. Adiabatic perturbations associated with mass scales larger than this are dominated by gravitational forces, while smaller mass scales are pressure dominated and vibrate like sound waves. After recombination, the radiation pressure is of no importance and the Jeans length is determined by the standard gas pressure $p = nkT$.

Adiabatic fluctuations are also influenced by dissipative phenomena. The photons and baryons are coupled via photon – electron collisions prior to recombination, but that coupling is not perfect. If we look at a patch of universe where the density is higher than average, but where the entropy per baryon is the same as everywhere else (an adiabatic fluctuation), it is possible for the photons to leak out by random walking their way to the boundary of the patch. If the patch is very large that will take a long time and the fluctuation will remain essentially adiabatic. For small enough patches the photon diffusion time may be short enough that the photons all leak out within one cosmic expansion time, leaving an isothermal perturbation whose amplitude is much less than the adiabatic perturbation it came from.

This process of "damping" adiabatic fluctuations by photon diffusion is particularly important during the very brief period when the universe is changing from a totally ionised photon – electron plasma to a neutral hydrogen gas in which the photons play little part. The recombination in fact takes about ten per cent of the cosmic expansion time, and that is long enough for severe attenuation of primordial adiabatic perturbations. The mass scale associated with that attentuation is on the order of

$$M_s \sim 2 \times 10^{12} (\Omega h^2)^{-5/4} M_\odot \qquad (49)$$

Clearly, if the source of cosmic structure were primordial adiabatic perturbations, the first structures to form would have scales more closely related to galaxy clusters than to galaxies themselves. All structure on scales smaller than clusters would be wiped out at the end of the fireball phase of the cosmic expansion. This is the essence of the "pancake theory" for galaxy formation in its original version involving no exotic particles. Galaxies have to form as a consequence of the collapse of gaseous protoclusters, though precisely how that is achieved is not yet fully worked out. The clusters are able to collapse to form large cool gaseous sheets which fragment to form galaxies. The flat gaseous sheets give rise to the name "pancake". There is certainly no observational constraint as to whether galaxies or clusters formed first.

What is interesting is that theories for the origin of the primordial fluctuations almost always produce adiabatic fluctuations. This is simply because the mechanism usually depends on microphysical processes. That is an attractive plus for the pancake theory. An important argument against the pancake theory is that if the universe consisted only of baryonic matter and photons, then we would have Ω_o = 0.1, and the fluctuations at the recombination epoch needed to produce the structure we see now would violate the stringent observational limits on the small scale isotropy of the microwave background radiation. The way out is to argue that in fact $\Omega_o = 1$, and introduce some dark nonbaryonic matter into the universe. This changes the details of the pancake theory since, as we shall see, the fundamental lengthscales are not then determined by photon diffusion.

The isothermal fluctuation theory gives a totally different picture of galaxy formation, though the question there is "where did these perturbations come from?". Isothermal perturbations are not affected by either gravitation nor by diffusion processes and so whatever the universe was born with emerges at recombination, unscathed by the physical processes of the fireball. The details of what happens depend on the spectrum of inhomogeneity (and there is no reason to believe that should be the Harrison - Zel'dovich spectrum). The amplitude and shape of the spectrum can be adjusted to give a variety of quite plausible schemes in which generally speaking the smallest scales collapse first to form subgalactic structures which coalesce together to build galaxies and clusters hierachically.

As unattractive as the isothermal picture may seem from the point of view of the arbitrariness of the initial conditions, if there is no exotic matter in the universe that is the idea we should have to accept (unless it were proved that the estimates of the microwave background anisotropies in the pancake theory had been wrongly calculated). The theory would be far more acceptable if we could think of sensible ways of making isothermal perturbations at an early phase of the expansion of the universe.

CHARACTERISTIC MASS SCALES IN THE UNIVERSE

It has become apparent in the past few years that the cosmic plasma may in fact be more complex than we had originally supposed and that its physics may be dominated by low mass particles like massive neutrinos or gravitinos. This adds additional phenomena and their associated length scales into the problem and we potentially have an embarrassment of riches!

Suppose that in addition to the photons and baryons, the universe is populated by a species of collisionless particle of mass m_+. These particles could be massive neutrinos, gravitinos, axions or whatever kind of exotica appeals to one's taste. There is an important time, t_+, when the temperature falls to a value low enough that the motion of these particles becomes nonrelativistic:

$$kT_+ \sim m_+ c^2 \qquad (50)$$

At times when the temperature is above T_+ the particles move around almost at the speed of light and in a cosmic expansion time can travel a distance ct, the horizon scale. Once the temperature drops below this value, the particle velocities fall and the particles can only diffuse a distance $d < ct$. The cosmic mass scale associated with the free streaming length scale of these particles reaches a maximum at t_+. This maximum mass, M_+ (the mass of baryons in a sphere whose radius is the maximum free streaming lengthscale), is of considerable importance. (There are a number of technical points of importance, for example whether the universe is still radiation dominated at the time t_+. That depends on the mass m_+ and the contribution of baryons to ρ.)

Now consider the evolution (following the period of inflation if there was one) of a spherical adiabatic density perturbation of mass M. (We fix on adiabatic modes for the reasons stated earlier: we can understand how tese originate and the resultant theory looks rather good). At first the perturbation is larger than the horizon (in the sense that its mass $M > M_H$) and it grows in amplitude under the influence of gravitational forces.

When it comes within its horizon it may be the case that its mass is smaller than the mass M_t, and so the collisionless particles will stream out at close to the speed of light leaving only a perturbation in the baryons and photons. That perturbation is subject to viscous forces until and during the recombination period. If on the other hand the perturbation is larger than M_t when it enters the horizon, it retains its collisionless component (since the diffusion time is slow under those circumstances). The collisionless component contributes to the gravitational forces acting on the perturbation. Whether these forces are successful in promoting the gravitational amplification of the perturbation amplitude depends on whether or not falls in the radiation dominated era.

For a universe that would be dominated by massive neutrinos of mass m_ν eV, the characteristic damping mass is calculated to be

$$M_\nu \sim 4 \times 10^{15} \left(\frac{m_\nu}{30 \, eV}\right)^{-2} M_\odot \tag{51}$$

The contribution Ω_ν of such neutrinos to the present mass density of the universe is calculated to be

$$\Omega_\nu h^2 \simeq 0.31 \left(\frac{m_\nu}{30 \, eV}\right) \tag{52}$$

(This can be seen approximately by noting that while the neutrinos are relativistic, their energy density is almost the same as the energy density in the radiation field and falls off as l^{-4}. Once they become nonrelativistic, $t > t_\nu$, their energy density falls off only as l^3 until the present epoch.) If the neutrinos are massive enough that they make a substantial contribution to Ω_o ($m_\nu \sim 100$eV) the scale M_ν is typical of clusters of galaxies rather than galaxies. Hence in a universe dominated by massive neutrinos galaxies have to form by a secondary process. This is the modern version of the pancake theory. There is the possibility of measuring the neutrino rest mass in a laboratory experiment and so verifying the basis of the theory.

In the massive neutrino dominated universe, the baryons cluster after recombination by falling into the potential wells of scale M_ν created by the neutrinos. The gravitational potential is presumably smooth enough that the baryons flow into the well, forming a planar shock structure (a "pancake") deep within the well. It is the fragmentation of this shock that is supposed to lead to galaxies, though the precise details of this are still unknown and it is difficult to confront such a scheme with observational data on galaxies.

The test of a massive neutrino based theory must come from looking at the properties of the clustering induced by the neutrino fluctuations. Given the complex nonlinear evolution that takes place after recombination the only way to handle this is by direct numerical simulation of the gravitational clustering of a gas of massive neutrinos. A number of such simulations have been done and it is found that when the present epoch is reached the neutrinos are clustered on lengthscales on the order of

$$\lambda_\nu \sim 40 \left(\frac{m_\nu}{30 eV} \right)^{-1} \text{Mpc}. \tag{53}$$

Since the typical galaxy clustering lengthscale is only ~ 8Mpc., it looks as though we may have a problem unless we adopt an unreasonably high mass for the particles, or unless we try to argue that the estimates of the Hubble parameter are seriously in error. Note however that the simulations follow the neutrinos, not the baryons, so a direct comparison with observed clustering implicitly contains assumptions about the relative distribution of the baryonic component relative to the neutrinos. There is in addition an assumption about the efficiency of galaxy formation being uniform and not biassed towards the places where clusters are forming. It is not clear that there would not in fact be such a bias in the galaxy formation process: galaxies may form first in the cluster volumes and by their influence on the surrounding cosmic gas might somehow turn off galaxy formation elsewhere. We shall return to "biassed" galaxy formation below.

What is clear is that numerical simulations of the galaxy and cluster forming process cannot, without considerable care, be used to eliminate hypotheses for the formation of galaxies. They do nevertheless help to understand the salient features of a particular theory and point the way to what areas are in need of close attention.

The evidence for a neutrino mass is somewhat controversial, but even if the neutrino mass proves too small to be of cosmological relevance, it is possible that other species of "exotic" particle may play the crucial role. An important alternative suggestion is that the present mass density of the universe is dominated by a particle like a massive gravitino with a mass on the order of $m_g \sim$ 1keV. The characteristic mass in that case becomes

$$M_g \sim 10^{12} \left(\frac{m_g}{1 \text{ keV}}\right)^{-2} M_\odot \qquad (54)$$

which resembles the mass of a healthy galaxy.

By going to particles of greater mass, like axions, we can evidently produce structure of arbitrarily small scale. We pick on the gravitino as an example of "cold dark matter", where the word "cold" refers to the fact that such particles would at the present time have extremely low kinetic energy (unlike the massive neutrinos). The situation with cold dark matter is in some ways easier to handle than with massive neutrinos since structure is built up hierachically on all scale of relevance. We do not have to appeal to a secondary process to form galaxies. Moreover, given the amplitude of the primordial cold dark matter fluctuations and the spectrum (which is presumed to be of the Harrison Zel'dovich type) the whole theory can be worked through. Of course there are always the complications of biassed galaxy formation to handle, but no one suggested that galaxy formation would be an easy problem to solve!

THE CONDENSATION OF GALAXIES

(The problem of the relative distribution of dark and luminous matter is not discussed in textbooks or in any review that has appeared to date)

One of the central issues facing us is the problem that the mass in the universe may not be distributed like the light. This would have to be the case if we lived in an $\Omega_o = 1$ Einstein - de Sitter universe with zero cosmological constant, otherwise it would be difficult to understand the results of the cosmic virial theorem discussed earlier. This means that galaxy formation would have to be biassed in favour of places like galaxy clusters so that they enhance the apparent inhomogeneity of the universe on scales of up to 10Mpc.

In the pancake type pictures for galaxy formation where clusters form first and the galaxies form by some secondary process (like the fragmentation of a cold baryonic shock in the cluster potential well) the bias of galaxy formation in favour of the cluster environment is very strong. In fact, it is difficult to imagine how a system like the Local Group of galaxies could have formed in such a picture. The difficulties arise when we look at models for galaxy and cluster formation in which the smallest (galactic or sub-galactic) scales fragment out first, since then the clusters hardly exist and it is difficult to imagine that they have a strong influence on the galaxy formation process.

There are a number of ways in which this biassing might occur. The most straightforward is if there is for some reason a threshold amplitude for the density fluctuations, below which the fluctuation fails to form a bound luminous system. It is evident that under such circumstances a fluctuation of galaxy size sitting within a protocluster is more likely to be above this threshold than an identical fluctuation sitting in a relative "hole" in the density distribution. Thus galaxy formation is biassed in favour of those places where the clusters will form. The question arises as to why there should be such a threshold.

If the universe is open and Ω_o = 0.1 or 0.2, there is an automatic threshold imposed at recombination:

$$\left(\frac{\rho_c - \bar{\rho}}{\bar{\rho}}\right)_{rec} \equiv \left(\frac{\delta\rho}{\rho}\right)_{thresh} = \frac{1}{1 + Z_{rec}}\left(\Omega_o^{-1} - 1\right), \qquad (55)$$

where Z_{rec} is the redshift of the recombination epoch ($\sim$ 1000). (Here $\bar{\rho}$ is the mean density of the universe at recombination and ρ_c is the critical density at that time). Thus there is an automatic bias in open universes, even though there is no demand for it from the cosmic virial theorem as there is in the Ω_o = 1 case. (The existence of this bias may explain the observation that there is a substantial difference in the amplitudes of the galaxy - galaxy and cluster - cluster correlation functions on scales   10Mpc.)

If Ω_o = 1, then this natural threshold does not exist and we have to search for a more complicated mechanism. We can still take advantage of the fact that fluctuations are more likely to be of higher total amplitude in proto-cluster environments: that will make those galaxies collapse before their counterparts that are not in the cluster environment. If we then appeal to the possibility that the forming galaxies may emit radiation and so heat up the rest of the universe, it is possible for the first formed galaxies (which lie in the cluster environment) to quench the formation of galaxies elsewhere. The details of this scenario pose an interesting problem in cosmology because we have to beware of constraints imposed by observations on the thermal history of the intergalactic medium. Nevertheless, it may work.

The existence of possible biasses in galaxy formation has an impact on our interpretation of the gravitational N-body simulations of the universe. The dynamics may be dealt with correctly, but there is as yet no means of properly identifying in these models which mass

points represent luminous galaxies. This in turn affects the
estimation of the correlation functon amplitudes in the models and the
application of the cosmic virial theorem. It behoves us to be careful
therefore in using N-body models to reject particular theories of
galaxy formation on the grounds that a simulation fails to produce the
correct structure or the correct lengthscales.

CAN WE UNDERSTAND THE EVOLUTION OF THE UNIVERSE?

It is clear that we have a number of possibilities for
understanding the universe as we see it now. It is true that we have
yet to explain the amplitude of the primeval perturbations that led to
the galaxies and clusters we see now, but nevertheless we can see how
things must have happened and it is "only a matter" of being given the
right theory for the nature of matter at the start of the universe.
The key question is perhaps whether we can know when we have the right
theory.

Let us take this in simple steps. Firstly, can we really know
anything about the earliest moments of the universe? The big success
here is of course the explanation of the cosmic baryon – antibaryon
asymmetry: we can allow ourselves to be optimistic and we could claim
that we at least know something as far back as the time when the
temperature was around 10^{16} GeV. It is difficult to be optimistic
about knowing what went on before that. A calculation of the shape
and amplitude of the initial spectrun of density fluctuations, based
on some theory for those epochs, is potentially testable. However, it
seems that all reasonable theories yield a "Harrison – Zel'dovich"
spectrum, so the only test lies in the amplitude and the theories at
our disposal generally have sufficient unknown parameters to fix that.
(Admittedly to cries of "fine tuning!", but it may be that the density
fluctuations were what they were just because those parameters had
those particular values. We would then have to wait for a theory for
the values of the coupling constants of nature!).

What would be really interesting is if we could show that the present structure arose from a spectrum of inhomogeneities that was not the Harrison - Zeldovich spectrum. That would constrain the parameters in theories of matter at ultra-high energy and we may even have to resort to explaining the origin of the spectrum as a consequence of happenings at or before the Planck time when the notion of "causality" ceased to make sense. We are not without hope there either, there may be ways of doing "quantum cosmology" that allow a proper answer to the question of the amplitude of fluctuations generated by quantum gravity processes before the Planck time. But that is another story.

OUTSTANDING MYSTERIES

We have come a long way with cosmology as a branch of the physical science since the discovery some 20 years ago of the cosmic background radiation, though there are still large numbers of unanswered questions. We should not be discouraged by the fact that all is not yet understood, rather we should be optimistic on the grounds that, with the recent advances in high energy physics, we have a possibility of addressing the issues on a scientific basis. If inflation did play a role in our distant past, why does the universe seem to be open? Is there any evidence for substantial amounts of non baryonic material in the universe? Did galaxies form from primeval adiabatic fluctuations (which are the obvious consequence of processes that generate inhomogeneity subsequent to the Planck time)? If not, and if there are no exotic non baryonic particles, how did galaxies form, and in particular where did the non adiabatic perturbations come from?

The problems are not confined to the first instants of the universe's existence. The process of galaxy formation itself might be so complex as to render the interpretation of what we see very difficult. We should be particularly aware of the difficulties posed if galaxy formation were systematically biassed towards the regions where clusters of galaxies were forming. That is a state of affairs that will improve only with a greater effort on the observational side.

The ultimate goal must be to explain the properties of our universe in as much detail as possible. In driving towards this goal we shall have inputs from high energy physics, from observations of galaxies themselves, and from vast numerical simulations of cosmogonic processes.

ACKNOWLEDGEMENTS

We wish to express our appreciation to Jose Luis Sanz, and those who helped him organise this series of lectures, for the invitation to participate in this school.

REFERENCES

REVIEW ARTICLES

Brandenburger,R.H. (1985), Rev. Mod. Phys., 57, 1.

Efststhiou, G and Silk, J. (1983) Fund. Cosmic Phys.

Jones, B.J.T. (1976), Rev. Mod. Phys., **48**, 107.

Linde, A.D. (1984), Rep. Prog. Physics, **47**, 925.

CONFERENCE PROCEEDINGS AND "SUMMER SCHOOLS":

Abell, G.O. and Chincarini, G. "The Early Evolution of the Universe and its Present Structure" (IAU Symposium No.104, Kolymbari, Crete. Reidel, 1984).

Audouze, J. and Tran Thanh Van, J. "The Birth of the Universe" (Proceedings of the 17th. Rencontre de Moriond. Editions Frontieres 1982).

Audouze, J. and Tran Thanh Van, J. "Formation and Evolution of Galaxies and Large Structures in the Universe" (3rd. Moriond Astrophysics Meeting. NATO ASI. Reidel, 1984)

Gerbal, D. and Mazure, A. "Clustering in the Universe" (Proceedings of a European Colloquium held at the Observatoire de Meudon, Paris. Editions Frontieres, 1983)

Gunn, J.E., Longair, M.S. and Rees, M.J. "Observational Cosmology" (10th. Advanced Course of the Swiss Society for Astronomy and Astrophysics, Saas Fee 1978)

Jones, B.J.T. and Jones, J.E. "The Origin and Evolution of Galaxies" (Proc. 7th. Course of the International School on Cosmology and Gravitation: NATO ASI. Reidel 1981)

BOOKS:

Michael Berry: "Principles of Cosmology and Gravitation". (Cambridge University Press, 1976)

Frank Close: "The Cosmic Onion". (Heinemann Educational Books, 1983)

J.E. Dodd: "The Ideas of Particle Physics". (Cambridge University Press, 1984)

David Layzer: "Constructing the Universe". (Scientific American Books, 1984)

P.J.E. Peebles: "Physical Cosmology". (Princeton, 1971)

P.J.E. Peebles: "The Large Scale Structure of the Universe". (Princeton, 1980)

D.J. Raine: "The Isotropic Universe". (Adam Hilger, 1981)

J. Silk: "The Big Bang". (Freeman, 1980)

S. Weinberg: "The First Three Minutes". (Basic Books, 1977)

S. Weinberg: "Gravitation and Cosmology". (John Wiley, 1972)

CLEAN RELATIVISTIC BINARIES: SOME OBSERVATIONAL CONSEQUENCES OF THE CM VELOCITY

Ramon Lapiedra[*]

Dpment. de Mecanica i Astronomia. Facultat de Matematiques
Burjassot (Valencia)
SPAIN

ABSTRACT

We consider a point-like binary slightly relativistic and look for long-time solutions of the equations of motion, at the PN approximation, in a quasi-inertial general frame. From these solutions we obtain a formula giving the times of passage by the periastron positions. In this formula some modulated terms appear which go to zero in the CM frame. Some other effects tied to the velocity of the CM frame are discussed.

INTRODUCTION

We are going to consider point-like (and so clean) relativistic binaries, but I will not do a review on the subject. Rather we will present in this lecture some specific points concerning the PN approximation. Why to revisit in the eighties such an standard topic? There are two reasons for it: the first one is that we are going to treat the dynamical systems in an arbitrary quasi-inertial frame, not just a frame where the velocity, V, of the Newtonian CM of the

---

[*]This lecture stands mainly on a reaserch work made with M. Portilla.The idea of searching for long time solutions to the equations of motion is due to him. Some other results have been obtained in collaboration of we, both, with J.L. Sanz.

binary is zero; the second one is that we search for "long-time" solutions of the equations of motion and not for "short-time" solutions as it is done usually (we will precise these terms later).

As long as the first reason is concerned (to study the binary in a general quasi-inertial frame) it is very often implicitly assumed, that what is essential is to make the study in the rest frame moving to a new quasi-inertial frame being a trivial matter. So, for example, some people assume erronously that once the relative orbital period of the binary (i.e., the time ellapsed between two consecutive periastrons) is known in the rest frame, one can easily to know the corresponding period in another quasi-inertial frame by simple calculating the time dilation implied by the boost conecting both frames. In fact there is nothing, since in order to calculate the relative orbital period in the new frame from the period in the old frame, one ought to know which are the two new events which are now the two corresponding periastrons (defined for example as positions of minimal radial relative coordinade) and this is not more easy than to calculate things in a general quasi-inertial frame from the very begining. This is just which we have done in this work.

2.- THE PN SYSTEM

Let it be a point-like binary of masses m_a, $a = 1,2$. Consider a quasi-inertial frame, i.e., a frame which coordinates (t,x) are nearly-Lorentzian (one where the metric is $g_{\alpha\beta} = \eta_{\alpha\beta} + h_{\alpha\beta}$ with $|h_{\alpha\beta}| \ll 1$ everywhere and zero far away from the sources) and satisfy the harmonic condition. Let $\vec{x}_1, \vec{x}_2$ be the position coordinates of the two stars and $\vec{v}_1, \vec{v}_2$, its velocities. Then, use the new variables $(\vec{x}, \vec{X}, \vec{v}, \vec{V})$ instead of the old ones $(\vec{x}_1, \vec{x}_1, \vec{v}_1, \vec{v}_1)$, as defined by

$$\vec{x} = \vec{x}_1 - \vec{x}_2 \quad , \quad \vec{X} = \frac{m_1 \vec{x}_1 + m_2 \vec{x}_2}{m} \quad , \quad m = m_1 + m_2 \tag{1}$$

$$\vec{v} = \vec{v}_1 - \vec{v}_2 \quad , \quad \vec{V} = \frac{m_1 \vec{v}_1 + m_2 \vec{v}_2}{m} \quad . \tag{2}$$

In the new variables the equations of motion read out, in the PN approximation

$$\frac{d\vec{x}}{dt} = \vec{v} \quad , \quad \frac{d\vec{v}}{dt} = -\frac{Gm}{r^3}\vec{x} + \vec{R} + O(v^6) \tag{3}$$

$$\frac{d\vec{X}}{dt} = \vec{V} \quad , \quad \frac{d\vec{V}}{dt} = \vec{S} + O(v^6) . \tag{4}$$

Here $Gm/r \sim v^2$, with G the gravitational constant.

For R we have the $O(v^4)$ expression

$$\vec{R} = \frac{G}{r^3}\left[2\frac{G(m_1 m_2 + 2m^2)}{r} + \frac{3\mu}{2r^2}(\vec{x}\cdot\vec{r})^2 - (3\mu+m)v^2 + m V^2 + \right. \tag{5}$$

$$\left. + 2 m_{21}\vec{v}\cdot\vec{V} + \frac{3m}{2r^2}(\vec{x}\cdot\vec{V})^2\right]\vec{x} + \frac{G}{r^3}\left[(4m-2\mu)\vec{x}\cdot\vec{v} + m_{21}\vec{x}\cdot\vec{V}\right]\vec{v} ,$$

$m_{21} = m_2 - m_1$

and a similar expression for $\vec{S}$ which we do not write here (see Portilla and Lapiedra 1984).

The PN motion contained in eqs. (3), (4) have been studied by Wagoner and Will (76), Epstein (77) and Spyrou (1981) only in the case $\vec{V} = 0$. Moreover their results are only valid for times not to much long, i.e., of order $O(1/v)$. Notice that the PPN corrections to the accelerations even if small can produce noticiable effects over times long enough.

3. THE NEWTONIAN SOLUTION

Let us recall briefly the solution of the equations of Newton for the relative motion

$$\frac{d\vec{x}}{dt} = \vec{v} \quad ; \quad \frac{d\vec{v}}{dt} = -\frac{Gm}{r^3}\vec{x} \ . \tag{6}$$

This solution is

$$\vec{x}(l) = a(1 - e \cos \xi)\vec{u} \ , \tag{7}$$

$$\xi - e \sin \xi = l \ , \tag{8}$$

$$\vec{u} = \left[\cos \Omega \cos(\omega+\varphi) - \sin \Omega \sin(\omega+\varphi) \cos i\right]\vec{e}_1 +$$
$$\left[\sin \Omega \cos(\omega+\varphi) + \cos \Omega \sin(\omega+\varphi) \cos i\right]\vec{e}_2 +$$
$$\sin(\omega+\varphi)\sin i \ \vec{e}_3 \ , \tag{9}$$

$$tg\frac{\varphi}{2} = \left(\frac{1+e}{1-e}\right)^{1/2} tg\frac{\xi}{2} \ . \tag{10}$$

Here a, e, Ω, i and ω are the constants of motion, nemed respectively semi-major axis, eccentricity, position of the line nodes, inclination of the orbital plane and position of the periastron. On the other hand $\vec{e}_1$, $\vec{e}_2$, $\vec{e}_3$ are three orthonormal vectors of an inertial frame. Also φ is the angle between the relative vector position $\vec{x}$ and the direction of the periastron. (see the figure below). Finally, l is the mean anomaly, i.e., $l = \frac{2\pi}{P_{bn}}(t - T)$

with P_{bn} the orbital Newtonian period and T the first periastron epoch, which only depends on the origin of time and the auxiliary variable ξ is defined by equation (8).

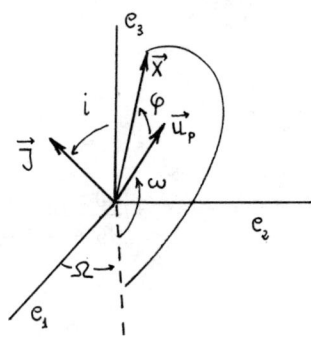

FIG.- This shows the elements of the orbit $(\vec{u}_p, \vec{J}, \Omega, \omega, i, \varphi)$

4. THE PN SYSTEM: THE METHOD OF THE VARIATION OF THE CONSTANTS

Let us make the change of variables $(\vec{x},\vec{v}) \rightarrow (c_\alpha, l)$, with $\alpha = 1...5$, c_α standing for the five first integrals of the Newtonian equations (6) a, e, Ω, i and ω, respectively. Then, equations (6) become

$$\frac{dc_\alpha}{dt} = 0 \quad , \quad \frac{dl}{dt} = n \quad ; \quad n = \left(\frac{Gm}{a^3}\right)^{1/2} \quad (11)$$

Now, consider the PN equations (3), (4). Let us write them by making the change of variables $(\vec{x},\vec{v},\vec{X},\vec{V}) \rightarrow (c_\alpha, l, \vec{X}, \vec{V})$, according to what is called the method of variation of the constants frequently used in Celestial Mechanics. (Brower and Clemence 1961) After some calculation, one obtains, modules PPN terms,

$$\frac{da}{dt} = \frac{2a^2}{Gm_1m_2} DE \quad , \quad \frac{de}{dt} = \frac{P}{Gm_1m_2 e} DE - \frac{\vec{J}\cdot D\vec{J}}{Gm_1m_2 \mu a e} \quad , \quad \frac{d\omega}{dt} = \frac{\vec{N}\cdot D\vec{A}}{Gm_1m_2 e} \qquad (12)$$

$$\frac{di}{dt} = 0 \quad , \quad \frac{d\Omega}{dt} = 0 \qquad (13)$$

$$\frac{d\ell}{dt} = \frac{2\pi}{P_{bn}} + \frac{2\pi\mu p a}{Gm_1m_2 P_n eJ}\left(-2r + \frac{P}{e}\cos\varphi\right)\frac{DE}{\sin\varphi} - \frac{2\pi p \cot g\varphi}{Gm_1m_2 e^2 J P_n}\vec{J}\cdot D\vec{J} \quad , \qquad (14)$$

$$\frac{d\vec{X}}{dt} = \vec{V} \quad , \quad \frac{d\vec{V}}{dt} = \frac{1}{m} D\vec{P} \quad , \qquad (15)$$

where E, $\vec{P}$ and $\vec{J}$ are the Newtonian relative energy, momentum and relative angular momentum. On the other hand μ is the reduced mass $\mu = m_1 m_2 / m$, $p = a(1-e^2)$ and D is the differential operator

$$D \equiv \vec{R}\cdot\frac{\partial}{\partial \vec{r}} + \vec{S}\cdot\frac{\partial}{\partial \vec{v}} \qquad (16)$$

Finally $\vec{N} = \vec{J}\wedge\vec{A}/JA$ where $\vec{A}$ is the Runge-Lenz vector, i.e., $\vec{A} = Gm_1m_2 e \vec{u}_p$ and $\vec{u}_p$ is a unit vector along the major semi-axis of the Newtonian ellipse relative to $\vec{x}$, pointing toward the periastron.

Consider equs. (12) and (15). Take mean values over the Newtonian trajectories. It is well known that one does not find any change in the mean for $\underline{a}$, $\underline{e}$ and $\vec{V}$, while from the change of ω one finds the standard advance of the periastron. These results can be seen to be independent of the fact that V be zero or not, as long as the time intervals which are considered are short enough (it is certainly true for times of order $0(1/v)$, i.e., of some few orbital periods). Nevertheless, as we will see next, another kind of analysis is needed when longer times, i.e., $0(1/v^3)$ times, are involved. (These times can be easily seen to be of the order of the apsidal period of the binary) are involved. In this case, the mean anormaly, l, is corrected with oscillations of long period associated to the motion of the CM of the binary. The reason for this new fact is that, in according with what has been pointed out at the end of the Section 1, when times larg enough are considered, one must device a method to be sure that, when treating the PN system, the PPN corrections to the accelerations can be actually neglected. Thus, when using the standard slow motion approximation, we have $|\vec{x}_{PN}(t) - \vec{x}_N(t)| \sim |a_{PN}(0) - a_N(0)| t^2 \sim 0(v^4) t^2$, where $\vec{x}_{PN}(t)$ and $\vec{x}_N(t)$ are the standard PN and Newtonian solutions and $\vec{a}_{PN}$ and $\vec{a}_N$ the accelerations corresponding to the same initial conditions, the point stands for the time derivation and t is a not too long time. So, under these conditions, both solutions do not differ too much. For example, if t is of the order of the orbital period of the binary, P_b, i.e., $t \sim 0(1/v)$, then we have $|\vec{x}_{PN}(t) - \vec{x}_N(t)| \sim 0(v^2)$, that is $|\vec{x}_{PN}(t) - \vec{x}_N(t)|$ is an small quantity. On the contrary, if we consider a time interval of order $0(1/v^3)$, which are the time intervals we are going to consider in the present work, we have $|\vec{x}_{PN}(t) - \vec{x}_N(t)| \sim 0(1/v^2)$, and so for these long times, we cannot be sure that $\vec{x}_N(t)$ and $\vec{x}_{PN}(t)$ are succesive approximations of the exact solution x(t).

In the following Section we consider an approximation method, called the method of averaging, valid for $0(1/v^3)$ times, i.e., for "long times", but not for "very long times".

5. THE METHOD OF AVERAGING

Let us consider the PN system (12)-(15) and let us write it in the more compact form

$$\dot{l} = \gamma(a) + f(I, l), \qquad (17)$$

$$\dot{I} = g(I, l), \qquad (18)$$

where I stand for any of the six variables $(a, e, \omega, \vec{V}/\varepsilon)$, $\gamma(a) = n/\varepsilon$, and f and g are appropiate periodic functions in l of period 2π, whose expression can be derived by simply comparing the eqs. (17), (18) with eqs. (12)-(15). The parameter $\varepsilon \sim O(v)$ has been introduced in order that γ be order $O(1)$ and f and g order $O(v^2)$, and finally the time has been rescaled: $t \to \tau = \varepsilon t$ (so in (17),(18), $\dot{\sim} \equiv \frac{d\sim}{d\tau}$). According to the method of averaging as it is described by Arnold (1978) let us make the change of variables $(I, l) \to (P, \lambda)$

$$P = I + h(I, l), \qquad (19)$$

$$\lambda = l + j(I, l), \qquad (20)$$

where h and j are $O(v^2)$ functions to be determined in order to simplify the integration of system (12)-(15). In the new variables, the differential system (17),(18) becomes

$$\dot{\lambda} = \gamma(P) + f(P, \lambda) + \gamma \frac{\partial j(P, \lambda)}{\partial \lambda} - \frac{\partial \gamma}{\partial P} \cdot h(P, \lambda) + O(v^4), \quad (21)$$

$$\dot{P} = g(P, \lambda) + \gamma \frac{\partial h(P, \lambda)}{\partial \lambda} + O(v^4). \qquad (22)$$

Now, define the averages

$$\bar{f} = \frac{1}{2\pi}\int_0^{2\pi} f\, d\lambda \quad , \quad \bar{g} = \frac{1}{2\pi}\int_0^{2\pi} g\, d\lambda \, , \tag{23}$$

and the same for any other quatity depending on P, λ (these integrals are taken along the Newtonian orbits, i.e., Keeping P constant). Let us define the oscillatory parts $\tilde{f}$ and $\tilde{g}$, of f and g, respectively as $\tilde{f} = f - \bar{f}$, $\tilde{g} = g - \bar{g}$, and choose the functions h and j of the change of variables (19) (20), to be

$$h(P,\lambda) = -\frac{1}{\gamma}\int_0^\lambda \tilde{g}(P,\lambda)\, d\lambda \, , \tag{24}$$

$$j(P,\lambda) = -\frac{1}{\gamma}\int_0^\lambda \left(\tilde{f} - \frac{\partial \gamma}{\partial P}\cdot \tilde{h}\right) d\lambda \, . \tag{25}$$

Then, the system (21), (22) turns into

$$\dot{\lambda} = \gamma(P) + \bar{f}(P) - \frac{\partial \gamma}{\partial P}\cdot \bar{h}(P) + \mathcal{O}(v^4) \, , \tag{26}$$

$$\dot{P} = \bar{g}(P) + \mathcal{O}(v^4) \, . \tag{27}$$

Notice that the second members do not depend on λ any more, which is a great simplification.

Now, one associates to the system (17), (18), wich we want to integrate in some approximate way, the so-called PN averaged system

$$\dot{\bar{\ell}} = \gamma(\bar{I}) + \bar{f}(\bar{I}) - \frac{\partial \gamma}{\partial \bar{I}}\bar{h}(\bar{I}) \, , \tag{28}$$

$$\dot{\bar{I}} = \bar{g}(\bar{I}) \qquad (29)$$

for the new variables l, I (the bars do not mean here averaged values, so $\bar{l},\bar{I}$ simply stand for the solutions of the system).

Now, since the evolution in time of (λ,P) and ($\bar{l},\bar{I}$), from some given values, is uniquely determined by systems (26), (27) and (28), (29), respectively, one cannot guarantee, in general, that $\bar{l},\bar{I}$ will remain close to λ,P, for any time, even if in the origin they are very close. Let us next discuss how long can be the time intervals considered in order to be sure that $|\bar{l}-\lambda| \sim 0(v^2)$ and $|\bar{I}-P| \sim 0(v^2)$: as long as this be the case the averaged system (28), (29) differs from the original system (26), (27) by terms of order $0(v^4)$. Then, for neighbouring initial conditions $|\lambda(0)-\bar{l}(0)| \sim 0(v^2)$, $|P(0)-\bar{I}(o)| \sim 0(v^2)$, we will have $|\lambda(\tau)-l(\tau)| \sim 0(v^2)$, $|P(\tau)-\bar{I}(\tau)| \sim 0(v^2)$, for rescaled times, τ, such as $\tau \sim 0(1/v^2)$, that is for times $t \sim 0(1/v^3)$, which are $1/v^2$ times greater than the orbital period of the binary. This is which we call here a "long time". Hence, since l,I differ from λ,P by terms of order $0(v^2)$, we have found an averaged solution $\bar{l}(t)$, $\bar{I}(t)$ such that $|l(\tau)-\bar{l}(\tau)| \sim 0(v^2)$, $|I(\tau)-\bar{I}(\tau)| \sim 0(v^2)$ up to times,t, of order $1/v^3$ if we take $\bar{l}(0) = l(0)$, $\bar{I}(0) = I(0)$.

It could be thought, at first sight, that the averaged solution $\bar{l}(\tau)$, $\bar{I}(\tau)$, would be closer to the solutions $l(\tau)$, $I(\tau)$, of the original PN system, for shorter times. Actually this is not true since $|I(t)-P(\tau)|$ is already $0(v^2)$ and so we cannot approach $I(\tau)$ with $\bar{I}(\tau)$ by more than $0(v^2)$ terms. Nevertheless, since we have
$I = P-h(I,l) = P-h(P,\lambda)+0(v^4)$, $l = \lambda-j(I,l) = \lambda-j(P,\lambda)+0(v^4)$, we can consider which is called the improved solution, $I_{imp}(\tau)$, $l_{imp}(\tau)$, i.e.,

$$I_{imp}(t) = \bar{I}(t) - h(\bar{I},\bar{\ell}), \quad \ell_{imp}(t) = \bar{\ell}(t) - j(\bar{I},\bar{\ell}), \qquad (30)$$

which for times of the order of one orbital period, i.e., order $O(1/v)$, differs from the exact solution by terms of order $O(v^4)$.

On the other hand, $h(P,\lambda)$ and $j(P,\lambda)$ are periodic functions of λ of period 2π. Then, modulo $O(v^4)$, the average value of I_{imp} and ℓ_{imp} coincides with $\bar{I}$ and $\bar{\ell}$, respectively.

6. THE SOLUTION OF THE PN AVERAGED SYSTEM

Let it be the averaged PN system (28), (29). Let us write the collective variable $\bar{I}$ in its explicit form $\bar{I} \equiv (a, e, \omega, \vec{V}/\varepsilon)$ and restore the physical time $t = \tau/\varepsilon$. We have

$$\frac{d\bar{a}}{dt} = 0, \quad \frac{d\bar{e}}{dt} = 0, \quad \frac{d\vec{V}}{dt} = 0$$

$$\frac{d\bar{\omega}}{dt} = \frac{3 G m \bar{n}}{\bar{p}}, \quad \bar{p} = \bar{a}(1-\bar{e}^2), \quad \bar{n}^2 \bar{a}^3 = Gm, \qquad (31)$$

$$\frac{d\bar{\ell}}{dt} = \bar{n}\left[1 - \frac{3\bar{a}}{G m_1 m_2} E^*(\bar{I},0) + \frac{1}{4}\vec{V}^2 + \frac{G(\mu - 15m)}{8\bar{a}}\right],$$

where

$$\frac{3\bar{a}}{G m_1 m_2} E^*(\bar{I},0) = \frac{3G}{2\bar{p}}\left[\mu + 4m + (6m - 3\mu)\bar{e}\right] + \frac{3G(3m - 9\mu)}{8\bar{a}} +$$

$$+ \frac{3}{4}\frac{5+3\bar{e}}{1-\bar{e}}\vec{V}^2 + \frac{3 G m_{21} \bar{J}(1+\bar{e})\vec{V}_\perp}{G m_1 m_2 \bar{p}} \sin(\beta - \bar{\omega}) - \frac{3\bar{e}\,\vec{V}_\perp^2}{2(1-\bar{e})}\cos^2(\beta - \bar{\omega})$$

and where $\vec{V}_\perp$ is the projection of $\vec{V}_\perp$ over the orbital plane, $\overline{V}_\perp$ its modulus and β the angle between $\vec{V}_\perp$ and the line of nodes.

Let us take for $\bar{a}, \bar{e}, \bar{w}, \vec{V}$ and $\bar{l}$ the initial conditions $a_0, e_0, w_0, \vec{V}_0, l_0 = 0$, respectively, at $t = T$. Then, the integration of (31) gives

$$\bar{a}(t) = a_0 \;,\; \bar{e}(t) = e_0 \;,\; \vec{\overline{V}}(t) = \vec{V}_0 \;,\; \overline{w}(t) = \dot{w}_0(t-T) + w_0 \qquad (32)$$

where $\dot{w}_0 = 3Gmn_0/P_0$ with $n_0^2 a_0^3 = Gm$ and $p_0 = a_0(1-l_0^2)$. For l one finds

$$\bar{l}(t) = \frac{2\pi}{\bar{P}_b}(t-T) + \xi(t) \qquad (33)$$

where

$$\frac{2\pi}{\bar{P}_b} = n_0 \left[1 + \frac{1}{4} \vec{V}^2 + \frac{G(\mu - 15m)}{8} - \frac{3G}{2\bar{p}}\left(\mu + 4m + e_0(6m - 8\mu)\right) - \frac{3G(3m - 9\mu)}{8 a_0} - \frac{3}{4}\frac{5 + 3e_0}{1 - e_0} V_\perp^2 \right] - \frac{1}{2} \delta_2 \dot{w}_0 \;, \qquad (34)$$

$$\xi(t) = \delta_1 (\cos\alpha(t) - \cos\alpha_0) + \frac{\delta_2}{4}(\sin 2\alpha(t) - \sin 2\alpha_0) \;, \qquad (35)$$

with

$$\delta_1 = -\left(\frac{a_0}{Gm}\right)^{1/2}(1+e_0)^{3/2}(1-e_0)^{1/2}\frac{m_{\perp l}}{m} V_{\perp 0} \;, \qquad (36)$$

$$\delta_\ell = \frac{a_0 e_0 (1+e_0)}{2 G m} V_{10}^2 , \qquad (37)$$

$$\alpha(t) = \alpha_0 - \dot{\omega}_0 (t - T) , \quad \alpha_0 = \beta - \omega_0 , \qquad (38)$$

According to eqs. (7)-(10) this solution can be given in the conventional way

$$\vec{x}(t) = a_0 (1 - e_0 \sin \xi) \vec{u} ,$$

$$\xi - e_0 \sin \xi = \bar{\ell}(t) ,$$

$$\vec{u} = [\cos \Omega \cos(\bar{\omega} + \varphi) - \sin \Omega \sin(\bar{\omega} + \varphi) \cos i] \vec{e}_1 +$$

$$+ [\sin \Omega \cos(\bar{\omega} + \varphi) + \cos \Omega \sin(\bar{\omega} + \varphi) \cos i] \vec{e}_2 + \qquad (39)$$

$$+ \sin(\bar{\omega} + \varphi) \sin i \, \vec{e}_3$$

$$tg \frac{\varphi}{2} = \left(\frac{1+e_0}{1-e_0}\right)^{1/2} tg \frac{\xi}{2} ,$$

$$\bar{\omega} = \omega_0 + \dot{\omega}_0 (t - T) ,$$

$$\bar{\ell}(t) = \frac{2\pi}{P_b} (t - T) + \zeta(t) .$$

It is worth to point here that, unlike the PN differential system (12)-(15) the averaged system (31) can be shown not to be Poincaré invariant (Portilla and Lapiedra 1984). That means that we cannot obtain an averaged solution in a general quasi-inertial frame ($\vec{V} \neq 0$) by boosting the averaged solution previously obtained in the CM frame ($\vec{V} = 0$). The equations of motion are Poincaré invariant but the method used to obtain a long taise solution is not.

The averaged long time solution (39) is seen to be periodic in the variable l, with period 2π . For l = 0, 2 ,...,2kπ , K being a positive entire number, we attain the periastron positions, defined as the configurations of minimal relative coordinte r. For K as great as $1/v^2$, we obtain the periastron epochs, t_k,

$$t_k = T + \kappa \bar{P}_b - \frac{1}{2\pi} \bar{P}_b \xi(t) + \theta(v) . \qquad (40)$$

(A similar expression can be obtained for the periastron positions).

The most relevant fact of the expression (40) is that the periastron epochs are not obtained by adding to the first epoch, t_0 = T, successive multiples of the same orbital period P_b, but some modulated terms, through $\xi(t)$, are also present. These modulated terms depend on cos $\alpha(t)$ and sin $2\alpha(t)$ (notice that $\alpha(t)$ is the angle formed by $\vec{V}_1$ and the periastron direction in the time t).

Now, it could seem that these oscillatory terms are in contradiction with the famous quadrupole radiation formula which gives a secular decreasing for the orbital period without any modulated terms. In fact there is nothing, since the quadrupole formula has been obtained in the framework of the successive approximations, i.e., for short times of the order of one orbital period. Then, it can be seen that if one tries to obtain a formula such as (40), valid for "long times", starting from the quadrupole formula, the errors accumulated when passing from the "short time" description to the "long time" description are greater than O(v) and so one cannot reach any conclusion. It is a trivial matter that the converse way -trying to move from (40) to a formula giving the variation of the orbital period in order to compare with the quadrupole formula- cannot be reached from eq. (40) because of the large error, $O(v)$, present there.

Notice that the oscillatory terms contained in eq. (40), through $\xi(t)$, go to zero for $V = 0$. In this case we obtain an expression for the periastron epoch, t_k, which has the same form that in the Newtonian case, i.e., $t_k = T + K P_b$, the only difference being that we have now $\bar{P}_b$, given by (5.4), instead of the orbital Newtonian period, P_{bn}. So, one could use formula (40), in principle, to derive the velocity of the CM of the binary.

Now, an expression such as (40) is a rather academic formula. Thus, in the interesting case of the binary pulsar PSR 1913+16, the times t_k are by no means observed, the only outstanding data being the arrival times of the successive pulses. So, one could doubt, if, in practice, any observational effect tied to the CM velocity could ever be present. In fact many authors, begenning with Blandford and Teukolsky (1976), emphasise, that, essentially, the boost which conects the CM frame of the binary with an arbitrary quasi-inertial frame, can only introduce constant (and so irrelevant) factors between both sequences of arrival times recorded by the corresponding two observes. And Schutz(1984) even claims that determining V from measuring arrival times would go against the principles of special relativity.

7. TIMING FORMULA, DECREASING OF THE ORBITAL PERIOD AND VELOCITY OF THE CM OF THE BINARY

In order to be convinced that the study made here of a binary as seen from a frame other than the CM frame is not an academic topic by itself let us consider an orbiting pulsar. The only outstanding observational data are the arrival times of the successive pulse to an inertial observer. Then, in order to compare the observations with the theory, one calculates first wich is called the timing formula, i.e., a formula wich gives the number, N, of pulses recorded, starting from an arbitrary given first pulse, as a function of the arrival time, t_N, of the N-th pulse. Blanford and Teukolsky (1976) calculated this timing formula in the case where the velocity, V, of the CM of the binary relatively to the observer, is zero. Then, Portilla and Sanz (1985) have calculated the same timing formula when $V \neq 0$. Some additional terms appear relatively to the case when $V = 0$. This new timing formula allows in principle for the determination of V, through the recording of the arrival times of the pulses to the observer. The additional terms appearing in the timing formula can be understood using only Special Relativity (Lapiedra, Portilla and Sanz 1985).

To finish with these final considerations concerning the observational effects tied to a non zero value of V, consider the well known formula giving the decreasing of the orbital period, P_b, of a binary, because of the quadrupole energy radiation, i.e.,

$$\langle \frac{dP_b}{dt} \rangle = - \text{positive constant} \times P_b^{5/3} \times \text{quadrupole flux} + O(v^6), \quad (42)$$

where the braket $\langle - \rangle$ means that $\frac{dP_b}{dt}$ has been averaged over the orbit, between two consecutive periastrons and V is the small typical velocity of the slow motion approximation. In this formula the velocity, V, of the CM of the binary does not appear. Nevertheless, in the most rigorous derivation of this formula, to our knowledge, that one of Damour 1983 , the calculation has only been made in the CM frame, i.e., for $V = 0$. Now, in Lapiedra, Portilla and Sanz

(1983), an equation similar to eq. (42) has been derived, in the general case $V \neq 0$, the only difference being that in the left side of the new equation we have $\left\langle \frac{dP_{bpn}}{dt} \right\rangle$ instead of $\left\langle \frac{dP_b}{dt} \right\rangle$, where P_{bpn} is the approximated value, up to the PN approximation, of the exact orbital period P_b, i.e.,

$$P_b = P_{bpn} + \Delta P_b \quad . \tag{43}$$

Here, ΔP_b means the PPN and higher corrections. So, ΔP_b is order $O(v^3)$. Then, we have

$$\left\langle \frac{dP_{bpn}}{dt} \right\rangle = \left\langle \frac{dP_b}{dt} \right\rangle - \left\langle \frac{d\Delta P_b}{dt} \right\rangle$$

and the question is, which is the order of $\left\langle \frac{d\Delta P_b}{dt} \right\rangle$? Since $\frac{d\Delta P_b}{dt}$ is order $O(v^4)$ and since the average of $\frac{d\Delta P_b}{dt}$ over one Newtonian orbit could be expected to be zero (the Newtonian orbit is a closed orbit) one would expect $\left\langle \frac{d\Delta P_b}{dt} \right\rangle$ to be order $O(v^6)$, which means that $\left\langle \frac{dP_{bpn}}{dt} \right\rangle$ and $\left\langle \frac{dP_b}{dt} \right\rangle$ would be different by terms of order $O(v^6)$. From this, one would think that in the above referred paper we have obtained eq. (42) for an arbitrary value of V.

Nevertheless this is not so, since, according to Portilla (1983), the above reasoning is not correct as we are going to show next:

First of all, notice that, once one has selected the pair of corresponding consecutive periastrons, P_b is determined by the particular orbit considered. Now, one particular orbit is deremined by the corresponding initial conditions, i.e., the initial relative vector separation and the initial velocities of the two point-like stars. Then, P_b is a function cannot change its value if we move its arguments along the given orbit, i.e., P_b is a constant of motion. But decreases in time because of the gravitational radiation of energy from the binary. Then, what happens is that the function P_b becomes discontinous for the periastron positions. As a consequence we cannot be sure that $\langle \frac{d \Delta P_b}{dt} \rangle$ be zero when we integrate over a Newtonian (and so closed) orbit. This means that we cannot be sure that $\langle \frac{d \Delta P_b}{dt} \rangle$ is order $O(v^6)$, as it has been suggested above. We only can be sure that $\langle \frac{d \Delta P_b}{dt} \rangle$ is at least order $O(v^4)$. So, the conclusion is that $\langle \frac{d P_b}{dt} \rangle$ and $\langle \frac{d P_{bpn}}{dt} \rangle$ differ by terms which are at least order $O(v^4)$ and so from the expression which we have obtained for $\langle \frac{d P_{bpn}}{dt} \rangle$ in the above referred paper, valid for $V \neq 0$, we cannot prove that eq. (42) be also valid for $V \neq 0$.

The final conclusion is that one could doubt if the standard quadrupole formula for the decreasing of the orbital period of a binary, has in fact have been proved when $V \neq 0$, i.e., when the velocity of the CM of the binary, relative to the observer is different from zero.

REFERENCES

1.- Arnold, V. 1976 "Méthodes Mathématiques de la Mécanique Classique". (Moscou. Ed. MIR)
2.- Blandford, R. Teukolsky, S. 1976, Ap.J, **205**, 580.
3.- Brower, D. and Clemence, G.M. 1961 "Methods of Celestial Mechanics" (New York. Academic Press)
4.- Epstein, R. 1977, Ap.J., **216**, 92.
5.- Lapiedra, R. Portilla, M. Sanz, J.L. 1985 (Submitted for publication)
6.- Portilla, M. 1983. Private communication.
7.- Portilla, M. Lapiedra, R. 1984. Ap.J. 15 Nov. **286**, 633.
8.- Portilla, M. Sanz, J.L. 1984 Submitted for publication.
9.- Schutz, B. 1984, Mon. Not. R. astr. Soc. **207**, 37.
10.- Spyrou, N. 1981, Gen. Rel. Grav. 13, 473 and 487.
11.- Wagoner, R.V. Will, C.M. 1976, Ap.J. **210**, 764.
12.- Damour,T.1983,in Proc. 10th Intern. Conf. on Gen. Rel. and Gravitation,ed.B.Bertotti,F.De Felice,and A.Pascolini (Dordrecht:Reidel).

RELATIVISTIC COSMOLOGICAL MODELS

M.A.H. MacCallum,
Theoretical Astronomy Unit,
School of Mathematical Sciences,
Queen Mary College,
Mile End Road,
LONDON E1 4NS, U.K.

ABSTRACT

The nature of the cosmological modelling problem is discussed in the first lecture. The available information is described, and the commonly-used ways of representing and simplifying this information are considered. In the second lecture the available classes of general-relativistic cosmological models are introduced; a few details of their structure and evolution are given, and some recent work on the spatially-homogeneous models is discussed. Finally, in the third lecture, the applications of relativistic models to discussions of the actual universe are described.

1. INTRODUCTION

My general intention in these lectures is
(a) to discuss the purpose and aims of modelling the whole universe and to mention the difficulties involved, (Lecture 1)
(b) to introduce the models which have been discussed up to now (Lecture 2), to explain the simplifying assumptions used, and to give some details of recent work on a few particular problems, and
(d) finally (Lecture 3) to describe the applications of the

models to the explanation or interpretation of actual cosmological data.

It should be stressed from the start that the models used can give only a crude representation of the realities of the universe, and are aimed, like most models in physics, only at simulating certain salient features rather than encompassing all details.

This written account follows my actual lectures fairly closely. I had originally planned to include more technical detail. This, as it turned out, would have been inappropriate for the audience, many of whom knew little relativity (though expert in other things). [In fact the first lecture as given included a thumbnail sketch of general relativity, but I have omitted that here since the reader, unlike the audience of a lecture, has the time to first study one of the many introductory texts on the subject.] Consequently, I had to omit some of my prepared material from the talks; although I did so with some regrets, I have not changed the balance in this written version by restoring more than those small parts of it which seemed essential for coherence and consistency. Instead, I have done my best to provide adequate references to the papers where the details can be found. The standard (Friedman-Robertson-Walker) cosmologies were introduced in Prof. Jones's lectures, and I refer to them for that material.

Inevitably, parts of these lectures are closely related to various earlier reviews and lectures by myself and others [1-3], and sections 2.3 and 2.4 in particular follow previous lectures of mine rather closely.

1.1 The Real Universe

The first problem we face is that the real universe is actually quite complicated, despite our best efforts to pretend that it is not. Bondi [4] pointed out that we like to model both the microscopic and macroscopic scales as if they possessed a simplicity not seen on our ordinary

everyday scale, where there is a high degree of
differentiation, i.e. that "complexity is a feature of
phenomena on our scale only". So let us start by thinking
about what we actually see.

Hoyle [5] was the first, as far as I know, to give the
schematic presentation, depicted in Fig. 1, of the region
from which we have astronomical or geological data. From the
point of view of large-scale structure we are effectively
confined to information about our past light-cone. (For a
review of the interaction between cosmology and geophysics
see Wesson [6].) To make any deductions from the purely
descriptive information, we need some theory which can be
used to transform the latter into initial data on our past
light-cone (that even this step requires a theory has been
discussed by Ellis [7]) and then to predict, or postdict,
its evolution off the light-cone. In these lectures I shall
assume general relativity to be the appropriate theory (see
Will [8] and Coley [9] for discussions of alternatives).
Within this framework one can prove various results about
the problem of the evolution of data given on the past null
cone, which is a characteristic surface; see Friedrich [10].

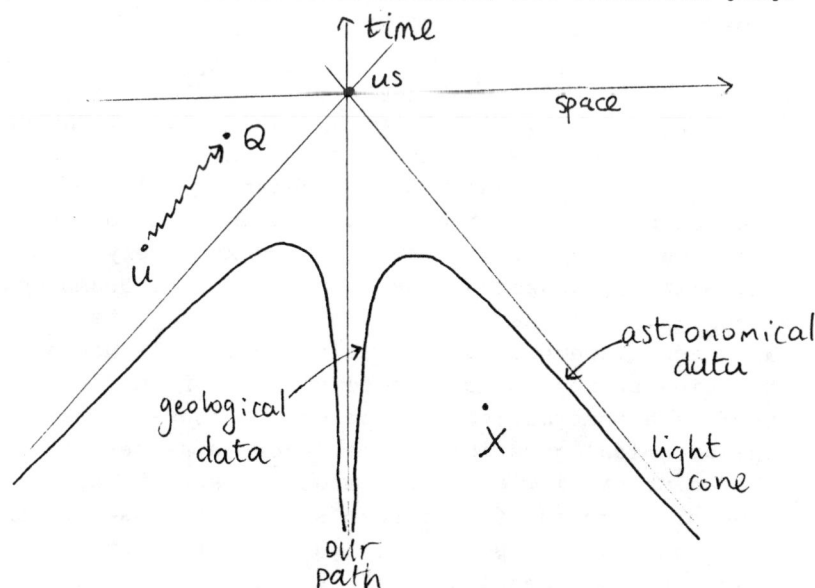

Fig. 1: The region from which we have cosmological
data; see the text for fuller explanantion

It is important to note that only the region within the past light cone could be determined completely from the data. We may be able to make exact statements about conditions at the point X in Fig. 1, but any statements about Q are more questionable since conditions there may be affected by signals unknown to us propagated from, for example, U.

In reality, our knowledge of the light-cone data is incomplete, as is our understanding of the physics of the evolution of celestial bodies. The difficulty is not so much ignorance of appropriate laws of gravity, electromagnetism, weak and strong interactions (though there has always been a tendency to suggest changing these as a way out of difficulties in using them), as our inability to apply them to the very complex systems with which we have to deal (e.g. radio galaxies) and our ignorance of the correct initial or boundary values to use. (In parenthesis, I would remark how very disappointed I was, as a student, to find that works with titles like 'Theory of radio galaxies' only explained how fast electrons and strong magnetic fields produce the radio emission, but not where those fields and electrons come from.)

The 'cosmographic' approach to cosmology, which starts, as I have just done, by considering light-cone data, has been gradually developed over many years by Dautcourt [11] and by Ellis and his colleagues [7,12]. There are both theoretical and practical problems in pursuing it rigorously, but it gives what I see as a very important guideline in assessing the significance of cosmological theories. (I do have doubts about some aspects of this approach, however, in particular because it hints that theories of physics can be constructed from data, which I feel is a most misleading notion; I discuss some philosophical questions of this sort elsewhere [13].) From the point of view of actual model construction, it is probably more helpful to take something closer to the view associated with Popper [14], in which a theory is considered scientifically meaningful if it could in principle be empirically proved to be false.

Let us now consider what data on the large-scale structure we have available.

(a) The most obvious thing is that the universe is <u>lumpy</u> — my density is different from that of the interstellar medium.

(b) By interpreting the well-known Hubble relation between the observed magnitudes and redshifts of galaxies we infer that the universe is <u>expanding</u>. The magnitude (flux received in a certain waveband) is interpreted as a measure of distance d, assuming all the galaxies considered have similar intrinsic brightness, and the redshift can be transformed into a velocity v by assuming the cause is a Doppler effect. The Hubble relation is then

$$d = v\tau$$

where the timescale τ, the Hubble time, is the same for all galaxies. The present estimates of τ lie in the range $10\text{-}20 \times 10^9$ years.

As Sciama [15] has pointed out, expansion also provides a resolution of the Olbers' paradox that the sky's darkness at night is inconsistent with a static uniform universe of infinite age with no overall evolution.

(c) The universe appears to be <u>evolving</u>. This is, of course, suggested by the expansion, but that was equally well-incorporated in steady-state theory. However, there is also the evidence of the radio source counts, which show that the density of the fainter (and hence more distant) sources was higher than the present nearby value, while for the very greatest distances (faintest sources) there is a decreased density.

It is important to note that this evolution is an evolution with affine parameter along the light cone, and can be as well accounted for by a model in which there is variation in space, e.g. one in which the surfaces of uniformity ('homogeneity') form spheres centred on our present position (Fig. 2b), as it is by the more conventional picture (Fig. 2a) in which the surfaces of

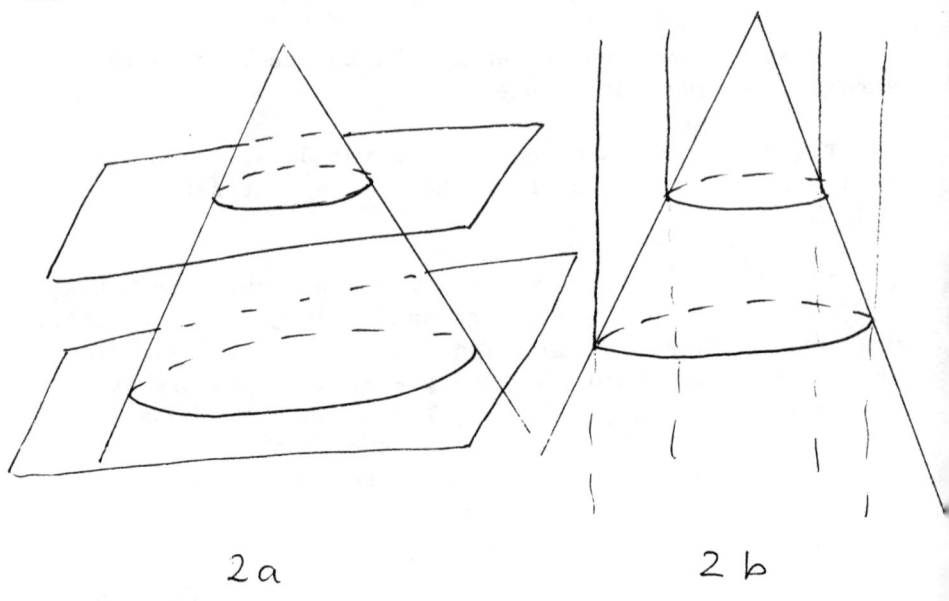

Fig. 2: The two ways of continuing homogeneous surfaces off the past light cone. Fig. 2a shows spacelike surfaces of homogeneity, Fig. 2b timelike surfaces.

homogeneity are spacelike and define surfaces of constant time. In the latter description we would say that the radio source counts are the result of an early phase in the universe during which the sources formed, a middle phase in which the source numbers and power are at a maximum, and a final phase during which they are gradually dying out.

The radio source count data is supplemented by the data on galaxy counts [16, 17] though this does not reach to quite such large redshifts.

(d) The data is nearly *isotropic*.

By isotropy is meant the equality of directions, i.e. that there is no observable or theoretical difference between two directions. [Beware of the fact that some relativists, especially in Latin countries (e.g. in French, 'isotrope'), use the term to mean lightlike.]

The evidence for isotropy comes from (i) the galaxy counts, where the main difficulty is to disentangle the effects of clustering, (ii) the Hubble relation, where any perturbation from isotropy is probably dominated by the supercluster, (iii) the radio source counts, (iv) the X-ray background (v) the cosmic ray background and (vi) the microwave background. For some details relevant to these tests see the lectures of Fabian, Peebles and Uson in this book.

Actually the microwave background shows a dipole anisotropy corresponding to a motion of our sun (and of our Galaxy) relative to this background. As Godel first pointed out, a prediction recently revived by Ellis and Baldwin [18], such a motion should lead to anisotropies in other data, in particular in galaxy and radio source counts.

There are also still some tantalising hints of anisotropy in the observations of radio sources, e.g. the Willson effect of alignments of radio sources [19] and the Birch effect of a relation between elongation direction and polarisation of high power radio double sources [20].

(e) The universe passed through a *hot dense phase*

There are two pieces of evidence for this. One is the chemical composition of the earth, meteorites, the solar system, more distant stars, and galaxies, our knowledge of which becomes less precise the larger the region we scan (and the further into the earth we probe). This composition is generally well accounted for by the ideas in the famous BBFH paper [21], with the exception of the abundances of the light elements relative to hydrogen, especially that of helium (deuterium and lithium being the most important of the other constituents). It appears that these can best be accounted for by fusion reactions in an initial fireball which produces roughly the observed abundances (see [22] for an introduction to the physics involved). [However, one should note that there are residual difficulties, e.g. the metal content of old stars is hard to reconcile with the standard picture unless there was an early generation of massive objects; see comments in the lectures of Carr and Jones in this book.]

The second piece of evidence is the well-known microwave background radiation, which appears to have an essentially black-body spectrum (see Uson's lectures for details of the microwave background) and is hard to account for other than as radiation escaping from a region of thermal equilibrium between photons and electrons. (The other explanations that have been attempted have not been successful.)

It is worth noting that while the microwave background can equally well fit either of the pictures 2a or 2b [23] the same is not true of the chemical composition, which implies that nearby matter has been processed in a hot dense phase, and that therefore the space-time region in which this occurred is in our past, separated in time, rather than at a great distance in space.

(f) There is some evidence of <u>homogeneity</u>

Here 'homogeneity' means that there are three-dimensional surfaces of uniformity, on which all points are equivalent, in four-dimensional spacetime. (Strictly, a homogeneous spacetime would be one in which all points in four dimensions were equivalent, but the evolution described under points (c) and (d) above implies that such models are irrelevant, and we can hence be a bit sloppy with terminology; a more precise term for the homogeneity we are concerned with now is 'hypersurface homogeneity'.) This definition is hard to distinguish, at first sight, from that of isotropy; however one can have spatially-homogeneous but anisotropic models (see Lecture 2) or spherically-symmetric models which depend on both radius and time (i.e. are isotropic about one point but not hypersurface-homogeneous). There are well-known relations between the concepts: in particular, a spacetime isotropic about every point must be spatially-homogeneous and have a metric of the

Robertson-Walker form. If field equations are used, further relations arise, e.g. the Birkhoff theorem and its generalisations which say that certain combinations of symmetry group and energy-momentum tensor imply additional symmetries (for details, see [24]).

The first problem in establishing homogeneity is that, like isotropy, it can only be true in some statistical average sense (as a consequence of the real universe being lumpy, see (a) above) and the required averaging is more difficult to achieve in tests of homogeneity than isotropy. In fact, there is no generally-accepted formal criterion for deciding whether data is compatible with cosmological homogeneity or not. It seems clear that such a definition is only sensible if the scales over which averaging is required are small compared with the size of the observable region and the length scales on which curvature is significant.

A more direct problem, however, arises from the point made by Figs. 1 and 2. To test spatial-homogeneity we must in principle check that conditions at all positions on some spatial (hyper)surface are the same. But we can only observe the intersection of such a surface with our past light-cone (as in Fig. 2a) and it is far from obvious that data compatible with spatial-homogeneity would be incompatible with either temporal-homogeneity (Fig.2b) or no hypersurface-homogeneity at all. Whichever way we continue the model off the past light-cone, we are making assumptions about evolution which may not be correct (the evolution of astronomical objects, with the possible exception of stars, is generally rather poorly understood). The lack of solid evidence for homogeneity may explain why elementary texts are prone to dwell on the isotropy, for which good evidence is available, while treating the spatial-homogeneity assumed in the standard models as if it were almost self-evident. One reason for this, in turn, may be that belief in homogeneity is itself of comparatively recent origin (e.g. until Baade's famous 1952 revision of the distance scale it appeared that ours was the uniquely largest galaxy in the universe) and we may be expressing the over-enthusiasm of the recently converted. The modern view, in sharp contrast to that up to comparatively recent times, is that models in which the Earth or solar system is in a special position in the universe are unacceptable.

So what evidence is there? First, the Hubble law suggests spatial-homogeneity, since only such a linear law could be true at all points of a homogeneous spatial surface. Indeed, it was the redshift-magnitude relation which proved difficult to fit with the temporally-homogeneous spherically-symmetric model [23]; see also [25].

The distribution of galaxies is hard to analyse, and conclusions from it are subject to the strictures above, but models of the clustering based on statistically-homogeneous distributions do seem to be compatible with the data (cf. Peebles' lectures in this book).

Lastly, the lack of any discernable small-scale angular variation in the microwave background limits the possible scale of the lumpiness.

(g) The evidence gives us several **parameters** of cosmological significance, based on the average density of matter ρ and the baryon number density B. The density parameter

$$\Omega = 8\pi G \rho \tau^2 / 3$$

is one, and the entropy per baryon, s, is another. s is essentially the ratio of photon number density to baryon number density, since most of the entropy of the universe is in the heat-bath of the (black-body) microwave background radiation, which also contains most of the photons. The observed lower limit on density (from the luminous matter) is that $\Omega > 0.01$ or so, while element formation work gives an upper limit to the baryon rest-mass density of $\Omega < 0.02$ or so. (There are very interesting possibilities of 'dark matter' in various forms which do not contribute either to observed luminous objects or to the baryon number density).

It should be noted that, with the exception of (a), all the statements above about the likely configuration of the universe rely on quite a bit of theoretical interpretation and are far from being straightforward statements of observed and incontrovertible fact.

1.2 Modelling The Observed Features

In these lectures I shall be concerned for the most part with exact models based on Einstein's field equations of general relativity. Thus I will be excluding quantum gravitational effects (which are expected to be significant only in the very very early universe). I will also be modelling the matter as a classical macroscopic fluid, within which framework one cannot really account for the parameters mentioned under (g) above. Explanations for these features, even the fundamental point that the universe appears not to contain any significant amounts of antimatter, are nowadays sought in terms of grand unified theories of particle physics, which may also drastically alter the evolution of cosmological models in their earliest phases (between the quantum gravitational phase and the 'lepton' era), e.g. the 'inflationary universe' picture; for reviews of these see [26].

In view of the (very understandable) fashionability of studies which blend elementary particle physics and cosmology, it may seem irrelevantly old-fashioned to talk about models based on classical physics. However, as my colleague Michael Rowan-Robinson once remarked, while cosmologists apparently believe they fully understand the first few minutes of the universe's life, the remaining 10 billion years remain a trifle mysterious. In fact since he made that comment the standard view about both the early and middle evolution has undergone several changes and both seem more controversial now. Since quantum gravity is expected to play a part only at extremely small scales, it is still very useful to have a thorough understanding of the classical models, which can be expected to be the appropriate background for almost all the physical processes of cosmological interest.

The study of exact solutions in general relativity has tended to be regarded as using techniques which lack mathematical interest to obtain results which lack physical interest. The first of these objections is I hope answered by the work in [24] and I will restrict my comments here to the second.

There are really only three ways to try to understand a physical theory. One is to prove results about the general nature of its solutions. In the case of relativity, a good example is given by the singularity theorems; another is the available results about general thermodynamics and kinetic theory in relativity. The difficulty with such approaches lies in the mathematical intractability of the problems. (See [27] for a recent review of the singularity theorems which highlights some unsolved problems.) For the case of cosmology the closest one has to such an approach is probably the formulation of fluid-dynamical cosmological modelling by Ellis [28]; see also [29].

The second is to use approximate solutions (which themselves start from an exact solution, however trivial). In the case of relativity, there are substantial difficulties in finding and justifying suitable approximate schemes for such a non-linear theory (the vacuum Einstein equations, in terms of the components of the metric, can be considered as of eighth order in the variables). The main applications of approximation methods, among which I include numerical methods, have been to the problems of galaxy formation and of the early evolution asymptotic to the initial singularity [30].

The third possibility, of approximating the physical situation but taking an exact solution of the governing equations, is the one most commonly used in cosmology, and is the approach with which these lectures are concerned. One important virtue of exact solutions is that they produce predictions of observable relations for comparison with the data and enable studies of astrophysical processes in a known gravitational field to be made; in these respects, they can offer more than general theorems or approximate solutions. In order to apply this approach, one must make simplifications in one's representation of the data (whose validity can be eventually checked by comparing theoretical predictions with observation). How is this actually done?

To model the features described above, it is usual to ignore (a), the lumpiness of the universe, completely. It is hardly surprising that subsequent attempts to re-introduce it (the modelling of galaxy formation processes) have not yet been entirely successful. It seems to me that very important advances in our understanding might arise if we could find a way of describing realistically lumpy universes; perhaps numerical methods will offer the best hope.

Features (b) and (c) will be adequately modelled by a hypersurface-homogeneous solution, but not by a homogeneous space-time.

Feature (d), the isotropy, strongly suggests a spherically symmetric model, or at least one close to spherical symmetry now.

The hot dense phase (e) suggests a model with high-density (and thus high-temperature) regions in our past. It would not be essential for very distant regions to have a similar history, since we have little evidence of their chemical composition, but our belief in an anti-geocentric view makes big-bang theories seem much more plausible than alternatives which also satisfy the singularity theorems but only have singularities on some rather than all of the matter worldlines.

Taken together with the preceding points, (f) suggests that spatially-homogeneous models should be favoured, although restrictions on spatial gradients of physical quantities would probably suffice for compatibility with observation. An objection to such models is that the observed degree of homogeneity is assumed rather than explained, and this was an important motivation of the chaotic cosmology programme [31], which was intended to consider arbitrarily large anisotropies and inhomogeneities and try to reduce them to observed levels by dissipation. In practice this investigation was (curiously) limited to the behaviour of anisotropy and horizons in spatially homogeneous models.

Summarising, the observational evidence suggests the standard (spatially-homogeneous and isotropic) big-bang models but by no means forces them on us. The actual constraints seem to be the degree of present-day homogeneity and isotropy, the presence of the background radiations, the chemical abundances, and, of course, the lumpiness.

1.3 The difficulties in making models

There are a number of difficulties in making cosmological models. The most obvious is the complexity of the Einstein field equations. If the curvature terms on the left hand side of

$$R_{ab} - \tfrac{1}{2} R\, g_{ab} + \Lambda\, g_{ab} = \kappa T_{ab}$$

where Λ is the cosmological constant and κ the gravitational coupling constant, are evaluated in a general system of coordinates they contain (according to Alan Held) 13,280 terms. [Here I am taking the convention that the signature of the metric is +2, Latin indices $a,b,..$ run from 1 to 4, where 4 usually denotes a time coordinate, Greek indices run from 1 to 3, $\kappa = 8\pi G/c^4$, where G is Newton's gravitational constant, and the signs of the Riemann and Ricci tensors are fixed by

$$v_{a;bc} - v_{a;cb} = - R^d{}_{abc} v_d$$

$$R_{ab} \equiv R^c{}_{acb}.$$

Skewing over a pair of indices is indicated by enclosure in square brackets, and symmetrisation by round brackets, while vectors are sometimes denoted by underlining rather than index notation.]

Thus solving the Einstein equations without some simplification is a practical impossibility. In any event, it would require complete specification of the data on the past light-cone even before facing the problem of evolution of the equations off the light-cone. Although it has been pointed out that in principle the data obtainable is just adequate, we can hardly expect to have complete data in practice. Therefore we attempt to find models which give a balance between mathematical tractability and gross misrepresentation of the main features of the universe, and enable us to make some sense of the latter.

The way to study increased complexity of models seems to me to be to increase the geometrical complexity. Retaining simple geometry but treating more complicated forms of the matter content for which the field equations can be exactly solved seems to me much less interesting, except in those rare cases where the more complicated energy-momentum is given a good physical motivation and realistic form, rather then just being cobbled together to create an exact solution for the problem.

However, even with relatively simple forms of energy momentum (e.g. a perfect fluid) one has to give a reasonable prescription for the behaviour under compression or expansion, i.e. one must specify the response of the matter to the geometry (while the Einstein field equations specify the response of the geometry to the matter). In the simplest case this means modelling the transition between the radiation-dominated state of the early universe to the effectively pressureless present-day state.

The next, and often most difficult step (and one frequently ignored by those interested in solving the Einstein equations for cosmological models) is to obtain the observable predictions; this means in particular solving the geodesic equations so that the behaviour of light rays can be understood. It is not at all hard to give exact solutions of Einstein's equations known in terms of simple 'elementary functions' for which the geodesic equations cannot be analytically solved.

Finally the early behaviour of a model is that asymptotic to a singularity, and our understanding of this problem seems to me to still be inadequate (though I suspect that Belinskii et al. [30] might not agree); in particular we do not yet even have a satisfactory definition of the boundary points to be attached to the space-time, at which the singularities actually occur [27].

2. THE AVAILABLE GENERAL-RELATIVISTIC COSMOLOGICAL MODELS

2.1 The energy-momentum content

To solve the Einstein field equations in a meaningful way, as Synge [32] first pointed out, it is essential to specify the two sides independently (rather than use the equations as a definition of the energy-momentum). So we want to give some reasonable specification of the cosmological energy-momentum. The most commonly used assumption is that of a perfect fluid

$$T_{ab} = \rho u_a u_b + p h_{ab}, \quad h_{ab} \equiv g_{ab} + u_a u_b$$

where u^a is the four-velocity of the fluid ($u^a u_a = -1$ in units in which $c = 1$), ρ is the density and p is the pressure of the fluid (as measured in the reference frame of an observer co-moving with the fluid). The equation of state is frequently taken to be $p = (\gamma - 1)\rho$, where $\gamma = 1/3$ for the early universe and $\gamma = 0$ now; the case of constant γ can be regarded as a relativistic version of the perfect gas laws [28].

In the standard Friedman-Robertson-Walker (FRW) models, the total energy-momentum must take the above form. However, this does not imply [33] that the matter actually is a perfect fluid or has a γ-law equation of state, because the same total energy-momentum can be made up in several ways (which is in a way a reworking of Synge's point that one needs more than the Einstein equations to fix the behaviour of matter in a realistic fashion).

The usual reason for a fluid approximation in terrestrial physics is that the number of particles per unit volume is so great that it makes sense to average over them. Even for scales of 1 mm the numbers are still enormous. The same argument looks convincing for intergalactic gas, where the rarefaction is offset by the large volume. However, it is far from clear that the same is true of the 'gas' of galaxies, where the number of particles in an averaging volume is comparatively small (perhaps 100 or 1000); a proper examination of this problem would be worthwhile (this has also been remarked by Heller [34]).

A second problem for the fluid approximation is that it cannot be strictly true where there is expansion, even if the expansion is isotropic [35], and indeed may break down completely [36]. To improve the treatment one should introduce both bulk and shear viscosity, but most attempts to do so analytically (rather than numerically) involve viscous coefficients which are mathematically convenient rather than physically motivated. One should also beware of the fact that if the energy-density is varying in the co-moving frame (referred to as the 'tilted' case)

$$\rho_{,a} h^{ab} \neq 0$$

then in reality one would expect some energy transport, for example heat conduction, which departs from the perfect fluid form. Again this can be difficult to model in a reasonable way analytically; perhaps the best attempt so far is [37]. For some details of the viscous terms and their effects see [38-43].

Finally it is worth mentioning that the effect of inflation primarily arises because the particle physics predicts a non-zero Lorentz-invariant energy-momentum which is the result of some supercooling in a phase transition. Because of the Lorentz invariance this just has the form of a cosmological constant term in the equations (which is a special case of a fluid with $\rho = -p$).

Having discussed the possible form of the energy-momentum, one must discuss its evolution. To do so, and for other purposes, it is useful, as in Newtonian fluid dynamics, to analyze the relative motion of neighbouring elements of the fluid. (Unfortunately, many modern physics courses never discuss continuum mechanics, which can be a serious handicap for students in learning cosmology.) It should be noted that this is a purely kinematic analysis, in no way involving the forces actually present, and indeed can be applied to any timelike vector field, whether or not it is the velocity of a fluid. We write

$$u_{a;b} = \Theta_{ab} + \omega_{ab} - \dot{u}_a u_b$$

where the first term on the right is symmetric and the second term skew, while both are projected into the comoving rest space, i.e.

$$\theta_{ab} = \theta_{ba}, \quad \omega_{ab} = -\omega_{ba}, \quad u^b \theta_{bc} = 0 = u^b \omega_{bc}$$

and $\dot{u}_a$ is defined by

$$\dot{u}_a \equiv u_{a;b} u^b.$$

Since this last quantity represents the amount by which the fluid flow lines fail to be geodesic, it is the acceleration (due to non-gravitational forces). By considering the effect on neighbouring worldlines, and by comparison with the Newtonian equations (see e.g. [28-9]) one can identify ω as the vorticity, giving the local angular velocity of fluid elements about an axis defined by

$$\omega_{ab} v^b = 0, \quad h^a{}_b v^b = v^a$$

while θ_{ab} can be further decomposed into trace and traceless parts

$$\theta_{ab} = \sigma_{ab} + \tfrac{1}{3} \theta h_{ab}, \quad \sigma^c{}_c = 0.$$

The trace θ gives the fractional rate of change of volume, called the expansion, while the traceless part is the shear. One can interpret the shear as causing expansions or contractions along the axes given by

$$\sigma^a{}_b v^b = \lambda v^b$$

where $\lambda > 0$ for an expansion axis, $\lambda < 0$ for a contraction axis, and the net effect is a distortion of shape without change of volume. One can define magnitudes of the vorticity and shear by

$$2\omega^2 = \omega_{ab} \omega^{ab}, \quad 2\sigma^2 = \sigma_{ab} \sigma^{ab}.$$

and one can define an averaged length scale (up to a constant) by

$$\theta = 3\dot{\ell}/\ell$$

The purpose of introducing these kinematic quantities is to give a succinct way of describing the qualitative features of a motion and to introduce quantities which can be used in the macroscopic equations of state for the viscous forces exerted by imperfect fluids. They also provide ways of making simplifications which are a sort of cross between restrictions on the energy-momentum and the geometry. Sufficiently strong restrictions on the kinematics can (for a perfect fluid) be sufficient to severely restrict or even force the choice of models.

2.2 Simplifying the geometry

To simplify the Einstein field equations one essentially does one or both of two things: either one reduces the number of independent variables, or one reduces the number of dependent variables. The general case, whose solution, even if it could be expressed in some formal way, is hardly likely to be usable in cosmological discussions, has 4 independent variables and 10 dependent variables (the metric coefficients) subject to arbitrary coordinate changes.

The reduction of independent variables implies the introduction of symmetry. Symmetry (e.g. rotational symmetry) may also imply reduction of the dependent variables, but this can be accomplished in other ways. At one time it was frequently done by rather ad hoc ansatzes on the metric components, but more recently the trend has been for simplifying assumptions to be expressed in a coordinate-free way. One way to do this is to insist on some special algebraic conditions on the Riemann tensor (which since we are dealing now with perfect fluids, will in fact be conditions on the Weyl tensor) or on its derivatives. (The general theory of the 'equivalence problem' tells us that any non-global condition can be reduced to a formulation in terms of the Riemann tensor and its derivatives; see [44] for the general theory and the Szekeres models below for an example of how a restriction in coordinate terms can be re-derived as conditions stated in a coordinate-free way).

Let us now consider the classes of cosmological model which have been studied in the literature, in terms of increasing symmetry.

(1) Only two types of model with no symmetry group at all have been studied. One is the class of Szekeres models

$$ds^2 = A^2 dx^2 + B^2(dy^2 + dz^2) - dt^2$$

where A and B may be functions of all four coordinates. The models were originally derived [45] just from this ansatz on the metric form, but were subsequently characterised in terms of some of their many special invariant properties [46]. For example, they have a type D Weyl tensor and conformally flat spatial sections. They contain a perfect fluid moving along the time coordinate lines (whose geodesic character implies that the pressure is dependent on time alone); in general this fluid has density dependent on spatial position. Hence, except for the case of dust, no γ-law equation of state is possible without the simultaneous introduction of symmetries. Attempts to generalise or modify these models by substituting different assumptions (e.g. a different energy-momentum tensor), which might make them more realistic as cosmologies, tend to either produce an insoluble problem or introduce some symmetry into the models, so leading to member(s) of the more specialised classes of models below which have symmetry. For example the Szekeres models include spherically symmetric and Robertson-Walker metrics as special cases.

(2) The second class without symmetry considered as cosmological models are the conformally flat perfect fluid solutions of Stephani, which generalise the Friedman-Robertson-Walker solutions. These have been considered by Krasinski [47].

(3) The next step is to impose a single symmetry, reducing the Einstein equations to partial differential equations in three, rather than four, variables. A few such solutions are explicitly known [24], but have not, as far as I know, been applied in cosmology.

(4) Next one has the solutions with two ignorable coordinates. These are divided into three cases; models with two non-commuting symmetries, models with two commuting symmetries, and models with an additional rotational symmetry. The first of these has not really been considered yet in the literature. A special case of the second, in which the metric takes the block diagonal form, has been discussed quite a lot (see [48] and Dr. Verdaguer's lectures for some references). The metric is

$$ds^2 = \epsilon f_{AB} dx^A dx^B + \delta e^{2\gamma}[(dx^3)^2 - \epsilon(dx^4)^2]/f$$

where A and B take values 1 and 2 and in matrix form f_{AB} is

$$\begin{bmatrix} f & -fw \\ -fw & fw^2 + \epsilon(x^4)^2/f \end{bmatrix}.$$

The indicators ϵ and δ take values ± 1 and the functions f, w and γ depend only on the essential variables x^3 and x^4. If $\delta = -\epsilon = 1$ we have the stationary axisymmetric case, which has been extensively investigated. In the $\epsilon = 1$ case the two Killing vectors are both spacelike, and there are two possibilities, depending on whether the determinant of f_{AB} has spacelike ($\delta = -1$) or timelike ($\delta = 1$) gradient (The third possibility, that of a null gradient, requires a different form of the metric.) It is the case $\epsilon = \delta = 1$ which is normally regarded as the one of interest in cosmology, as discussed in Dr. Verdaguer's lectures. It is worth noting, incidentally, that the name 'cylindrical' often applied to these models, is misleading, in that no requirement that one of the ignorable spacelike coordinates is periodic is necessary. (Other names used for these models are plane, which is also misleading in that it suggests an extra rotational symmetry, planar and pseudoplanar.)

(5) If a rotational symmetry is present, and there are two non-ignorable variables, we obtain metrics of the form

$$ds^2 = C^2(dx^2 + \epsilon \Sigma^2(x) dy^2) + A^2 dr^2 - \epsilon B^2 dt^2$$

where $\epsilon = \pm 1$, A B and C depend on r and t only, and

$$\Sigma(x) \equiv \begin{Bmatrix} \sin x \\ x \\ \sinh x \end{Bmatrix} \begin{Bmatrix} \text{spherical} \\ \text{plane} \\ \text{pseudospherical} \end{Bmatrix} \quad \text{if } \epsilon = 1$$

The plane and pseudospherical cases are special cases of the metrics with a group with two generators (respectively a commuting and non-commuting pair); the spherical case is not. Like the metrics considered under (4), these models are governed by partial differential equations in two variables, but have fewer dependent variables than the previous case.

(6) Up to now I have treated 'symmetry' as equivalent to the concept of an isometry, a transformation under which length is preserved. However, it suffices, in order to simplify the equations, to impose other weaker types of symmetry. One such is 'self-similarity' in which the dependence on one of the variables is completely known (as an exponential in that variable). The consequence is that the number of independent variables in the differential equations is in effect reduced. Models with two isometries (as discussed in (4) above) and an additional self-similarity, in particular 'spatially self-similar' models, have been discussed by a number of authors [49-55].

(7) The next step is to replace the self-similarity with an isometry. As in the previous case this implies that the equations reduce to ordinary differential equations. The symmetry group can act on either spacelike or timelike surfaces; the latter have not been much used in cosmology, but have been studied by Harness [56]. In either case the three-parameter symmetry group can be classified by the Bianchi(-Behr) type. This is done by considering the operators generating the symmetry in the same way as is done in quantum mechanics; we write

$$[\xi_\alpha, \xi_\beta] = C^\gamma{}_{\alpha\beta}\xi_\gamma \quad , \quad \xi_\alpha = \xi_\alpha{}^i \frac{i\partial}{\partial x^i}$$

Writing out the distinct choices for α and β in a cyclic order, the coefficients appear as a 3×3 matrix, and we use this to classify the different groups, first taking its symmetric part $N_{\alpha\beta}$ and a vector A^γ equivalent (in the standard way) to its skew part. The Jacobi identity then gives

$$N_{\alpha\beta}A^\beta = 0$$

and after diagonalising $N_{\alpha\beta}$, rescaling the basis vectors to

remove unimportant constants, and, if necessary, re-
numbering the axes, one can arrive at the possibilities
shown in the Table [57-8].

Class	Type	N_1	N_2	N_3	A		
A	I	0	0	0	0		
	II	1	0	0	0		
	VI_0	0	1	-1	0		
	VII_0	0	1	1	0		
	VIII	-1	1	1	0		
	IX	1	1	1	0		
B	V	0	0	0	1		
	IV	0	0	1	1		
	VI_h	0	1	-1	$\sqrt{(-h)}$	h < 0	
	[III	0	1	-1	1	same as VI_{-1}]	
	VII_h	0	1	1	$\sqrt{h}$	h > 0	

Table The Bianchi classification in terms of the
values of the eigenvalues of the matrix $\underline{N}$ and
the length of the vector $\underline{A}$.

For fuller details of this classification see [3, 24,
57-9]. Although it is a classification of groups (more
exactly, of Lie algebras) it is commonly used as a
classification of the spatially-homogeneous cosmologies on
which those groups occur as symmetry groups. It is possible
to prove, see references above, that one must be able to put
the metric of such models in the form

$$ds^2 = -dt^2 + \gamma_{\alpha\beta}(t)e^{\alpha}{}_{\mu}e^{\beta}{}_{\nu}dx^{\mu}dx^{\nu} \qquad (1)$$

where the vectors $e_{\alpha}{}^{\mu}$ have the same commutators as $\xi_{\alpha}{}^{\mu}$,
and $e^{\alpha}{}_{\nu}$ is the matrix inverse of $e_{\beta}{}^{\mu}$ and depends only on
the spatial coordinates. It is worth noting that all Bianchi
types except VIII and IX have two commuting generators
forming a subgroup, and hence can be considered to be
special cases of the metrics considered under (4) above.

(8) One can add to the metrics of (7) above additional
isotropies, leading to metrics which are still governed by
ordinary differential equations, but in fewer dependent

variables. A full list of possibilities appears in [60], but the cases of most interest to cosmology are those in which the extra isotropy is a rotational symmetry at every point; these are the locally rotationally symmetric (LRS) cases. Alternatively one can add an additional translation symmetry to the metrics of (5); this yields one useful possibility which is not a special case of (7) above, namely

$$ds^2 = -\epsilon dt^2 + \epsilon A^2(t)dx^2 + B^2(t)(d\theta^2 + \sin^2\theta\, d\varphi^2)$$

of which the spatially homogeneous case $\epsilon = 1$ is referred to as the Kantowski-Sachs class [61]. The case $\epsilon = -1$ describes static spherically symmetric metrics (they include the Schwarzschild metric, for which the region inside the black hole horizon is the vacuum Kantowski-Sachs metric).

(9) Finally adding the full spatial rotation group, to obtain an isotropic model, one arrives at the Friedman-Robertson-Walker metric form

$$ds^2 = -dt^2 + \ell^2(t)(dr^2 + \Sigma^2(r)[d\theta^2 + \sin^2\theta\, d\varphi^2])$$

which is the one used to obtain the standard cosmological models. Its evolution is governed by ordinary differential equations for the only remaining undetermined function ℓ. (These were discussed by Prof. Jones.)

(10) One can go one step further and impose symmetry throughout all space and time. Such homogeneous spacetimes were used as 'steady-state' models, but play no part in current discussions because they are considered to be incompatible with the observed expansion, evolution, and hot dense phase properties expected.

I had originally intended to give details of the three classes of model which have been most intensively studied, i.e. the FRW models, the Bianchi models, and the models of case (4). However, time prevented this; Prof. Jones covered the FRW case while Prof. Verdaguer dealt with the examples of type (4). Accordingly, I restricted myself to some remarks about perturbations in FRW models and about recent developments in study of Bianchi models. In both cases these echo (rather closely) what I have said elsewhere [2, 62].

2.3 Perturbations of the FRW models

The purpose of this section is to draw attention to the non-triviality of the handling of gauge choice when discussing perturbations in general relativity. A variety of discrepant calculations have appeared in the literature for the rate of growth (or decay) of density perturbations in FRW models. The results are given physical significance by their use for models of the early stages of galaxy formation; consequently it is useful to resolve the disagreements in quoted results. Recently a series of papers have shown exactly how these discrepancies, and some errors in physical predictions, arise from (mis-) handling of the gauge problem [63-67]. The details are quite lengthy so I shall aim only to explain how the problem arises.

Suppose we have a model of the real universe with no symmetries and some lumpiness. In treating it as a perturbed FRW model we are comparing it with a more unrealistic, smoother, model. The comparison is usually described by setting up coordinates in both the models so that points with the same coordinates in the two spaces are compared. The difference between physical quantities at these points (e.g. density) is then ascribed to the perturbation. The difficulty is that a priori one has no way of deciding precisely how the comparison should be made. If a point P in the perturbed model is compared with Q in the FRW model (see Fig. 3), the density perturbation will be different from that obtained if P is compared to R. Yet both may be small enough to reasonably be considered perturbations. The choice of one among the many possible identifications is called the choice of gauge (because the formulae that arise look similar to those encountered when a gauge transformation in electromagnetism, or one of the more recent non-abelian gauge theories of particle physics, is made).

The objective is to find results which are independent of this gauge choice, since it clearly cannot be of physical significance. To do so the gauge change can be described as a coordinate transformation in just one of the two spaces

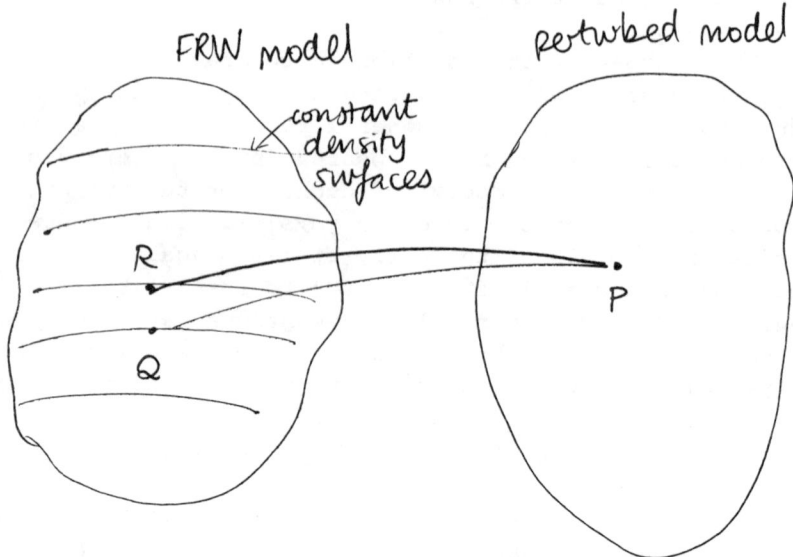

Fig. 3: The correspondence between a perturbed FRW and a FRW model

(if the coordinates in both spaces undergo the same change, the identification does not alter). While this is computationally convenient, it is conceptually confusing. The gauge problem in handling perturbations is **not** the problem of the arbitrariness of coordinates in general relativity, but the problem of indeterminacy of the identification of two distinct space-time models.

Quantities which are constant in the FRW model (in particular, those which are zero) must give rise to gauge invariant perturbations, since the identification cannot affect the difference in values at the compared points (see e.g. [68]). However the density of FRW models is not such a variable; thus the gauge choice does affect the result for the perturbation. Press and Vishniac [63] showed, for example, that the 1/t mode previously discussed was a pure gauge mode. Bardeen [64] has approached the problem by constructing combinations of variables such that the perturbations are gauge invariant; see also [65-67]. There

are certain standard gauge choices which preserve one or
another of the special properties of the FRW metrics, and
the effects of these choices are described in, for example,
[67].

2.4 Spatially homogeneous models

These have been, apart from the FRW models, the most
intensively studied cases; see [1, 3, 59] for reviews. In
the last few years there have been three new (related)
approaches to the study of the spatially-homogeneous models
using methods from modern dynamics or the qualitative theory
of differential equations. These have each been recently
reviewed by their originators, and are mentioned in a
previous review of mine [2]. They are Jantzen's study [69]
of the true degrees of freedom and dynamics for each Bianchi
type, the qualitative study of the evolution using a
representation on a compactified space of variables due to
S.P. Novikov and Bogoyavlenskii [70], and the study of the
properties of Poincare return maps applied in particular to
Bianchi IX models (the famous "Mixmaster universes") by
Barrow [71-72].

Jantzen begins by casting the metric into a canonical
form. This is achieved by first putting the structure
constants in a canonical form as in the Table. The second
step is to write the metric in the form

$$\gamma_{\alpha\beta} = S_\alpha{}^\epsilon \, \gamma'_{\epsilon\delta} \, S_\beta{}^\delta \qquad (2)$$

where $\gamma'_{\epsilon\delta}$ is a diagonal metric and $S_\alpha{}^\epsilon$ is a time-dependent
automorphism of the Lie algebra of the symmetry group, i.e.
the vectors $S_\alpha{}^\beta e_\beta{}^\mu$ have the same commutators as $e_\alpha{}^\mu$.
(Variants of this idea had appeared in earlier work of
several authors.) This technique leads to a separation of
true degrees of freedom from gauge terms [73], and to
clarification of the Hamiltonian and the possible choices of
time variable for the system. It results in a description of
the evolution for all Bianchi types in terms of a
generalisation of the potentials for the cases where

$C^{\alpha}{}_{\beta\alpha} = 0$ (Class A) and the matter content is a fluid at rest in the coordinates of (1), which were given in [74]; the generalisation to matter in relative motion is like that given for type IX by Ryan [75]. The time derivatives of the automorphisms appearing in (2) behave in a way similar to that of angular momenta in a problem involving central forces, and the equations are partially decoupled [77-78].

The diagonal matrix γ' is parametrised by

$$\gamma' = \exp \underline{\beta}, \quad \underline{\beta} = \beta^0 \underline{e}_0 + \beta^+ \underline{e}_+ + \beta^- \underline{e}_-$$

$$\underline{e}_0 = \text{diag}(1, 1, 1), \quad \underline{e}_+ = \text{diag}(1, 1, -2),$$

$$\underline{e}_- = \sqrt{3} \,\text{diag}(1, -1, 0).$$

The additional variables required to describe a perfect fluid matter content can be introduced [69, 78] in a way adapted to the constants of the fluid motion. As an example of the final diagrams of potential in the β plane I reproduce the diagrams for the gravitational potential for Bianchi types II, VI, VII, VIII and IX in Fig. 4 [69, 74-76]. The double-headed arrows indicate the direction of force down exponential potential walls analogous to that of Bianchi type II, while the potential wells show corner channels of two kinds illustrated by the Bianchi VI and VII diagrams. Additional potential terms arise from (a) the centrifugal potentials involving the automorphism velocities and (b) the tilt potentials arising from the motion of the matter relative to the obvious rest-spaces of (1). The centrifugal potential barriers may or may not be penetrable. There is a wealth of detail required for a full discussion of the possible cases arising from the different Bianchi types and behaviour of the matter content for which I can only refer the reader to Jantzen's paper [69].

Before describing the remaining approaches, I give the Bianchi I vacuum (Kasner) solution, which is

$$dt^2 = -dt^2 + t^{2p_1} dx^2 + t^{2p_2} dy^2 + t^{2p_3} dz^2,$$

where $p_1 + p_2 + p_3 = 1 = p_1^2 + p_2^2 + p_3^2$.

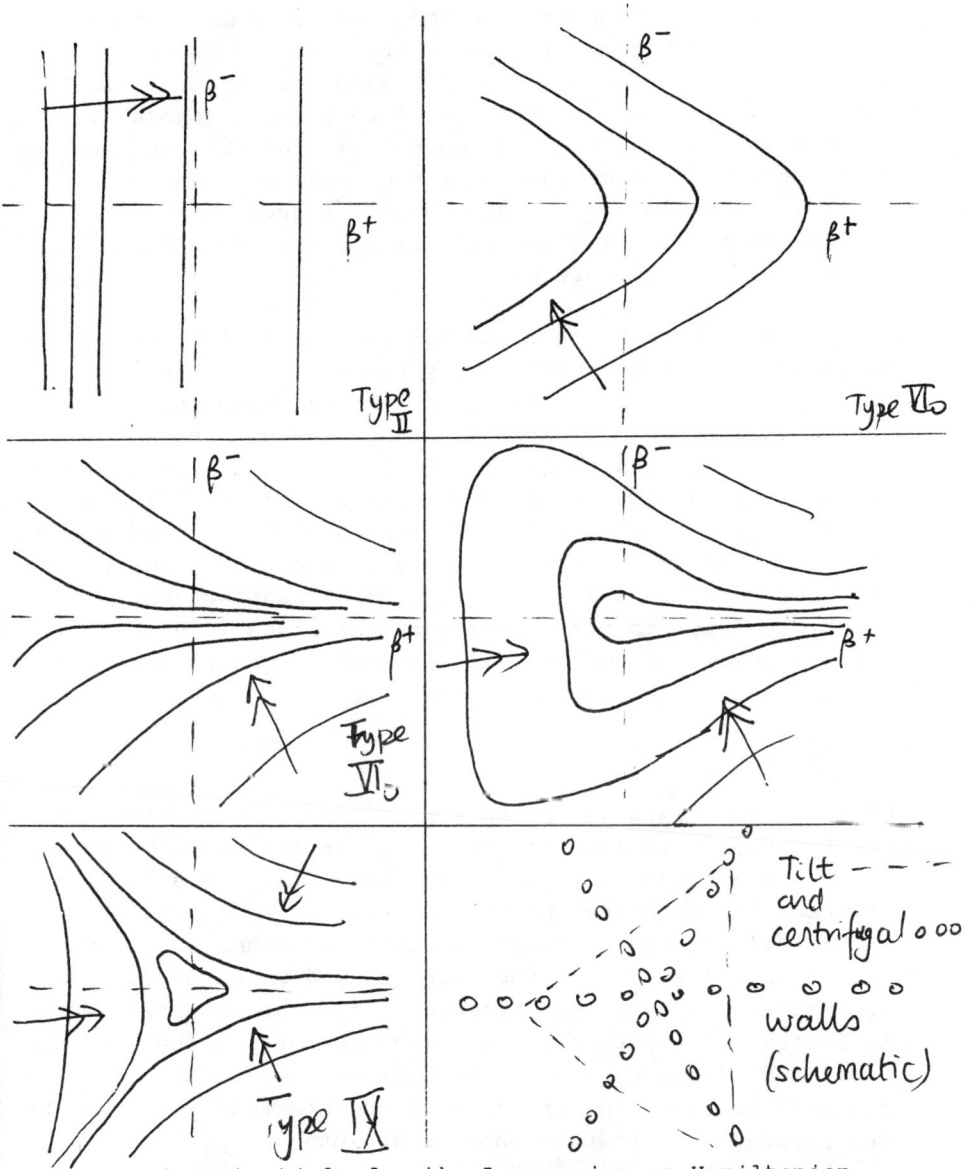

Fig. 4: Potentials for the Lagrangian or Hamiltonian forms of the Einstein equations arising from the spatial curvature in various Bianchi types; for explanation see text.

This is the exact solution if the potential for the motion
is zero. It is an approximate solution for small potential
in all the cases shown in Fig. 4 (for the vacuum case,
anyway), and the evolution of a Bianchi IX universe can be
considered as a series of Kasner-like periods joined by
bounces off the walls of the potential. The BLK treatment of
inhomogeneous metrics says that they follow a similar
oscillatory evolution. These points are necessary to
understanding the remarks that follow; more detail can be
found in [1, 3, 30, 62, 69].

The work of Novikov and Bogoyavlenskii [70] uses
somewhat different methods. Very crudely, it can be
expressed as the following recipe. First choose new
variables for the ordinary differential equations so that
the space of values is compact (and the singularities are on
the boundary of the manifold of configurations). Find a
function (essentially a scaled version of the expansion rate
of the universe) monotone on the solution curves and use it
to prove that, in the time direction of contraction, the
solutions approach the boundary of the space (i.e. the
singularities). Show that the differential equations take
the form of an autonomous system, i.e.

$$\frac{d\underline{x}}{dt} = \underline{f}(\underline{x})$$

where $\underline{x}$ represents the n-tuple of dependent variables, and
the function $\underline{f}$ is analytic in $\underline{x}$ both on the solution curves
and on the boundary. These facts imply that the solutions
approach the boundary and that as they do so the solution
curves will differ arbitrarily little from the solution
curves of the system on the boundary and will follow them
arbitrarily closely (as a result of the analyticity). Thus
to follow the asymptotic evolution one has only to find the
behaviour of the curves in the boundary, which can be
achieved by examining the critical points (where $\underline{f} = 0$) and
the separatrices (curves whose endpoints are critical
points) on the boundary. By these means, Novikov and
Bogoyavlenskii were able to show that the BLK [30]
approximations gave correct results for general type IX
Bianchi models. They also showed that as the models evolved
away from the big-bang the typical behaviour ceased to be of

the oscillatory BLK type and approached instead one of a
small number of possible power-law behaviours (i.e.
behaviours in which the $\gamma'_{\alpha\beta}$ are simply powers of t).

In the Bianchi IX models, they were also able to
characterise typical states at the maximum of expansion.
These ideas have been extended to other Bianchi types but
the treatment in [70] is rather briefer and lacks full
details. The possible power-law behaviours for Bianchi
models have been investigated by Wainwright [79].

The two approaches so far described are not unrelated.
Bogoyavlenskii explicitly uses the Hamiltonian character of
the system, and his work and that of his colleague
Peresetskii (reported in [70]) uses a parametrisation
related to Jantzen's. However, Bogoyavlenskii does not
reduce the system to the true degrees of freedom in the
configuration space before performing his analysis.
Recently, Rosquist [116] has combined the two ideas in a
study of Bianchi VI metrics containing perfect fluid.

Barrow's work [71-72], which develops in certain
respects that of BLK for the homogeneous case, considers
primarily Bianchi IX models. The method is to approximate
the potential shown for type IX in Fig. 4 as a triangular
well with infinite vertical walls and to parametrise the
evolution by the angle at which the curve representing the
evolution of the universe meets the wall on successive
bounces. In the cases with fluid, additional variables are
required [72, 80]. The evolution is then characterised by
the succession of values of these quantities, related by
recurrence relations representing the effect of collisions
with the potential walls. This is essentially the same as
the BKL technique. Barrow (in collaboration with D.
Chernoff) has studied the recurrence in the manner of modern
studies of iterated maps arising as Poincare return maps in
dynamical systems. Although the system is deterministic, the
solutions are very sensitively dependent on initial data and
exhibit the phenomenon of chaotic behaviour in that any
given solution wanders throughout the phase space and
departs by an arbitrarily large amount from an initially

neighbouring trajectory. Because the probability density preserved by the return map can be uniquely determined, the degree of chaos can be expressed quantitatively. Recently, numerical simulations by Zardecki [81] have been used to check this analysis, although they show that white noise tends to reduce the chaos. (I find it interesting that the correction introduced by going back to the true potential is not expected to affect the result although Barrow ascribes the chaos to the behaviour at the corners of the potential well which is precisely where the approximation is worst.)

3. APPLICATIONS TO COSMOLOGICAL PROBLEMS

The aim of this section is to review briefly the application to real cosmological problems of the models introduced in the previous lecture. Each application has many technical details, so this can be no more than a cursory and superficial survey. The various topics are arranged roughly in increasing order of their 'exoticness', i.e. of increasing remoteness from terrestrial conditions either in astronomical distance, time or density. In addition to the topics mentioned here which can be treated purely classically one also of course has the effects of quantum processes to consider. Quantum field theory in curved space can produce particle creation which can in turn lead to the removal of anisotropy. Grand unified theories of physics can predict baryon asymmetry and the entropy per baryon, while the inflationary theories give explanations for the observed homogeneity (though perhaps not the isotropy [101]). I have not attempted to cover these (very large) issues here, partly because I do not feel competent to do so and partly because an adequate discussion would take many pages (see e.g. [26]).

It may be noted here that the advantage of exact solutions over perturbation theory for some of the problems mentioned below lies essentially in the fact that they incorporate the full non-linearity of the theory.

3.1 Local observations

It is at first a surprise to realise that something more than a smoothed-out model is necessary to understand observations of galaxies and quasars correctly. The reason is that in many models, especially those of low average density, there is a strong density contrast between galaxies and the intergalactic medium. A beam of light from a distant galaxy is focussed by the matter within the beam, not by the averaged density. Clearly we see most galaxies not through other galaxies but only through the more diffuse intergalactic medium (conversely, when the beam does pass sufficiently close to massive objects image-doubling may arise, as it is thought may happen for some quasars). This effect is really just an example of the light-bending test of relativity. To model its effect on the interpretation of observed magnitude-redshift relations, two models have been used. One is the "Swiss cheese" models, in which spherical regions of an FRW space are replaced by a Schwarschild solution; this has been studied in [82-83]. The other model used is McVittie's treatment, also for spherical density inhomogeneities [84]. Newman [85] found that there were significant differences between the two models in their predictions. Further study of this problem might be useful.

3.2 Clustering and voids

As well as the local observations, inhomogeneous spherically symmetric models have long been used to model gravitational collapse, for example of galaxies. More recently they have been used to represent the behaviour of large-scale, but not collapsing, density inhomogeneities. Kantowski [86] modelled the Coma cluster with a "Swiss cheese" model. Wesson used spherical models to represent hierarchical cosmologies [87], and other work on modelling the Virgo supercluster and the correlation function of galaxies, interpreted in terms of a hierarchical model [88-90] has been done. Most recently, Maeda et al. [91] have used spherical models to study the possible dynamics of the voids found in the spatial distribution of galaxies.

3.3 Galaxy formation

At one time it was hoped that anisotropic models might have a galaxy collapse rate faster than that in the Friedman models. The calculations of Perko et al. [92] showed that this is not the case for Bianchi I models. Spherical collapse has frequently been used to model the galaxy formation process (see e.g. [93, 55]). Recently Goode [67] showed that certain of the Szekeres evolve exactly according to the perturbation equations for the FRW universes.

3.4 Microwave background

Because the isotropy of the microwave background is established to very high precision (see Uson's lectures) it has been a very useful constraint on possible models, and conversely the possibility of detecting very small deviations has stimulated many people to work on the possible distortions of the background. The consideration of the angular distribution was begun by Thorne [94] and continued by many authors (see e.g. [1] for references). The most recent work on this line has been by Barrow et al. who have used a combination of Bianchi models as a representation of the perturbed FRW universes [95]; these models demonstrate both quadrupole distributions (in Bianchi type I for example, cf. [94, 38]) and 'hot spot' behaviour (in Bianchi type V, for example; cf. [100]). It seems to me surprising that one can correctly represent inhomogeneous models by homogeneous ones (the calculations have been criticised [96], and the situation is not yet clear). Another approach has been to use spherical models to study the passage of the microwave background through density enhancements (e.g. [83, 97]).

As well as the angular distortions, non-FRW models can affect the spectrum and the polarisation of the background. The effect on the spectrum is probably small compared with other astrophysical effects, and does not seem to have been discussed since [99], but the polarisation has recently been considered in [98], where the effects of propagation through Bianchi models was considered. It was found that some models display a handedness and that the rotation of the plane of

polarisation can be significant. This effect is on large
angular scales. The effect of inhomogeneity on the
polarisation on small angular scales has been considered by
Anile.

3.5 Element formation

The primordial abundances of light elements place
constraints on the anisotropy present during the element
formation epoch. Calculations by several authors [94, 102-
103, for example] dealt with spatially-homogeneous models
close to FRW models. The result was very stringent limits.
More recently, these calculations were improved by the
consideration of the anisotropy in the distribution
functions of particles, which had previously been
ignored,and the consequent effects on reaction rates [104-
105]. The result seems to be that helium abundance will
decrease with the introduction of anisotropy.

A radically different result has been found for some
anisotropic models [106-107], in which the usual FRW results
are reproduced in a highly anisotropic situation. The trick
is that in these models the equation governing the rate of
volume expansion, which is the most important parameter in
element formation, contains both shear and spatial
curvature, and it is possible to arrange matters so that
these balance one another.

3.6 Primordial black holes

Exact solutions have been used to model the formation
of black holes in the early universe [108] and to discuss
the effect of anisotropy on this process [109]. The "Swiss
cheese" model has recently been studied with a view to its
global structure, and the amusing result was found that it
is possible to add two different exterior universes to a
Schwarzschild black hole in such a way that it is possible
to send a message (sufficiently early in the expansion) from
one to the other, but not to receive a reply [110].

3.7 Gravitational waves

The metrics with two commuting Killing vectors provide us with exact models of gravitational waves in expanding universes. These have been quite widely discussed in the recent literature, but as this is the subject of Dr. Verdaguer's lectures I refer the reader to his contribution to this volume.

3.8 Isotropy and homogeneity

Misner's articles [31, 38] prompted a large body of work on classical processes to bring about the observed homogeneity and isotropy. The discussions of the effects of dissipation in reducing anisotropy were detailed and technical, but the essential conclusions were that no classical process governed by non-singular differential equations could remove arbitrary amounts of anisotropy, and that the horizon removal predicted for the "Mixmaster universes" only occurred with very small probability (thus failing to meet the objective of showing that the current state is independent of initial conditions at the big-bang.This work was surveyed in [1] and there has been little new since then; [38-40, 43] are relevant articles.

3.9 The far future

It is of some interest to consider the (unobservable) future. The reason, apart from the fun of the speculation, is that models which are isotropic now but will not be so in the far future are clearly less plausible than those in which the present isotropy and homogeneity is a stable property. Bianchi models show that in general this is not the case [74, 111-112] and more recently Wainwright has shown [115] that certain models do indeed fulfil Siklos' prediction that a plane wave state is the appropriate final one [77]. The conclusion is that the demand for a long-term FRW state is a selection principle on models.

3.10 The initial singularity

This problem has already been alluded to earlier. The

work of Belinskii et al. [30] which used the Bianchi IX
oscillatory models as a basis for studying inhomogeneous
models was strongly criticised by Barrow and Tipler [113].
It is difficult to produce a rigorous and conclusive
argument about the correctness of the BLK method [62]. It
has been shown to lead to generally correct results in the
homogeneous cases [70] although it was also found that the
oscillatory phase might only affect those very early times
when the calculations should in fact be replaced by a
quantised gravity theory, while the subsequent phase, in
which a classical treatment might be more relevant, is
apparently more likely to be close to a power-law solution
(for which see [79]). Futher investigation of the
astrophysics of these power-law models is indicated.

Inhomogeneous models may have physically unacceptable
singularities of a non big-bang character. However, Goode
and Wainwright [67, 114] have found that there may be FRW-
like singularities in Szekeres models. Wainwright has also
found solutions which, near the singularity, are in some
sense like a plane wave [115]. For further details of the
many papers on singularities in specific models, see [1-3].

3.11 Conclusion

I hope that the examples described above have shown
that the study of exact relativistic cosmological models,
although it may sometimes appear to be more mathematics than
physics, is not, and cannot be, divorced from the study of
important cosmological questions, including some of very
immediate relevance to the physical interpretation of
observations.

REFERENCES

1. MacCallum, M.A.H., 'Anisotropic and inhomogeneous
    cosmological models', in 'General relativity: an
    Einstein centenary survey' ed. S.W. Hawking and W.
    Israel, Cambridge University Press (1979)
2. MacCallum, M.A.H., 'Exact solutions in cosmology', in
    'Solutions of Einstein's equations: techniques

and results' (Proceedings of the international seminar on exact solutions of Einstein's equations, Retzbach 1983) ed. C. Hoenselaers and W. Dietz, Springer Lecture Notes in Physics, **205**, 334 (1984).
3. Ryan, M.P., jr., and Shepley, L.C., 'Homogeneous relativistic cosmologies', Princeton University Press Princeton, New Jersey (1975).
4. Bondi, H., 'Cosmology', Cambridge University Press (1952).
5. Hoyle, F., 'Cosmological theories of gravitational theories' in 'Evidence for gravitational theories' ed. C. Moller (Proceedings of the International School of Physics 'Enrico Fermi' course 20) 141, Academic Press, New York (1962).
6. Wesson, P., 'Cosmology and Geophysics', Adam Hilger, Bristol and Oxford University Press, New York (1978)
7. Ellis, G.F.R., 'Limits to verification in cosmology, Ann. N.Y. Acad. Sci., **336**, 130 (1980).
8. Will, C.M., 'Theory and experiment in gravitational physics', Cambridge University Press (1981).
9. Coley, A.A., 'Analysis of non-metric theories of gravity' 'I: Electromagnetism', Phys. Rev. **D27**, 728 (1983), 'II: The weak equivalence principle' Phys. Rev. **D28**, 1829 (1983), 'III: Summary of the analysis and its application to theories in the literature', Phys. Rev. **D28**, 1844 (1983).
10. Friedrich, H., 'On the regular and the asymptotic characteristic initial value problem for Einstein's vacuum field equations', Proc. Roy. Soc. **A375**, 169 (1981).
11. Dautcourt, G., 'The cosmological problem as initial value problem on the observer's past light-cone' ':geometry', J.Phys A.**16**, 3507 (1983), ':observations' Astr. Nachr. **304**, 153 (1983).
12. Ellis, G.F.R., Maartens, R. and Nel, S.D. 'Observational Cosmology I: Ideal cosmographic Observations' preprint, (1983).
    Nel, S.D., Stoeger, W.R., Ellis, G.F.R., Whitman, A.P., and Maartens, R. 'Observational cosmology II: Ideal Cosmological Observations' preprint (1983).
13. MacCallum, M.A.H., 'Some philosophical aspects of cosmology', (Lecture given at the Pontifical Academy

of Krakow), Philosophy of Nature (to appear).
14. Popper, K., 'The logic of scientific discovery' Hutchinson, London (1959).
15. Sciama, D.W. 'The unity of the universe', Faber, London (1959).
16. Ellis, R.S., 'Evolution of faint galaxies' in 'The origin and evolution of galaxies' ed. B.J.T. and J.E. Jones, (Nato Advanced Study Institute C 97), D. Reidel and co., Dordrecht (1983).
17. Sievers, A.W., Ellis, G.F.R., and Perry, J.J. 'Cosmological observations of galaxies: number counts' Mon. Not. R.A.S. 212, 197 (1985).
18. Ellis, G.F.R., and Baldwin, J.A., 'On the expected anisotropy of radio source counts', Mon. Not. R.A.S. 206, 377 (1984)
19. Sanders, R.H., 'Alignment of distant radio sources', Nature 309, 35 (1984).
20. Kendall, D.G., and Young, D.A., 'Indirectional statistics and the significance of an asymmetric distribution discovered by Birch', Mon. Not. R.A.S. 207, 637 (1984).
21. Burbidge, E.M., Burbidge, G.R., Fowler, W.A., and Hoyle, F. 'Synthesis of the elements in stars', Rev. Mod. Phys. 29, 547 (1957).
22. Peebles, P.J.E. 'Physical Cosmology', Princeton University Press (1975).
23. Ellis, G.F.R., Maartens, R. and Nel, S.D. 'The expansion of the universe', Mon. Not. R.A.S. 184, 439 (1978).
24. Kramer, D., Stephani, H., MacCallum, M., and Herlt, E., 'Exact solutions of Einstein's field equations', Deutscher Verlag der Wissenschaften, Berlin, and Cambridge University Press, Cambridge (1980) (also in Russian translation, ed. Yu. S. Vladimirov, Energoisdat, Moscow (1982)).
25. Partovi, M.H., and Mashoon,B., 'Toward verification of large-scale homogeneity in cosmology', Ap.J. 276, 4 (1984).
26. Gibbons, G.W., Hawking, S.W., and Siklos, S.T.C. (eds.) 'The very early universe', Cambridge University Press, Cambridge (1983).
27. Tipler, F.J., Clarke, C.J.S., and Ellis, G.F.R.

'Singularities and horizons - a review article', in 'General Relativity and Gravitation: one hundred years after the birth of Albert Einstein', $\underline{2}$, 97, ed. A. Held, Plenum, New York (1980).
28. Ellis, G.F.R., 'Relativistic cosmology', in 'General relativity and cosmology' (Proceedings of the International School of Physics 'Enrico Fermi', course 47) ed. R.K. Sachs, Academic Press, New York (1971).
29. MacCallum, M.A.H. 'Cosmological models from a geometric point of view' in 'Cargese lectures in physics, vol. 6' ed. E. Schatzman, Gordon and Breach, New York (1973).
30. Belinskii, V.A., Khalatnikov, I.M., and Lifshitz, E.M., 'A general solution of the Einstein equations with a time singularity' Adv. Phys. $\underline{31}$, 639 (1982).
31. Misner, C.W., 'Relativistic fluids in cosmology' in Colloques Internationaux de C.N.R.S. $\underline{220}$ (1969).
32. Synge, J.L., 'Relativity: the general theory', North Holland, Amsterdam (1960).
33. Coley, A.A., and Tupper, B.O.J., 'A new look at FRW cosmologies' Gen. Rel. Grav. $\underline{15}$, 977 (1983). 'An exact viscous fluid FRW cosmology' Phys. Lett. $\underline{A95}$, 347 (1983). 'Zero-curvature Friedman-Robertson-Lemaitre models as exact viscous magnetohydrodynamic cosmologies'. Astrophys. J. $\underline{271}$, 1 (1983).
Raychaudhuri, A.K. and Saha, S.K. 'Viscous fluid interpretation of electromagnetic fields I', J. Math. Phys. $\underline{22}$, 2237 (1981). 'II', J. Math. Phys. $\underline{23}$, 2554 (1982).
34. Heller, M. 'On the interpretative paradox in cosmology', Acta. Cosm. $\underline{2}$, 37 (1974).
35. Stewart, J.M., MacCallum, M.A.H., and Sciama, D.W., 'Thermodynamics and cosmology', Comments Astrophys. Space Phys. $\underline{2}$, 206 (1970).
36. Stewart, J.M. 'Non-equilibrium processes in the early universe' Mon. Not. R.A.S. $\underline{145}$, 347 (1969).
37. Bradley, J.M., and Sviestins, E. 'Some rotating, time-dependent Bianchi type VIII cosmologies with heat flow', Gen. Rel. Grav. $\underline{16}$, 1119 (1984).
38. Misner, C.W. 'The isotropy of the universe', Astrophys. J. $\underline{151}$, 431 (1968).
39. Doreshkevich, A.G., Zeldovich, Ya. B., and Novikov, I.

D., 'The kinetic theory of neutrinos in the anisotropic universe', Astrofizika $\underline{5}$, 539 (1969).
40. Stewart, J.M., 'Non-equilibrium relativistic kinetic theory', Springer Lecture Notes in Physics, vol. 10, Springer-Verlag (1971).
41. Stewart, J.M., and Israel, W. 'Progress in Relativistic Thermodynamics and electrodynamics of continuous media', in 'General Relativity and Gravitation: one hundred years after the birth of Albert Einstein', $\underline{2}$, 491, ed. A. Held, Plenum, New York (1980).
42. van Leeuwen, W.A., Polak, P.H., and de Groot, S.R. 'On relativistic kinetic gas theory IX: transport coefficients for systems of particles with arbitrary interaction', Physica $\underline{63}$, 65 (1973).
43. Caderni, N. 'Viscous dissipation and evolution of homogeneous cosmological models' in 'Physics of the Expanding Universe', ed. M. Demianski, Springer Lecture Notes in Physics vol. $\underline{109}$, 81 (1979).
44. Karlhede, A. 'A review of the geometric equivalence of metrics in general relativity' Gen. Rel. Grav. $\underline{12}$, 693 (1980).
45. Szekeres, P. 'A class of inhomogeneous cosmological models' Comm. math. phys. $\underline{41}$, 55 (1975).
46. Szafron, D.A., and Wainwright, J. 'A class of inhomogeneous cosmological perfect fluid cosmologies', J. Math. Phys. $\underline{18}$, 1608 (1977).
Wainwright, J. 'Characterisation of the Szekeres inhomogeneous cosmolgies as algebraically special solutions' J. Math. Phys. $\underline{18}$, 672 (1977).
47. Krasinski, A. 'On the global geometry of the Stephani universe' Gen. Rel. Grav. $\underline{15}$, 673 (1983).
48. Kitchingham, D.M. 'The use of generating techniques for space-times with two non-null commuting Killing vectors in vacuum and stiff perfect fluid cosmological models' Quant. Class. Grav. $\underline{1}$, 677 (1984).
49. Cahill, M.E., and Taub, A.H., 'Spherically symmetric similarity solutions of the Einstein field equations for a perfect fluid' Comm. math. phys. $\underline{21}$, 1 (1971).
50. Eardley, D.M. 'Self-similar spacetimes: geometry and dynamics' Comm. math. phys. $\underline{37}$, 287 (1974).
51. Jantzen, R.T., 'Variational of parameters in cosmology' Ann. Phys. (N.Y.) $\underline{127}$, 307 (1980).

52. Luminet, J.P. 'Spatially homothetic cosmological models' Gen. Rel. Grav. **9**, 673 (1978).
53. Henriksen, R.N., and Wesson, P.S. 'Self-similar space-times I: three solutions' Astrophys. Sp. Sci. **53**, 429 (1978), 'II: Perturbation scheme' Astrophys. Sp. Sci. **53**, 445 (1978).
54. Wu, Z.C. 'Self-similar cosmological models' Gen. Rel. Grav. **13**, 625 (1981) and J. China Univ. Sci. and Tech. **11**(2), 25 and **11**(3), 20 (1981).
55. Bogoyavlenskii, O.I., and Moschetti, G. 'The investigation of some self-similar solutions of Einstein's equations' J. Math. Phys. **23**, 1353 (1982).
56. Harness, R.S. 'Spacetimes homogeneous on timelike hypersurfaces' Ph. D. thesis, Queen Mary College, London and J. Phys. A **15**, 135 (1982).
57. Estabrook, F.B., Wahlquist, H.D. and Behr, C.G. 'Dyadic analysis of spatially homogeneous world models' J. Math. Phys. **9**, 497 (1968).
58. Ellis, G.F.R., and MacCallum. M.A.H. 'A class of homogeneous cosmological models' Comm. math. phys. **12**, 108 (1969).
59. MacCallum, M.A.H., 'The mathematics of anisotropic spatially homogeneous cosmologies' in 'Physics of the expanding universe' ed M. Demianski, Springer Lecture Notes in Physics vol. **109**, 1, Springer-Verlag, Berlin (1979).
60. MacCallum, M.A.H., 'Locally isotropic spacetimes with non-null homogeneous hypersurfaces' in 'Essays in general relativity: a festschrift for A.H. Taub' ed. F.J. Tipler, Academic Press, New York (1980).
61. Kantowski, R., and Sachs, R.K. 'Some spatially homogeneous anisotropic cosmological models' J. Math. Phys. **7**, 443 (1966).
    Kantowski, R. 'Some relativistic cosmological models' Ph. D. thesis, University of Texas (1966).
62. MacCallum, M.A.H. 'Relativistic cosmology for astrophysicists' in 'The origin and evolution of galaxies' ed. B.J.T. and J.E. Jones, (Nato Advanced Study Institute C 97), D. Reidel and co., Dordrecht (1983) and 'The origin and evolution of galaxies' ed. V. de Sabbata, World Scientific, Singapore (1982).
63. Press, W.H. and Vishniac, E.T. 'Tenacious myths about

cosmological perturbations larger than the horizon size' Astrophys. J. <u>239</u>, 1 (1980).
64. Bardeen, J.M. 'Gauge-invariant cosmological perturbations' Phys. Rev. <u>D22</u>, 1882 (1980).
65. Brandenberger, R., Kahn, R., and Press, W.H. 'Cosmological perturbations in the early universe' Phys. Rev. <u>D23</u>, 1809 (1983).
66. Abbott, L.F. and Wise, M.B. 'Gauge-invariant cosmological fluctuations of uncoupled fluids' Nucl. Phys. <u>B237</u>, 226 (1984).
67. Goode, S.W. 'Spatially inhomogeneous cosmologies and their relation with the FRW models' Ph. D. thesis, University of Waterloo (1983).
68. Stewart, J.M. and Walker, M. 'Perturbations of spacetimes in general relativity' Proc. Roy. Soc. Lond. <u>A341</u>, 49 (1974).
69. Jantzen, R.T. 'Spatially homogeneous dynamics: a unified picture' in 'Cosmology of the early universe' ed. L.Z. Fang and R. Ruffini (Adv. Series in Astron. and Astrophys. vol. 1, 233) World Scientific, Singapore (1984).
70. Bogoyavlenskii, O.I. 'Methods of the qualitative theory of dynamical systems in astrophysics and gas dynamics' (in Russian) Nauka, Moscow (1980).
71. Barrow, J.D. 'Chaotic behaviour in general relativity' Phys. Repts. <u>85</u>, 1 (1982).
72. Barrow, J.D. 'Chaotic behaviour and the Einstein equations' in 'Classical general relativity' ed. W.B. Bonnor, J.N. Islam and M.A.H. MacCallum, Cambridge University Press (1984).
73. Jantzen, R.T. 'The dynamical degrees of freedom in spatially homogeneous cosmologies' Comm. math. phys. <u>64</u>, 211 (1979).
74. MacCallum, M.A.H. 'A class of homogeneous cosmological models III: Asymptotic behaviour' Comm. math. phys. <u>20</u>, 57 (1971).
75. Ryan, M.P. jr., 'Qualitative cosmology: diagrammatic solutions for Bianchi type IX universes: I. the symmetric case' Ann. Phys. <u>65</u>, 506 (1971). 'II. the general case' Ann. Phys. <u>68</u>, 541 (1971).
76. Ryan, M.P. jr., 'Hamiltonian cosmology' Springer Lecture Notes in Physics, vol. 13, Springer-Verlag,

Berlin (1972).
77. Siklos, S.T.C. 'Field equations for spatially homogeneous spacetimes' Phys. Lett. A76, 19 (1980).
78. Jantzen, R.T. 'Perfect fluid sources for spatially homogeneous cosmologies' Ann. Phys. (N.Y.) 145, 378 (1983).
79. Wainwright, J. 'Power law singularities in orthogonal spatially homogeneous cosmologies' Gen. Rel. Grav. 16, 657 (1984).
80. Elskens, Y. 'Alternative descriptions of the discrete mixmaster universe' Phys. Rev. D28, 1033 (1983).
81. Zardecki, A. 'Modelling in chaotic relativity' Phys. Rev. D28, 1235 (1983).
82. Kantowski, R. 'Corrections in the luminosity relations of the homogeneous Friedmann models' Astrophys. J. 155, 89 (1969).
83. Dyer, C.C. 'The gravitational perturbation of the cosmic background radiation by density concentrations' Mon. Not. R.A.S. 175, 249 (1976).
84. Newman, R.P.A.C., and McVittie, G.C. 'A point particle model universe' Gen. Rel. Grav. 14, 591 (1982).
85. Newman, R.P.A.C. 'Singular perturbations of the empty Robertson-Walker universes' Ph. D. thesis, University of Kent (1979).
86. Kantowski, R. 'The Coma cluster as a spherical inhomogeneity in relativistic dust' Astrophys. J. 155, 1023 (1969).
87. Wesson, P.S. 'Relativistic hierarchical cosmology III: comparison with observational data' Astrophys. Sp. Sci. 32, 315 (1975).
88. Fennelly, A.J. 'Anisotropy in the Hubble parameter and large-scale cosmological inhomogeneity' Mon. Not. R.A.S. 181, 121 (1977).
89. Mavrides, S., and Tarantola, A., 'Local supercluster and anomalous Hubble expansion' Gen. Rel. Grav. 8, 665 (1977).
90. Bonnor, W.B. 'A non-uniform cosmological model' Mon. Not. R.A.S. 159, 261 (1972).
91. Maeda, K., Sasaki, M. and Sato, H. 'Voids in the closed universe' Prog. Theor. Phys. 69, 89 (1983). Maeda, K., and Sato, H. 'Expansion of a thin shell around a void in expanding universe II' Prog. Theor.

Phys. 70, 1276 (1983).
92. Perko, T.E., Matzner, R.A., and Shepley, L.C. 'Galaxy formation in anisotropic cosmologies' Phys. Rev. D6, 969 (1972).
93. Tolman, R.C. 'Effect of inhomogeneity on cosmological models' Proc. Nat. Acad. Sci. (Wash.) 20, 169 (1934).
94. Thorne, K.S. 'Primordial element formation, primordial magnetic fields and the isotropy of the universe' Astrophys. J. 148, 51 (1967).
95. Barrow, J.D., Juszkiewicz, R. and Sonoda, D.H. 'The structure of the cosmic microwave background' Nature, 305, 397 (1983).
96. Lukash, V.N. and Novikov, I.D. 'The effect of spottiness in large-scale structure of the microwave background' Moscow preprint (1984).
97. Goicoechera, L.J., and Sanz, J.L. 'Evolution of spherically symmetric perturbations of Friedman models and the propagation of cosmic background radiation' Phys. Rev. D29, 607 (1984).
98. Tolman, B.W., and Matzner, R.A. 'Large scale anisotropies and polarisation of the microwave background in homogeneous cosmologies' Proc. Roy. Soc. London 392, 391 (1984).
99. Rasband, S.N. 'Expansion anisotropy and the spectrum of the cosmic background radiation' Astrophys. J. 170, 1 (1971).
100. Novikov, I.D. 'An expected anisotropy of the cosmological radio-radiation in homogeneous anisotropic models' Astr. Zh. 45, 538 (1968), trans. as Sov. Astr. A.J. 12, 427 (1968).
101. Barrow, J.D. 'Can inflation isotropise the universe?' Preprint (1984).
102. Barrow, J.D. 'Light elements and the isotropy of the universe' Mon. Not. R.A.S. 175, 359 (1976).
103. Olson, D.W. 'Helium production and limits on the anisotropy of the universe' Astrophys. J. 219, 777 (1978).
104. Rothman, T., and Matzner, R.A. 'Effects of anisotropy and dissipation on the primordial light isotope abundances' Phys. Rev. Lett 48, 1565 (1982).
Rothman, T. and Matzner, R.A. 'Nucleosynthesis in anisotropic cosmologies revisited' Phys. Rev. D30,

1649 (1984).
105. Juszkiewicz, R., Bajtlik, S., and Gorski, K. 'The helium abundance and the isotropy of the universe' Mon. Not. R.A.S. **204**, 63P (1983).
106. Matravers, D.R., Vogel, D.L. and Madsen, M.S. 'Helium formation in a Bianchi type V cosmological model with tilt' Class. Quant. Grav. **1**, 407 (1984).
107. Barrow, J.D. 'Helium formation in cosmologies with anisotropic curvature' Preprint (1984).
108. Lin, D.N.C., Carr, B.J. and Fall, S.M. 'The growth of primordial black holes in a universe with stiff equation of state' Mon. Not. R.A.S. **177**, 51 (1976).
109. Barrow, J.D., and Carr, B.J. 'Primordial black hole formation in an anisotropic universe' Mon. Not. R.A.S. **182**, 537 (1978).
110. Sussman, R.A. 'Conformal structure of a Schwarzschild black hole immersed in a Friedman universe' Gen. Rel. Grav. (to appear, 1985).
111. Collins, C.B. and Hawking, S.W. 'The rotation and distortion of the universe' Mon. Not. R.A.S. **162**, 307 (1972).
112. Barrow, J.D. and Tipler, F.J. 'Eternity is unstable' Nature **276**, 453 (1978).
113. Barrow, J.D. and Tipler, F.J. 'An analysis of the generic singularity studies by Belinskii, Lifshitz and Khalatnikov' Phys. Repts. **56**, 371 (1979).
114. Goode, S.W. and Wainwright, J., 'Singularities and evolution of the Szekeres cosmological models' Phys. Rev. **D26**, 3315 (1982).
115. Wainwright, J. 'A spatially homogeneous cosmological model with plane-wave singularity' Phys. Lett. **A99**, 301 (1983).
116. Rosquist, K. 'Regularised field equations for Bianchi type VI spatially homogeneous cosmology' Class. Quant. Grav. **1**, 81 (1984).

PRESENT AND FUTURE OF HIGH RESOLUTION RADIO OBSERVATIONS OF ACTIVE
GALACTIC NUCLEI

J.M. Marcaide

Max-Planck-Institut für Radioastronomie
Auf dem Hügel 69
D-5300 Bonn 1
Federal Republic of Germany

ABSTRACT

Using Cygnus A and 3C273 as examples, an introduction to radio interferometric observations of active galactic nuclei is made. Some aspects of present day research in extended and compact radio sources are presented. New instruments which are being presently built or planned and their possible scientific impact are discussed.

1. INTRODUCTION

The earliest indication that the nuclei of some galaxies are places of violent activity was provided by the work of Seyfert[1] on a number of spiral galaxies. He noted that in these galaxies the hydrogen emission lines are much wider than in normal galaxies and their widths, if interpreted in terms of Doppler motions of the ionized gas, correspond to velocities of several thousand km/s. The significance of Seyfert's findings was not realized for about 15 years until further evidence of activity accummulated as a result of the newly discovered radio galaxies. Many of these, like Cygnus A, shown in Figure 1, have double-lobe structures, that is, two enormous radio lobes straddling the suspected parent galaxy. From the early days it is believed that the radio luminosity of these lobes is due to emission via the synchrotron mechanism, which implies lobe energies of $\sim 10^{60}$erg in the form of weak magnetic fields and relativistic particles. It is also believed that the engine ultimately responsible for providing all these energies must be sitting in the central part of the parent galaxy and indeed, in most, if not all, radio galaxies one finds a - usually weak - radio core coinciding precisely with the visible galaxy.

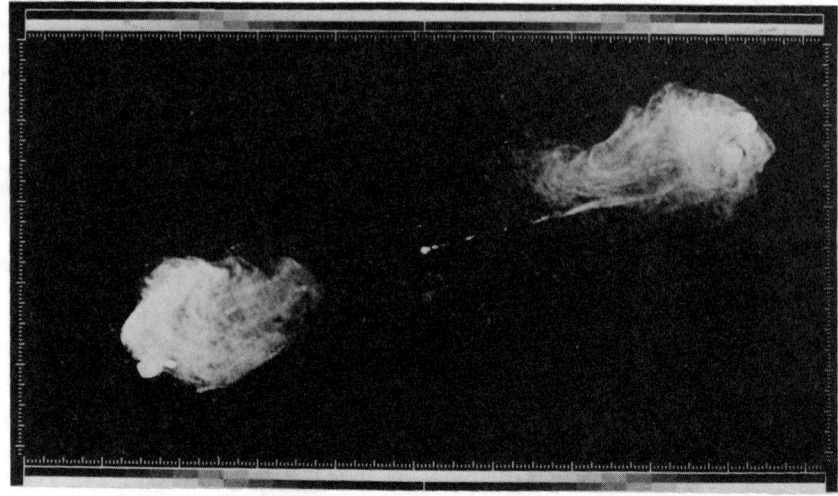

Figure 1: Radiograph of Cygnus A. Courtesy of J.W. Dreher and the NRAO. In reconstructing this image non-standard techniques have been used for enhancing the emission from the jet (and the counterjet?). Thus, there is no linear relationship between the brightness of the map features and the actual brightness of the source features. See also Perley, Dreher and Cowan[2].

There remain two outstanding questions to be answered: a) what is the nature of the central engine? and b) how is the energy transported from the central engine to the outmost parts of the radio source? Present observations can only provide general constraints on the nature of the central engine and hence mostly the question is under theoretical scrutiny[3][4]. As far as the second question is concerned, we do not have yet a satisfactory answer, but in the last years considerable progress has been made in both theoretical and observational grounds[3].

In this lecture, I should like to present results from some recent radio interferometric observations which might be helpful towards answering the questions posed above. I will not attempt to be complete. The latest review on radio observations of Active Galactic Nuclei (AGN) has been given by Porcas[5] and the interested reader is referred to this review and to the references therein. I should like to start with a short introduction into the instrumentation used in interferometry. Then, to lead the reader into the subject, I will use Cygnus A and 3C273 as observational prototypes of radio galaxies and quasars, respectively. I will finish by presenting an overview of new instruments which are presently being built or planned and by discussing their possible scientific impact.

2. INTERFEROMETERS

The radio telescopes (antennas) operate typically at cm wavelengths and their resolution is generally diffraction limited. This resolution is about λ/D radians, where λ is the wavelength of observation and D is

the maximum physical separation between the aperture elements of the
antenna (i.e. the diameter for a paraboloidal antenna). Generally, the
shape of the antennas is such that the signals reflected at each telescope panel (aperture element) are coherently added at the antenna
focal point. Conceptually, one can think of two of the antennas of an
interferometer as two aperture elements of a single huge telescope, of
diameter the separation of the antennas, whose focal point has been
replaced by another physical point, defined by electronic means, where
the signals from each antenna can be coherently added or compared.

The actual signal comparison is not in practice a signal-sum but a
signal cross-correlation, which provides the estimates of the two
fundamental observables called correlation coefficient and fringe phase,
respectively. As the earth rotates about its axis the relative position
of the antennas, given by the vector $\vec{D}/\lambda$, where now $\vec{D}$ is the vector
distance between the two antennas, changes in the radio source reference
frame. In other words, as the earth goes round, for as long as the
source remains above the horizon at both antennas, the interferometer
probes the structure of the radio source for all possible orientations.
The interferometer samples a complex function, the visibility function,
whose absolute value is the calibrated correlation coefficient - called
visibility amplitude - and whose argument is the visibility phase.

It can easily be demonstrated that the two-dimensional Fourier
transform of the radio source brightness distribution function, defined
on the sky plane, is the visibility function which is defined on the
so-called uv-plane. Thus, to reconstruct a source brightness distribution, usually referred to as source structure, it is necessary to
determine the visibility function. In practice the source structure is
reconstructed with limited information: With two antennas (i.e. one
interferometer) and one observation the visibility function is sampled
at a single point. Using this interferometer during the whole day the
visibility function can be sampled at a loci of points which happens
to be a part of an ellipse in the uv-plane. A second day of observation
will not provide, in principle, additional information. By adding a
third antenna we have an array of three interferometers. That is, the
visibility function can be sampled along three ellipses. In general,
the number of ellipses is $N(N-1)/2$ where N is the number of antennas.
The more antennas are available the better the visibility function will
be sampled. Since any practical interferometric array will be limited
in size as well as in number of antennas it will also have limited
resolution, as dictated by the largest separation between member antennas.
In other words, the brightness distribution will have to be reconstructed
in general under conditions of limited resolution and less than ideal
sampling.

Another consideration for the characteristics of any array will be
the method chosen to make the cross-correlation of the signals possible:
a) The signals from the different antennas, which can be separated by a
   few kilometers, can be brought together in real time to a common
   point for cross-correlation using cable, waveguide, optical fiber,
   etc.
b) Microwave links can be used instead of waveguide to bring the signals
   together in real time when the separations between the antennas are

less than a few hundreds of kilometers.
c) The signals can be recorded on magnetic medium along with very precise timing information and some time later the tapes recorded at the different antennas can be brought physically together to a special purpose correlator and the data streams cross-correlated. Presently, the earth size establishes the only limit in the separation of the antennas for using this method known as Very Long Baseline Interferometry (VLBI).

Table 1 presents the characteristics of three widely used arrays shown in Figure 2 which at present time realize each of the three possibilities mentioned above. All these arrays provide equal or better resolutions than the best optical seeing. In the case of VLBI its submilliarcsecond resolution provides direct structural information of the most compact regions known to exist in quasars and galaxies.

TABLE 1

Array	Number Antennas	Maximum Antenna Separation	Shortest Operational Wavelength	Maximum Operational Resolution
VLA[6]	27	21 km	1.3 cm	0".15
MERLIN[7]	6(5)	125	18 (6)	0".1
VLBI[8]	7*	10,000	1.3	0".00015

*This is a typical number. The antennas used in VLBI are not fully dedicated to the array. As many as 18 antennas have participated simultaneously in one occasion.

The price which VLBI has to pay for being able to use antennas arbitrarily removed from each other is that only the visibility amplitude, but not its phase, can be observed. The observed phase is the visibility phase plus a known geometric phase plus an unknown corruption-phase mainly due to the instrumentation and the propagation medium. The visibility phase cannot be recovered from the observed phase. Nevertheless, it can be easily shown that the sum of the observed interferometric phases over a closed loop of antennas exactly coincides with the sum of the visibility phases over that loop. This phase-sum is called closure phase[9]. Using the closure phase as constraint an iterative procedure[10] allows to determine a set of pseudo-visibility-phases which can be used with the fringe amplitudes to reconstruct the brightness distributions or maps.

Since the visibility phase is the only interferometric quantity which carries information on the precise sky position of a given radio source and since, as said above, this visibility phase cannot be recovered from the observed phase, it means that the maps reconstructed from VLBI cannot be localized on the sky to a precision comparable to the map resolution. Hence, maps reconstructed from different sets of data, be they observations at the same wavelength and different epoch or observations at different wavelengths and at the same epoch, cannot be rigorously registered.

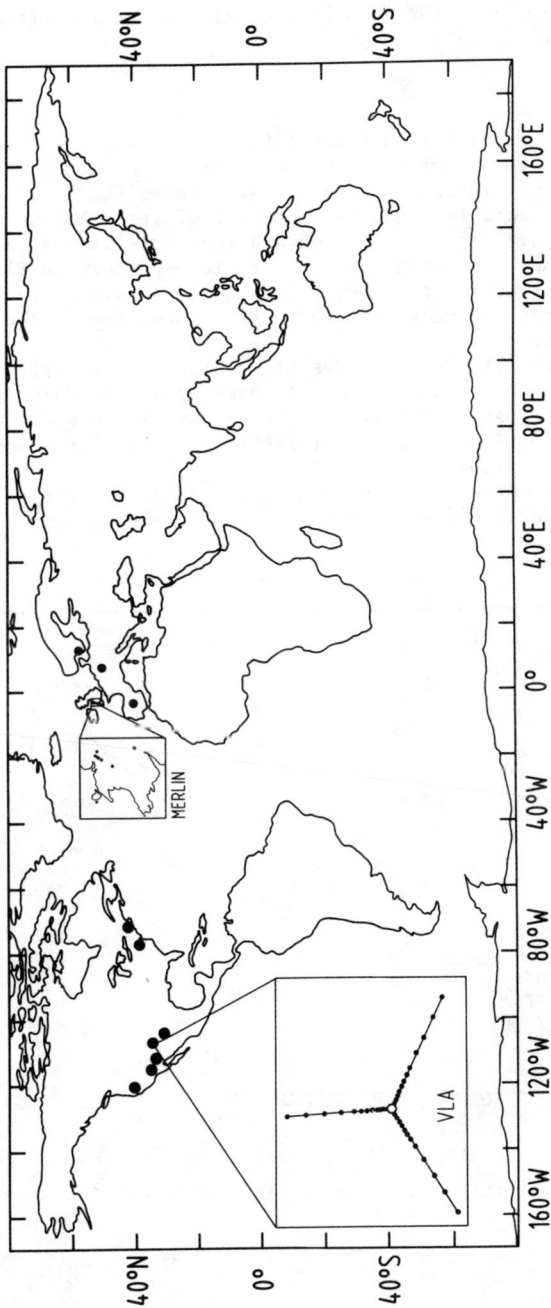

Figure 2: Three widely used interferometric arrays: VLA, MERLIN, and VLBI. See Table 1. Only some of the antennas regularly used in VLBI are indicated. Presently, nearly thirty antennas can be used for VLBI but they do not all have compatible instrumentation.

We have just outlined the simplest conceptual aspects of interferometry and synthesis mapping. We refer the reader to, for example, Thompson and D'Addario[11] for technical details about interferometry and image reconstruction.

3. CYGNUS A

Cygnus A is optically identified with a cD galaxy[12] of redshift 0.562 and lies within a luminous X-ray cluster[13]. At radio wavelengths it is one of the most luminous radio galaxies known ($L_R \sim 10^{45}$ erg s^{-1}), comparable to the radio loud quasars. Because of its relative proximity, its observed flux density is very high and therefore it has been extensively studied since the early days. The radio map shown in Figure 3 was the first detailed radio map ever made and displays already some important structural features which have since been found in other radio galaxies and quasars:

a) <u>lobes</u>, which are extended features of relatively low surface brightness. In the case of Cygnus A these lobes straddle symmetrically the parent galaxy. The lobes have steep spectral index ($\alpha \lesssim -0.5$, $S \propto \nu^\alpha$), characteristic of optically thin synchrotron radiation;

b) <u>nucleus</u>, coincident with the core of the galaxy. The nucleus has a flat spectral index ($\alpha \simeq 0$).

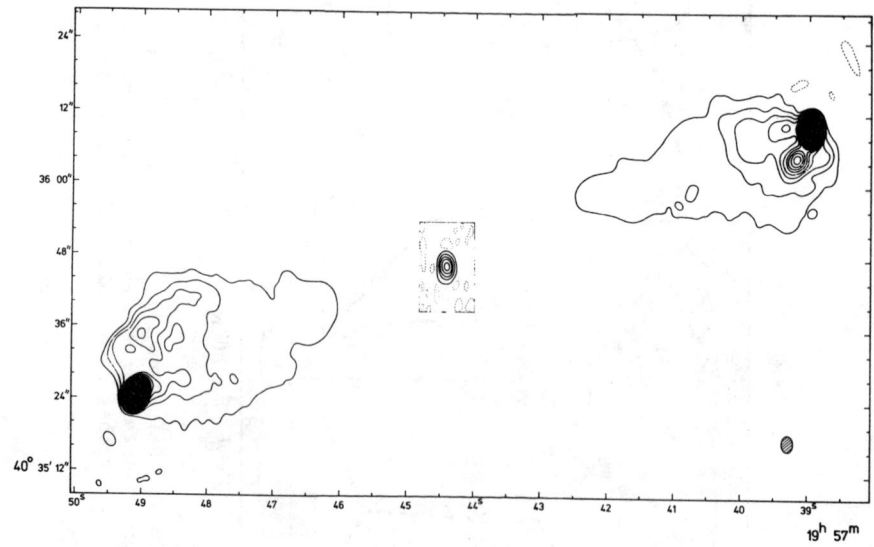

Figure 3: Arc second scale radio map of Cygnus A[14]

The discovery of the hot spots[14] located at ~ 60 kpc from the nucleus (use of $H_0 = 75$ km s^{-1} Mpc^{-1} gives a scale of 1 kpc per arcsecond) inspired the "twin-exhaust" model[15] in which energy produced by a central engine is continuously and efficiently channelled from the nucleus to the hot spots in highly collimated supersonic jets. In this model, the hot spots are shocks or working surfaces of the jet against the surrounding medium where the jets are decelerated thus causing efficient particle acceleration, which in its turn enhances the radiation. The "twin-exhaust" model predicted the existence of jets but these were for a long time not visible in Cygnus A. Recently[2] the production of detailed maps of extraordinary dynamic range has shown that indeed there is also a jet in Cygnus A and that it precisely points towards a hot spot. See Figure 1. Also in the last years it has been found that jets are very common in radio sources[16].

The highest resolution VLBI map of Cygnus A reproduced in Figure 4 shows that there exists a core which is elongated in the position angle of the jet[17]. A question which remains to be answered is why the core-jet structure is one-sided? Has it always been one-sided or is it only at this epoch? Similar questions have been posed many times because in recent years many sources with asymmetric structures have been found.

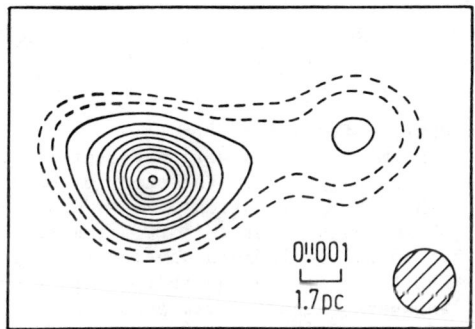

Figure 4: VLBI map of the core of Cygnus A[17]

The "twin-exhaust" and other similar models predict symmetric structures except for an apparent asymmetry introduced by the Doppler boosting of the radiation if the emitting material of the jet has relativistic speed and the jet is not perpendicular to the line of sight. As we shall see later this explanation is also attractive for other reasons. Nevertheless, the possibility of an alternating ejection from the central engine, a flip-flop mechanism, has also been advanced[18].

In Figure 5 we show the structure of Hercules A, a radio galaxy of similar global characteristics to Cygnus A. Notice nevertheless that many details are here very different: the jets are much more visible, the structure appears symmetric, the hot-spots are missing and instead shell structures in the lobes are apparent.

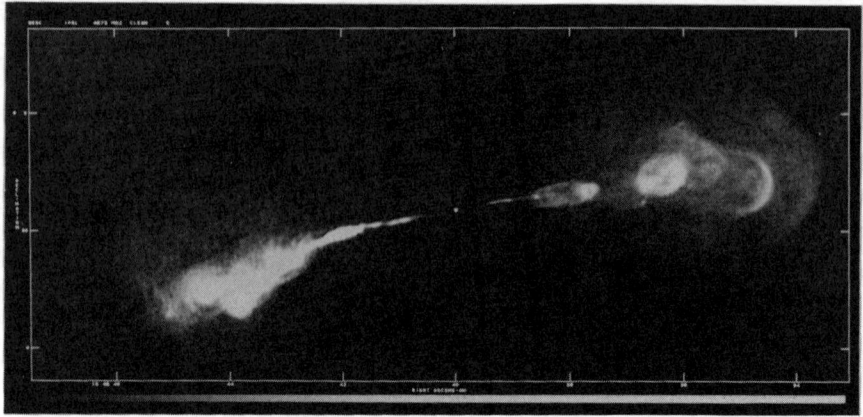

Figure 5: Radiograph of Hercules A. Courtesy of J.W. Dreher.

4. 3C273

With the determination of the redshift (z = 0.158) of this 13th magnitude object by M. Schmidt quasars were discovered[19]. 3C273 has played eversince 1963 a central role in the study of quasars. As is apparent in Figure 6 the radio and optical images display a conspicuous, well collimated jet, which reaches out to $\sim$ 20" from the quasar core, and whose first 12" are invisible in both wavelength ranges. It can also be seen in the MERLIN map of Fig. 6a that the radio structure of the jet shows a wiggling peak intensity rim and a jet cross section which increases outwards. These two characteristics of the radio jet are absent in the optical jet in Fig. 6b. Figure 7 shows a 6 cm VLA map[22] of the jet where the wiggles are also apparent. Superimposed on these intensity map we can see the polarization vectors for each region of the jet. Notice that at the end of the jet this vector has an abrupt change of 90 deg in orientation.

Furthermore, recent optical work[23] shows that there is no optical emission at the hot spot position at the outer end of the radio jet. At the region with both radio and optical emission the polarization vector is parallel in both wavelength ranges. That is, where the polarization angle of the radio jet has an abrupt change there is a hot spot and no optical counterpart. Clearly, this situation must be a consequence of physical processes taking place near the leading surface of the jet which is working against the intergalactic environment as it is moving out. A schematic of the situation is shown in Fig. 8.

237

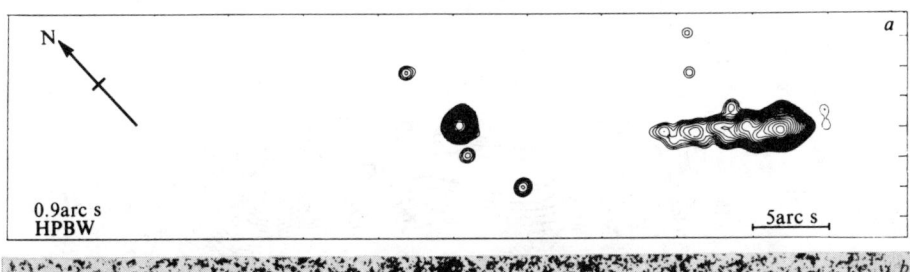
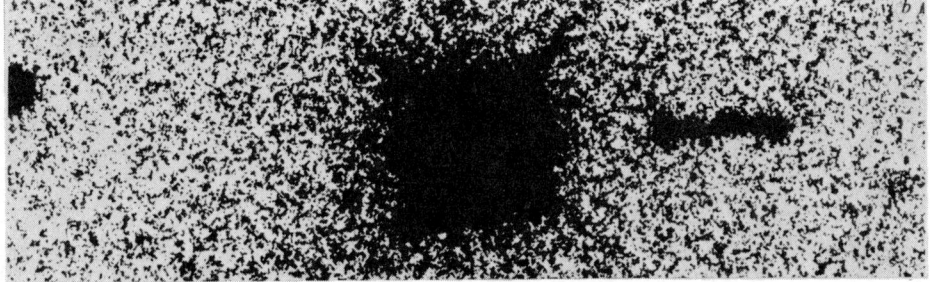

Figure 6: (a) optical picture of 3C273[20], (b) MERLIN 73 cm map of 3C273[21]

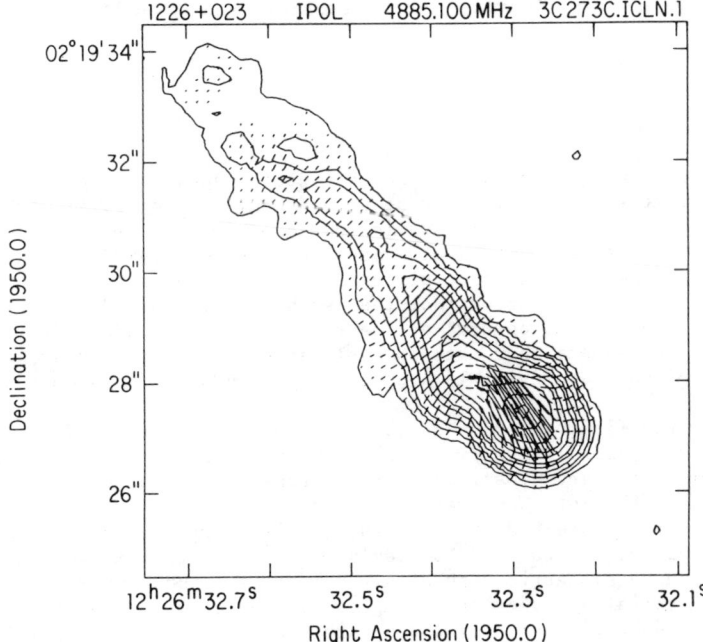

Figure 7: Arc second jet of 3C273. The dashes indicate the polarized flux.[22]

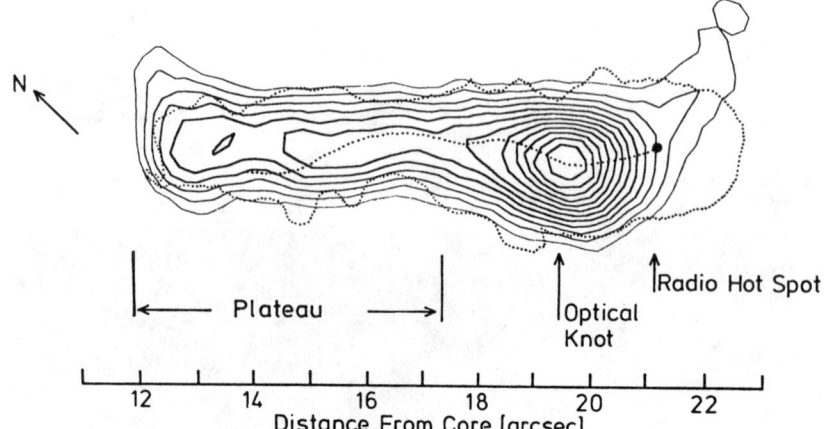

Figure 8: The optical map at .610μm (continuous line) of the 3C273 jet is shown superposed with an schematic of the map shown in Figure 7 (dotted line) where the outer and the ridge profiles are indicated. There is no optical counterpart to the radio hot spot. Courtesy of Meisenheimer and Röser[23]).

The existence of wiggles in the structures shown in Fig. 6a and Fig. 8 indicates that there must be some dynamical process at work. It has been suggested that hydrodynamic instabilities in the jet as it traverses the intergalactic medium may cause the wiggles[21]) or that these may be due to precession of the central source axis[22]). Both suggestions take for granted that the jet has been ejected from the source core, where the central engine is presumably located. Still there are two aspects that need an explanation. Firstly, why are the first 12" of the jet invisible and secondly, why don't we seen a counter-jet?

Maybe the first question can be answered as an effect of insufficient dynamic range presently available for the maps combined with a very efficient (lossless) energy transport mechanism for the first part of the jet. With regard to the second question, it is possible that the one-sidedness of the jet is intrinsic or perhaps it is only apparent due to the observer's Doppler favoritism. The jet emission can be symmetric with highly relativistic bulk motion of the emitting material and very small angles of the jet with the line of sight, thus enhancing greatly the apparent emission of the incoming jet

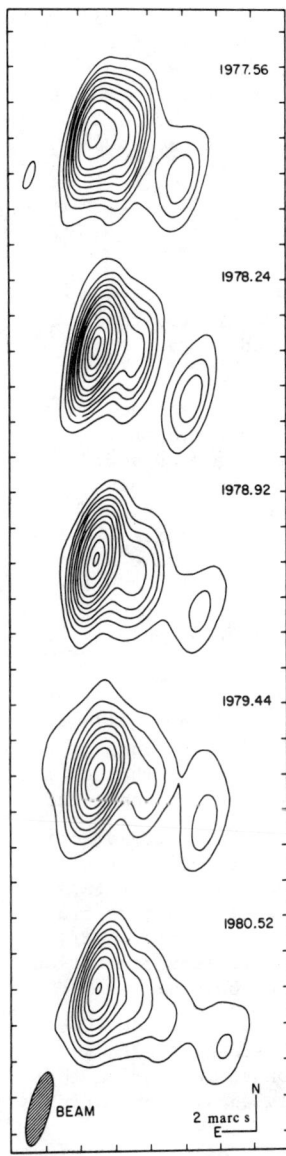

Figure 9: Superluminal expansion of the pc-scale structure of 3C273[25].

and decreasing that of the receding jet (as well as shortening it due to time delay effects). According to Perley[22] the present limits in dynamic range of the maps do not require too stringent limits on the relativistic speeds needed to account for the observed jet asymmetry.

The picture of the relativistic jet pointing directly at us receives strong support from a different set of observations. VLBI observations show that there is compact pc-sized structure at the quasar core. This pc-scale structure is elongated and although the elongation does not point exactly in the direction of the kpc-scale jet it is plausible that the latter is continuously fed by the pc-scale jet. Furthermore, it has been found[25] that the pc-scale structure changes in a way which suggests source feature motions with apparent speeds greater than the speed of light (superluminal motion[26]). There are 6 other active galactic nuclei where apparent superluminal motion has been observed. Figure 9 shows a sequence of maps of the pc-scale structure of 3C273. The size of the source has grown from 19 to 27 pc in 3 years implying apparent superluminal expansion of the source, which in turn implies apparent superluminal motion of the source features, since these seem to have been well identified from one epoch of observation to the next. But, which feature is moving and which one is stationary? Is the southwestmost component moving away from the northeastmost component, or vice-versa in Figure 9? The self-absorbed spectrum of the northeastmost component together with the orientation of the kpc-scale jet suggests that this component may be the stationary one and that the southwestmost component is moving away from it. But as we said in Section 2, because the VLBI maps are constructed using the closure phase condition, the precise positional information of the quasar maps has been lost and hence it is impossible to register maps from different epochs for determining the absolute proper motion of each of the quasar features.

I also want to point out that the situation in 3C273 is radically different from that in M87, a galaxy with a famous jet. The jet of M87 has seemingly periodic structure in well defined clumps both in the visible[27] and in the radio[28] and these structural details have a one-to-one correspondence in both wavelength ranges. Like in 3C273 a milli-arcsecond-scale jet has been found[29] in its core but here, in sharp contrast with 3C273, one finds that the expansion of the jet, if any, is <u>sub</u>-luminal.

5. EXTENDED AND COMPACT EXTRAGALACTIC RADIO SOURCES

In this section I present a very short review of some trends of present day radio interferometric research of extragalactic radio sources stressing relationships between extended and compact sources. Needless to say, the review is rather subjective.

Usually one speaks of extended extragalactic radio sources as those which show significant radio structure when mapped with lower than arcsecond resolution. Figs. 1 and 10 show two examples of such sources. Angular scales of several arcseconds correspond to linear scales of several kpc for typical extragalactic radio source distances. With superb instruments like the VLA it has been found that radio jets are very abundant in extragalactic radio sources. These jets take all sorts of imaginable shapes and a detailed study of them may allow a determination of the physical conditions in the jet and in the surrounding intergalactic medium. A review on radio jets has been written recently by Bridle and Perley[16].

VLBI is the only instrument capable of study extragalactic radio sources with pc-scale resolution. Kellermann and Pauliny-Toth[31], and more recently Porcas[5] have reviewed the research of these compact radio structures. Here again, jets are found to be commonplace although often they are barely seen[32] because the VLBI maps have small dynamic range.

According to Bridle and Perley, in a wide range of extragalactic sources kpc-scale jets often have properties which are well correlated with those of pc-scales. Thus, they say, it is reasonable to relate these properties to the fundamental process of energy transport from the cores to the lobes. The presence of jets supports continuous-flow source models and suggests that collimation and particle acceleration occur on kpc- and pc-scales in extragalactic radio sources.

One of the properties which are correlated between pc- and kpc-scales is that the asymmetric structures are same-sided. Are these asymmetries intrinsic? If not, is this correlation an indication that there exists a Doppler boosting mechanism which is equally operating on both size scales?

Rudnick and Edgar[18] pointed out that in many radio sources extended features at distances periodically increasing from the core appear at alternating sides of it and suggested that perhaps a "flip-flop" ejection mechanism is operating by which the direction of ejection changes by 180 deg between two consecutive ejections. It has been pointed out[16] that detailed studies of the hot spots in the lobes of sources with and without jets might help decide between Doppler boosting and "flip-flop" models of the jet sideness.

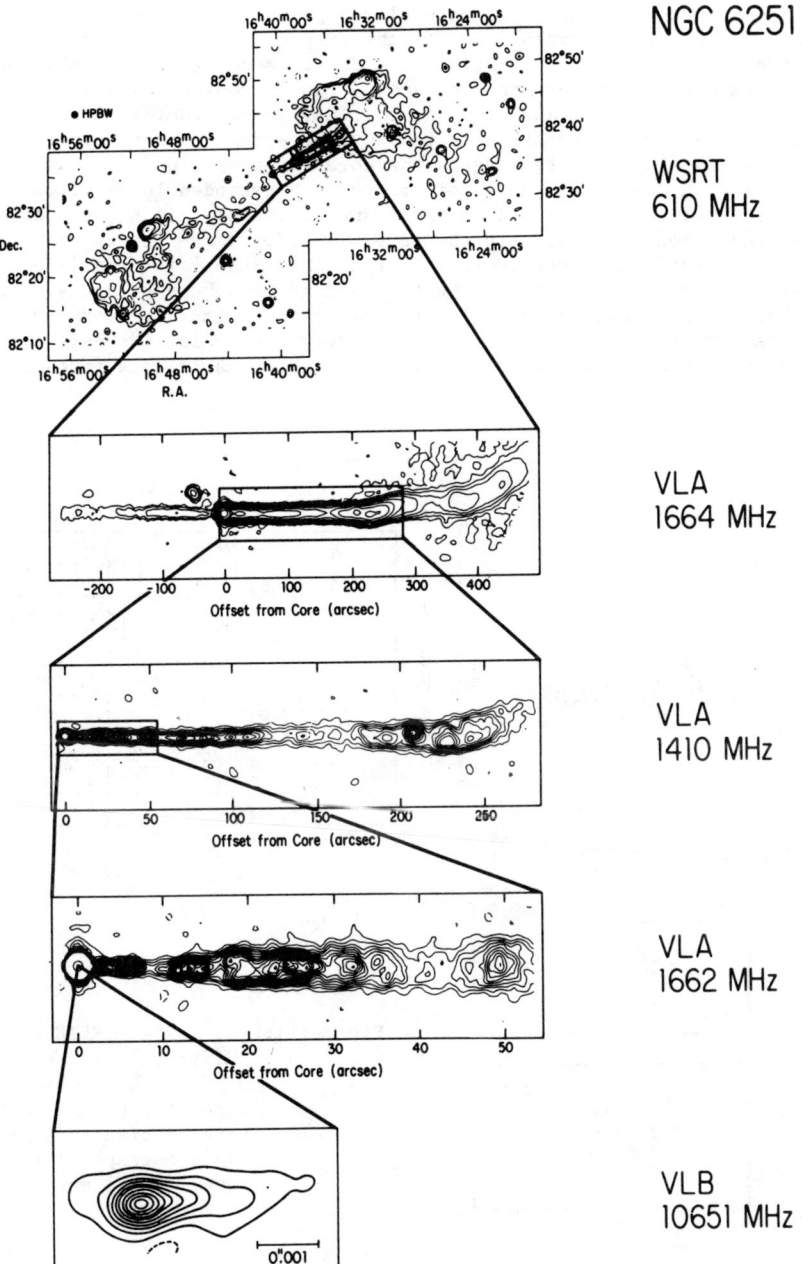

Figure 10: The radio structure of NGC6251 [16]).

The source asymmetry is naturally explained for a handful of sources where one needs to invoke close to the line of sight – and relativistic – motion for the emitting material in order to account for the observed superluminal motion. Are the cores of extended sources with very weak asymmetries therefore excluded as superluminal candidates? Porcas[34] initiated a very interesting approach towards testing this hypothesis by studying with VLBI techniques the cores of a well defined sample of quasars with weak cores and similar strength extended lobes which suggest that the latter are on the plane of the sky. It has turned out that the strongest source in the sample, 3C179, shown in Fig. 11 and first case studied, is a superluminal source! This surprising result can still be accommodated within the standard model scheme by invoking an ad hoc source geometry. However, if more superluminal sources were found in the sample, the situation would severely contradict the expectations from the standard model. As yet, the available results are inconclusive[35].

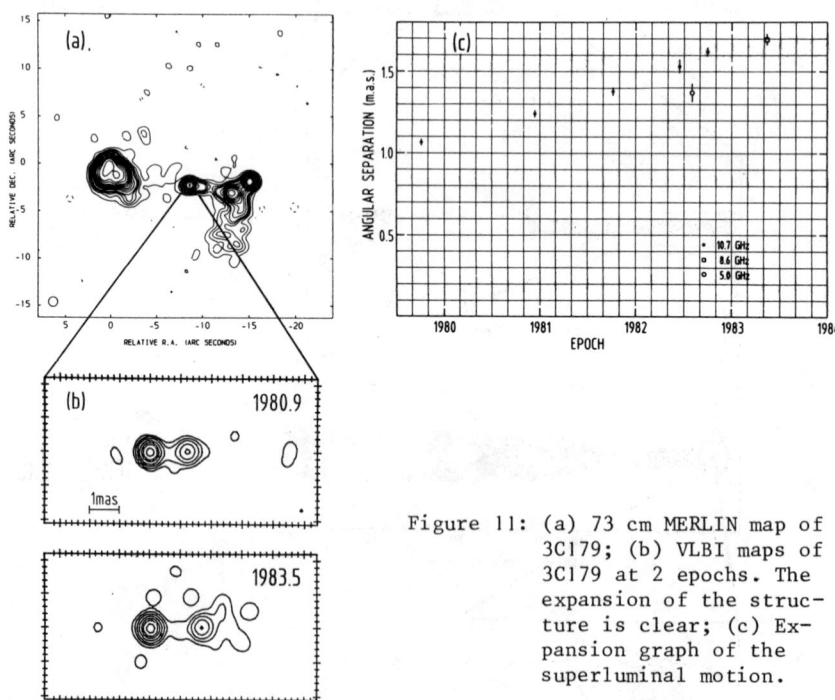

Figure 11: (a) 73 cm MERLIN map of 3C179; (b) VLBI maps of 3C179 at 2 epochs. The expansion of the structure is clear; (c) Expansion graph of the superluminal motion.

I want to briefly point out a morphological optical-radio connection which I find particularly interesting. There are now more than half-a-dozen cases known in which optical emission-line gas appears associated with the radio jet of extragalactic radio sources. What authors like Heckman, van Breugel, and Miley[36] find is that: a) the emission-line gas occurs along the boundaries of bright regions of radio emission; b) the radio source morphologies show abrupt bends in the vicinity of the emission-line gas; c) regions of enhanced optical and radio emission seem to be roughly associated; and d) the emission-line spectrum of the gas bears strong resemblance to that of the narrow-line region of active galactic nuclei. There are still many questions open on the origin of the emission line gas and its relationship with the kpc-scale radio structures but this line of comparative research looks to me a very promising one for figuring out the physics of the jets and the conditions of the dense, highly inhomogeneous ambient media which they go through in the circumnuclear gas.

There are types of information which are straightforward to obtain for kpc-scale structures but which are very difficult for pc-scale structures, namely polarization information and spectral information. The polarization information is very difficult to obtain with present VLBI arrays because the available telescopes have all different characteristics, the antenna polarization parameters are not well determined and often they are single polarization channel receivers. The spectral information cannot rigorously be obtained because, as already mentioned several times, the VLBI maps cannot be accurately registered since much of the absolute positional information is lost by being forced to use the closure phase in the map reconstruction. Also, as already mentioned, this same lack of registrability prevents generally being able to say which component in a superluminally expanding source is stationary.

Recently, the first map of the polarized emission of a compact source has been published[37]. The registration problem for spectral-index map making has been solved in a special case[38] and the stationarity of one source feature with respect to an external reference radio source has been shown in two cases[39,40].

The gross features in the first polarization map shown in Fig. 12 are probably correct in spite of the experimental difficulties the observers had to overcome. This map and other polarization maps[41] of the compact sources show that the polarization in the core is small and that the polarization angle changes by 90 deg between core and jet. In polarization mapping in VLBI we are just witnessing the birth of a subfield with all the difficulties associated to any pioneering work.

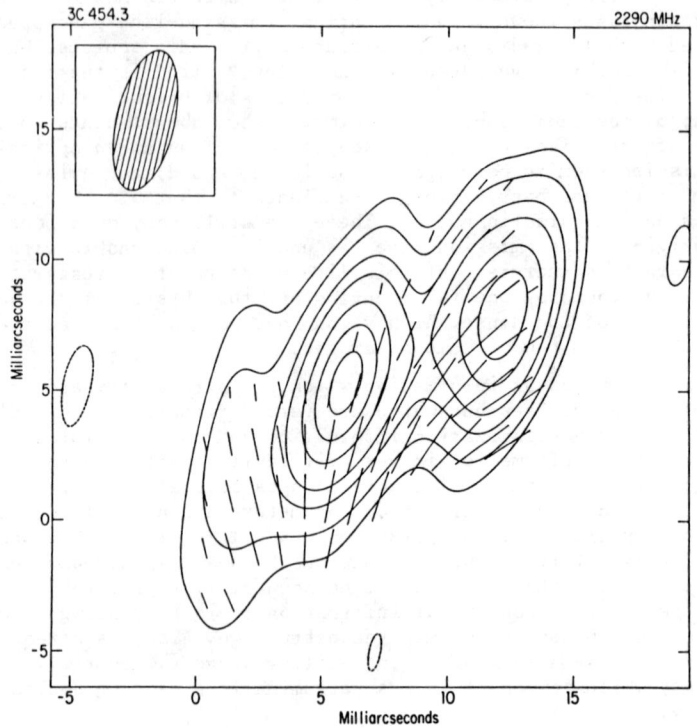

Figure 12: The first polarization map made in VLBI. The dashes indicate the polarized emission. The lengths of the dashes are proportional to the intensity of the polarized emission. The contour map corresponds to the total intensity emission.

In order to study the spectral-index distribution or the temporal evolution of source features it is essential to be able to correctly register maps of a radio source made from data at different wavelengths or from different epochs. In general the sky position of a VLBI map is not known to any precision comparable to the map resolution. However, in the cases where two or more compact radio sources appear very close on the sky one can be used as phase reference for the others and the relative locations of the sources can be determined to great accuracy. In such favorable situations, if one or more sources appear pointlike or with compact structures perpendicular to each other (most, if not all pc-scale structures are linear) then the registration of all the sources will be unique. One might think that either of the two requirements are difficult to meet in practice, but in fact the quasars in the pair 1038+528 A, B have quasi-perpendicular structures and the first member of the NRAO512/3C345 pair has a pointlike structure.

We made simultaneous dual-wavelength VLBI observations of the pair 1038+528 A, B and found[39] that we could determine the relative separation of the quasars with a precision better than 0".000004 at λ3.6 cm and 0".00001 at λ13 cm. A comparison of our result at λ3.6 cm with an earlier and less accurate one gave us an upperbound of less than 0".00002/year for the relative proper motion of the quasars. We also determined the relative registration of the maps made at the wavelengths of λ3.6 cm and 13 cm with an accuracy of 0".0001 and we therefore were able to obtain the spectral-index maps shown in Figure 13. It can be seen in Figure 13 that both sources have regions where the emission appears strongly self-absorbed ($\alpha \sim 1$, $S \alpha \nu^{\alpha}$). These 'core' regions are likely not far away from the centers of activity of the quasars. In both quasars the spectral-index decreases (the emission becomes optically thinner) along the directions of their jets in a smooth manner being the spectral-index uniform across the jet. For the quasar 1038+528 B the spectral-index gradient along the jet becomes extremely pronounced at its tip reaching very low values. These low values may indicate that in the tip-of-the-jet region the electron population is evolving without new input of energetic electrons, a possibility which questions the very same use of the term jet to denote an elongated structure with a self-absorbed core at the other end.

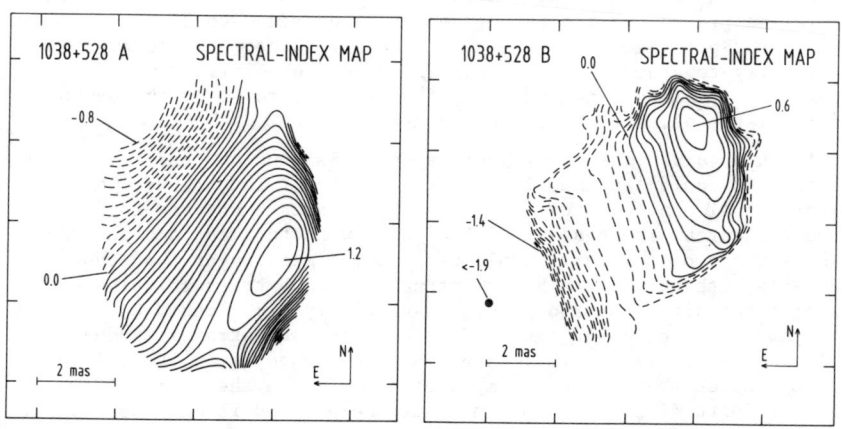

Figure 13: Spectral-index maps of the quasars 1038+528 A and B.

In the pair NRAO512/3C345 the relative proper motion limit, which has been obtained over a longer time-baseline, is similar to the one we cited above, at least in the right ascension direction. The superluminally expanding source 3C345 has a structure which extends mainly in the right ascension direction. Thus, it has been possible to establish[40]) that the component which appears to have the most inverted spectrum (although spectral-index maps are not yet available for this source) remains stationary and it is therefore more likely related to the central engine thought to be responsible of ejecting the other source components.

6. FUTURE PROSPECTS

Connected element interferometry is a mature observational subfield. In the case of the VLA, that array can still uncover a wealth of detail in the study of some key objects by combining data from several arrays, observing at new wavelengths, allowing long integrations, improving the already very good data processing algorithms, etc., but perhaps it is fair to say that it has already achieved peak production. By contrast, VLBI is still far from reaching this point. Until now the antennas used in VLBI have always been built for a general purpose. If by the end of this decade, a dedicated VLBI array (VLBA) is built, as it is now planned, the situation will change substantially. The present plan contemplates building 10 antennas, each of 25 m diameter, and locating them spread all over the US from Arecibo to Hawaii. When this project is completed we can hope to:
a) study unbiased samples of compact radio sources;
b) perform routine polarization VLBI observations;
c) perform routine multifrequency observations; and
d) have enhanced reliability in the operation and hence experimental repeatability, etc.

The drawback of the VLBA as a standalone instrument is its lack of sensitivity in comparison with arrays which include the 100 m telescope in Effelsberg, FRG, or the 64 m (soon to be upgraded to 70 m) telescopes of the Deep Space Network. Naturally, these telescopes could also work occasionally together with the VLBA when added sensitivity were essential to the project, but then it would have to be at the expense of some of the advantages we just pointed out.

It is possible that a project named QUASAT and now under study in ESA and NASA could see the light in the middle or late 1990s. The project QUASAT contemplates putting a 15 m diameter antenna in orbit around the earth. The baselines obtainable with this antenna and those on the surface of the earth could be 3 to 5 times longer than those obtainable with only earth based antennas. This extra resolution may look at first sight not too important but every experienced astronomer knows that his conclusion can, and often does, go astray if such a resolution factor is missing. To put a recent example in VLBI, I should like to mention that the new component found[42]) in the source 4C39.25 at a wavelength of 3.6 cm would not have been found if only arrays with half the resolution would have been available.

Not less important than adding resolution, which by the way can also be obtained by going to shorter wavelengths, is the fact that QUASAT would allow the mapping of sources in very short observation periods because the motion of the satellite would allow a fast filling of the uv-plane. Actually, fast mapping may be very important for the study of strong transient phenomena like star flares. QUASAT, in conjunction with the now planned Australia-array in Australia (I will not comment on this southern hemisphere development) would also allow VLBI studies of the southern sky which at high resolution is practically unexplored. Finally, the resolution obtainable with this antenna in orbit will likely make possible comparative studies with the optical emission from the Broad Line Region of active galactic nuclei.

A new area of research where breakthroughs are possible is mm-VLBI. Successful observations have been made at wavelengths of 3 and 7 mm. Wavelengths as short as 1 mm appear usable. Large improvements in receiver technology are necessary but they are forthcoming. With the likely upgrading of some antennas it appears that we can expect in the future a reasonably large array of antennas capable of operating at wavelengths as short as 3 mm. This tentative array, which is presented in Figure 14, should be able to produce, in about ten years, mm-images comparable in quality to the cm-images obtained nowadays with our best arrays. Naturally, then the beam will be as incredibly small as 50 micro-arcseconds, which will allow us to poke into the today inscrutable cores of the radio sources.

It has been shown[43] that for a sample which had been selected on the basis of its mm-emission excess, a strong correlation exists between the mm-radio emission and the X-ray emission. This correlation is probably an indication that the inverse-Compton mechanism is operating for the sources in this sample. If so, the X-ray photons may be the same mm-wavelength photons which have been up-scattered in Compton collisions by the electrons which produced them. Thus, one expects coordinated mm-wavelength VLBI studies and X-ray studies to be of much interest.

I will finish by briefly mentioning about two arrays which will provide a spatial resolution similar to that of the VLA but operating at millimeter and submillimeter wavelengths and which are presently being built at the Plateau de Bure (France) and planned by the Smithsonian Astrophysical Observatory, respectively. The instrument which is being built in France will consist of 3 (and possibly 4) 15 m antennas which can be moved on railway tracks. The maximum separation between the antennas will be $\sim$ 400 m and the antennas will be able to operate at wavelengths of $\sim$ 1 mm or shorter. The instrument planned in America is an array of six 6 m antennas, with maximum separation between the antennas of $\sim$150 m, and capable of operating to wavelengths as short as 0.35 mm.

These instruments, when and if built, will allow us to address questions which presently cannot be addressed. Let me just give you one simple example. There is good circumstantial evidence that the spectra of most quasars and AGN will peak in the millimeter-submillimeter range. The wavelength where this turnover occurs is a very important parameter in the models of these sources: From a knowledge of the turnover wavelength information about the energy mechanisms, the

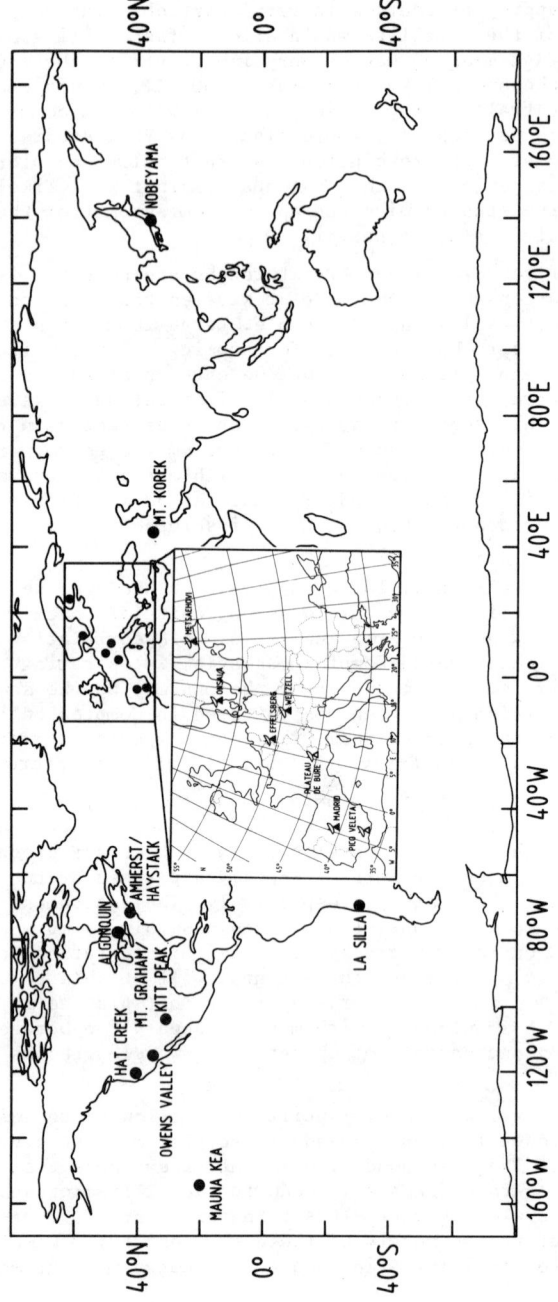

Figure 14: mm-telescopes with diameters larger than 10 m which eventually could work together as elements of a mm-VLBI array.

magnetic fields present, etc. can be derived. In this regard, these instruments would be extraordinary complements to the VLA and to VLBI. But still, I believe that instruments like these will make their best contributions in ways which we simply cannot anticipate.

7. ACKNOWLEDGEMENTS

In preparing this manuscript I have largely benefited from help from my colleagues Brian Corey, Marc Gorenstein, and Richard Porcas, to whom I wish to express my thanks. I also want to thank Rosel Bock for her skill and patience in typing the manuscript. This work was supported in part by the NASA-JPL Contract 956542.

8. REFERENCES

1) Seyfert, C.K., Astrophys. J. 97, 28 (1943)

2) Perley, R.A., Dreher, J.W. and Cowan, J.J., Astrophys. J. Lett. 285, L35 (1984)

3) Begelman, M.C., Blandford, R.D., Rees, M.J., Reviews of Modern Physics 56, 255 (1984)

4) Rees, M.J., Annual Review of Astron. and Astrophys. 22, 471 (1984)

5) Porcas, R.W., Workshop on Active Galactic Nuclei, Manchester University Press (1985) (in press)

6) Napier, P.J., Thompson, A.R., Ekers, R.D., Proc. IEEE 71, 1295 (1983)

7) Davies, J.G., Anderson, B., Morison, I., Nature 288, 64 (1980)

8) Rogers, A.E.E., Cappallo, R.J., Hinteregger, H.F., Levine, J.I., Nesman, E.F., Webber, J.C., Whitney, A.R., Clark, T.A., Ma, C., Ryan, J., Corey, B.E., Counselman, C.C., Herring, T.A., Shapiro, I.I., Knight, C.A., Shaffer, D.B., Vandenberg, N.R., Lacasse, R., Mauzy, R., Rayhrer, B., Schupler, B.R., Pigg, J.C., Science 219, 51 (1983)

9) Rogers, A.E.E., Hinteregger, H.F., Whitney, A.R., Counselman, C.C., Shapiro, I.I., Wittels, J.J., Klemperer, W.K., Warnock, W.W., Clark, T.A., Hutton, L.K., Marandino, G.E., Rönnäng, B.O., Rydbeck, O.E.H. and Niell, A.E., Astrophys. J. 193, 293 (1974)

10) Cotton, W.D., Astron. J. 84, 1122 (1979)

11) Thompson, A.R. and D'Addario, L.R., Synthesis Mapping, Proc. NRAO-VLA Workshop, Socorro, N. Mexico., Green Bank, W. Va: Natl. Radio Astron. Obs.

12) Spinrad, H. and Stauffer, J.R., Monthly Notices Roy. Astron. Soc. 200, 153 (1982)

13) Fabbiano, G., Doxsey, R.E., Johnston, U., Schwartz, D.A. and Schwartz, J., Astrophys. J. Lett. 230, L67 (1979)

14) Hargrave, P.J. and Ryle, M., Monthly Notices Roy. Astron. Soc. 166, 305 (1974)

15) Blandford, R.D. and Rees, M.J., Monthly Notices Roy. Astron. Soc. 169, 395 (1974)

16) Bridle, A.H. and Perley, R.A., Annual Rev. of Astron. and Astrophys. 22, 319 (1984)

17) Kellermann, K.I., Downes, A.J.B., Pauliny-Toth, I.I.K., Preuss, E., Shaffer, D.B., Witzel, A., Astron. and Astrophys. 97, L1 (1981)

18) Rudnick, L. and Edgar, B.K., Astrophys. J. 279, 74 (1984)

19) Schmidt, M., Nature 197, 1040 (1963)

20) Arp, H., Proc. ESO/ESA Workshop SP-162, 53 (1981)

21) Conway, R.G., Davis, R.J., Foley, A.R., Ray, T.P., Nature 294, 540 (1981)

22) Perley, R.A., Proc. IAU Symposium No. 110, 153 (1984)

23) Meisenheimer, K., Röser, H.-J., Proc. MPG Workshop, Garching bei München, p. 51 (1984)

25) Pearson, T.J., Unwin, S.C., Cohen, M.H., Linfield, R.P., Readhead, A.C.S., Seielstad, G.A., Simon, R.S., Walker, R.C., Nature 290, 365 (1981)

26) Porcas, R.W., Nature 302, 753 (1983)

27) Lelievre, G., Nieto, J.L., Proc. 24th Liège Int. Astrophys. Colloquium, 495 (1983)

28) Biretta, J.A., Owen, F.N., Hardee, P.E., Astrophys. J. 274, L27 (1983)

29) Schmitt, J.H.M.M., Reid, M.J., Astrophys. J. 289, 120 (1985)

31) Kellermann, K.I., Pauliny-Toth, I.I.K., Ann. Rev. of Astron. Astrophys. 19, 373 (1981)

32) Marcaide, J.M., Shapiro, I.I., Corey, B.E., Cotton, W.D., Gorenstein, M.V., Rogers, A.E.E., Romney, J.D., Schild, R.E., Bååth, L., Bartel, N., Cohen, N.L., Clark, T.A., Preston, R.A., Ratner, M.I. and Whitney, A.R., Astron. Astrophys. 142, 71 (1985)

34) Porcas, R.W., Nature 294, 47 (1981)

35) Zensus, J.A., Porcas, R.W., Proc. IAU Symposium No. 110, 163 (1984)

36) Heckman, T.M., van Breugel, W.J.M., Miley, G.K., Astrophys. J. 286, 509 (1984)

37) Cotton, W.D., Geldzahler, B.J., Marcaide, J.M., Shapiro, I.I., Sanroma, M., Rius, A., Astrophys. J. 286, 503 (1984)

38) Marcaide, J.M., Shapiro, I.I., Astrophys. J. 276, 56 (1984)
39) Marcaide, J.M., Shapiro, I.I., Astron. J. 88, 1133 (1983)
40) Bartel, N., Ratner, M.I., Shapiro, I.I., Herring, T.A., Corey, B.E., Proc. IAU Symposium No. 110, 113 (1984)
41) Roberts, D.H., Pottasch, R.I., Wardle, J.F.C., Rogers, A.E.E., Burke, B.F., Proc. IAU Symposium No. 110, 35 (1984)
42) Marcaide, J.M., Bartel, N., Gorenstein, M.V., Shapiro, I.I., Corey, B.E., Rogers, A.E.E., Webber, J.C., Clark, T.A., Romney, J.D., Preston, R.A., Nature (in press)
43) Owen, F.N., Helfand, D.J., Spangler, S.R., Astrophys. J. 250, L55 (1981)

THE HALO PUZZLE

P. J. E. Peebles

Joseph Henry Laboratories, Princeton University

Princeton, N. J. 08544

U.S.A.

ABSTRACT

Studies of the distributions of mass and starlight around galaxies have revealed some remarkable systematics that so far have resisted any easy interpretation. I review here some of the more interesting of these phenomena and the problems they seem to me to present to conventional models for the origin of the large-scale structure of the universe.

1. HALO PHENOMENA

The dark halo of a galaxy is the mass that must be presumed to be present on scales of tens of kpc around galaxies to account for stellar dynamics within conventional Newtonian mechanics. Remarkable features of these halos are their approximate uniformity from galaxy to galaxy, their symmetry, their large density variation within an individual galaxy, their continuity with the "fractal" halos revealed on larger scales by the clustering of galaxies, and of course their low luminosities. In this section I review some aspects of what is known about each of these points.

In spiral galaxies the streaming motion of the interstellar gas can be measured by the Doppler shift of optical and 21 cm radio emission lines. The symmetry of the Doppler shift shows that in many galaxies the streaming motion is very nearly circular. The circular velocity $v_c(r)$ is almost constant within a galaxy (but tends to increase slowly with increasing r).[1,2] This implies that the mass within radius r is

$$M(< r) = r v_c^2 / G \sim r. \qquad (1)$$

This equation assumes a spherical mass distribution, but the numbers for a flattened object are not much different. The mass density averaged over spherical shells is then

$$\rho(r) = v_c^2 / (4\pi \, G r^2) \sim r^{-2}. \qquad (2)$$

The starlight in spheriod and disc components generally decreases with increasing r significantly more rapidly than the mass density derived this way, so we conclude that the net mass is not distributed like the "seen" stars. We need a new component, a dark halo. (A way out, as discussed by Milgrom,[3] is to assume that the gravitational acceleration law changes from inverse square to $g \propto r^{-1}$ at large r. The only serious observational problem with this arises in cosmology, as is discussed by Felten.[4] It seems doubtful that the beautiful success of the standard nucleosynthesis theory described at this conference by Tenreiro would survive such a drastic change in gravity theory. But even ignoring such direct problems, I share the conventional view that it is best to concentrate attention on the standard laws of physics until we are driven from them.)

The correlations of the circular velocity v_c with morphological type and luminosity have been widely discussed as an important clue to the origin of galaxies, but perhaps equally interesting is the fact that v_c scatters so little, particularly at the large v_c end. In bright spirals v_c usually is in the range 150 to 300 km s^{-1}. The largest observed values, for galaxies like the Sombrero Galaxy, NGC 4594, are $v_c \sim 400$ km s^{-1}. We can translate v_c into the mass M_{10}

contained within a sphere of fixed radius, 10 kpc. We see that among bright spirals M_{10} generally is between 5×10^{10} M_O and 2×10^{11} M_O, and only rarely exceeds 4×10^{11} M_O. In effect, there is a fairly rigid constraint on the amount of mass that can accumulate around a spiral galaxy.

Gravitational accelerations in elliptical galaxies generally are more poorly known than for spirals because it requires a measurement of the broadening of the stellar lines by the random motions of the stars, which is much harder than the measurement of the mean shift of the stronger emission lines in spirals. In a few cases, however, as Fabian discusses at this meeting, ellipticals are seen to be surrounded by pools of plasma, and the gravitational field can be estimated from the condition that it must hold the plasma in hydrostatic equilibrium.

Fabricant and Gorenstein[5] used this method to obtain a beautiful measurement of the gravitational field around the giant elliptical galaxy M87 in the Virgo cluster. The measurement is based on X-ray observations. The X-rays are thought to come from a pool of plasma around the galaxy because the spectrum agrees with thermal bremsstrahlung plus emission lines from the elements one would expect to be common if the plasma had roughly solar composition. The plasma is thought to be in hydrostatic equilibrium or close to it because otherwise the plasma would disperse or collapse in a Hubble time. Consistent with this, the X-ray surface brightness contours are strikingly smooth and symmetrical around the galaxy, as if the gas were well relaxed. As the X-ray contours are very nearly circular it is reasonable to assume the plasma pool is spherically symmetric, and so there is little ambiguity in deriving the plasma density $n(r)$ as a function of distance r from the galaxy. The temperature run $T(r)$ is more difficult but the evidence from the X-ray spectrum as a function of projected radius is that $T(r)$ is nearly constant. The product $n(r)T(r)$ fixes the plasma pressure. The pressure gradient is the force per unit volume of the plasma which must be balanced by the gravitational force $g\rho_p$ per unit volume, where $\rho_p = nm$ is the plasma mass per unit volume. This yields the gravitational acceleration g, and hence the toal mass $M(<r)$. The result at $r = 400$ kpc is $M \sim 6 \times 10^{13}\ M_\odot$. This is the largest known mass for a single galaxy. (The cD galaxies found in the centers of some clusters are more luminous and so very possibly are even more massive.) The mass around M87 scales roughly as $M(<r) \propto r$, as is the case for a spiral with a flat rotation curve $v_c(r)$. At $r = 10$ kpc, $M(<r)$ amounts to about $M_{10} \sim 6 \times 10^{11}\ M_\odot$, about 5 times the mass of our galaxy at the same radius, and about comparable to M_{10} for the most massive known spirals. This indicates that the constraint on M_{10} we found for spirals applies also to ellipticals, though we might be inclined to put the mass at which the constraint becomes rigid at a somewhat larger value.

The striking symmetry of the X-ray map of M87 indicates that the gravitational field is very nearly radial and hence that the mass is fairly smoothly distributed, the density dropping off roughly as $\rho \sim r^{-2}$. The starlight is similarly smoothly distributed but with a steeper density run, the luminosity per unit volume in optical and infrared varying roughly as $j(r) \propto r^{-3}$.[6,7] There is no discernible edge to the galaxy. The r^{-3} starlight density run is measured to $r \sim 30$ kpc. Oemler[8] detected the galaxy to about three times that radius. A remarkable photograph obtained by Arp[9] shows the optical surface brightness continuing to decrease in a smooth way with increasing distance from the center of the galaxy to at least $r \sim 200$ kpc. This is half the radius to which Fabricant and Gorenstein mapped the mass from the X-ray data. Arp's optical measurement is exceedingly difficult and as far as I know no one has attempted to repeat it. However, a strong argument for its reality is the fact that the position angle of Arp's elliptical surface brightness contours lines up with the position angle of the X-ray contours and with the long axis of the much smaller optical image seen on the Palomar sky survey print. At $r = 100$ kpc, where the optical surface brightness can be fairly reliably estimated, we find that the optical luminosity per unit volume is $j \sim 10^{-6}\ L_\odot\ pc^{-3}$, which is some 5 orders of magnitude fainter than j in our local stellar neighborhood.

Continuity of the colors and of the surface brightness run around M87 from the central parts, where stellar lines are seen, to the halo suggests the faint emission in the outer regions is starlight. But that leads to an interesting puzzle: how did stars form at $r \sim 100$ to 200 kpc from M87, where conditions are so different from the star forming regions we can observe?

What is the dominant form of the mass in a galay? Again M87 provides a beautiful illustration of the problems. As we have just seen, the densities of starlight and of mass vary with radius in distinctly different ways. Thus if the net starmass per unit of starlight (the mass-to-light ratio) were constant the dominant mass component could not be stars. As neutral and ionized hydrogen are observationally ruled out we are led to consider new forms of matter, like primeval black holes, massive neutrinos, axions, or the supersymmetric partners of ordinary elementary particles.[10] We pay a lot of attention to such exotic ideas because if any should prove correct it would be such a great discovery. On the other hand, it should be noted that our assumption of constant starmass per unit of starlight is somewhat daring given the great range of ambient physical conditions in M87. It certainly would not be surprising if the stars seen at $r \sim 100$ kpc, where the luminosity density is $\sim 10^{-5}$ times the value in the solar neighborhood, formed in a different way from the stars seen at $r \sim 1$ kpc, where the density is comparable to the solar neighborhood. And that could well lead to a substantial difference in the frequency distribution dn/dm of star masses and hence the mass-to-light ratio.

The distribution of star masses is constrained by color measurements, but the limits are not as restrictive as one would like. Boughn and Saulson[7] showed that the ratio of blue to infrared (2.2 μ) surface brightness of M87 is very nearly constant at $B - K \cong 4$ at $r \lesssim 30$ kpc. This is sufficient to show that the net mass density at $r \sim 30$ kpc is not provided by stars like VB10, which is among the least massive of known stars. But it is more difficult to constrain a possible systematic variation of the stellar mass distribution with radius, such as might accompany a variation with r in the mass-to-light ratio of the stellar component. A conventional model for the distribution of stellar masses near the main sequence turnoff is the power law

$$dn/dm \propto m^{-(1 + x)}. \tag{3}$$

The spectrum synthesis models of Tinsley and Gunn[11] show that if $x \lesssim 2$ the blue-infrared color B-K is dominated by red giants (and has a value consistent with what is observed in M87). This means that broad-band colors like B-K are insensitive to the shape of dn/dm within quite broad limits, so that the Boughn-Saulson observation that B-K is nearly constant in M87 implies only a very broad constraint on the possible variation of dn/dm with r. It will be noted also that at $x = 2$ the mass density

$$\rho = \int dm \; m \; dn/dm \;, \tag{4}$$

diverges as m^{-1}. This means that if the power law were truncated at the mass of Jupiter the mass density from objects between the cutoff and the threshold for nuclear burning would be ~ 30 times the mass from the seen stars between threshold and the turnoff from the main sequence.

Spectral features in the infrared have been used as an aid to distinguishing between the light from red giants and from main sequence stars.[11] It will be interesting to see whether this method can improve the constraint on the possible variation of x with r in M87.

Let us return to the distribution of mass around a galaxy and consider how far from the galaxy the mass concentration might be traced. At $r \gtrsim 300$ kpc a galaxy generally has bright neighbors, and we can use the relative velocities of the galaxies as a measure of the mass concentration. The traditional problem in doing this has been that some of the apparent neighbors will be accidentals with substantial cosmological redshift differences. As the relative velocity dispersion $\langle \Delta v^2 \rangle$ is sensitive to the high velocity tail of the distribution it is easy to introduce a considerable underestimate or overestimate of $\langle \Delta v^2 \rangle$ through a relatively modest error in the correction for accidentals. However, with increased sample size it has become apparent that the distribution of Δv is quite close to exponential.[12] Thus one can find the dispersion by fitting the observed distribution of Δv to an exponential plus a constant term (to represent the accidentals). The results say that $\sigma = \langle \Delta v^2 \rangle^{1/2}$ increases slowly with increasing separation r, ranging from $\sigma = 200 \pm 20$ km s^{-1} at $r \sim 10$ kpc to $\sigma = 350 \pm 50$ km s^{-1} at $r \sim 1$ Mpc.[12,13,14] The former separation is comparable to the scales probed by the circular motions of gas in discs of galaxies, and the result, $\sigma = 200$, is about what we would expect from the observed circular velocities. The observation that $\sigma(r)$ is nearly flat resembles the observation that rotation curves in spirals are nearly flat, and the interpretation is the same: the mean value $M(< r)$ of the mass found within distance r of a galaxy scales as

$$M(< r) \sim r^{1.2}. \tag{5}$$

This applies to at least $r \sim 1$ Mpc, beyond which the uncertainty in the estimates of $\sigma(r)$ from the available data becomes too large.

The result $M(< r) \sim r^{1.2}$ can be compared to two other observations. First, we have seen that at $r \sim 10$ to 100 kpc the internal dynamics of galaxies often reveals a smooth mass distribution that scales with r in much the same way. Second, studies of the statistics of the distributions of galaxies show that the mean number of bright galaxies in excess of random within distance r of a galaxy varies as $\delta N(< r) \propto r^{1.2}$.[15] This is consistent with the result $M(< r) \propto r^{1.2}$ if we assume that on large scales mass clusters like galaxies. The scaling law for galaxy clustering, $\delta N(< r) \propto r^{1.2}$, applies from $r \sim 10$ kpc to at least $r \sim 10$ Mpc, where it is lost in the noise.

The curious thing about these observations that a single power law connects the smooth "monolithic halos" found around many galaxies with the clumpy "fractal halos" representing the large-scale clustering of galaxies. Might this continuity indicate that monolithic and fractal halos formed by related processes? Let us turn now to a discussion of such questions.

2. MODELS FOR THE ORIGIN OF HALOS

The smooth rotation curves of spirals and the smooth plasma pools seen around some ellipticals show that in these objects the mass distribution is a fairly smooth function of radius, $\rho \sim r^{-2}$. A terrestrial analogy would be the smooth flanks of a volcano. To continue the analogy, we might compare the snowy peak of a volcano to the starlight in the central parts of a dark halo. Yet another point of similarity is the character of the distribution of elevations. There are volcanos higher than 7 km, but they are rare and their abundance drops rapidly with increasing elevation. As we have discussed, a similar situation applies to the amount of mass found within a sphere of radius 10 kpc centered on a galaxy. In a "typical" galaxy like the Milky Way, $M_{10} \sim 10^{11}$ Solar masses. The known extremes on the high mass end are a factor of 5 to 10 larger than this, indicating that the abundance of galaxies as a function of mass M_{10} is rapidly decreasing at $M_{10} \sim 5 \times 10^{11}$ Solar masses.

The final part of the picture of the large-scale distribution does not have such a close analogy: the dark halos of galaxies appear in clusters, and the clusters are so arranged that the mean mass density averaged over spherical shells continues to vary about as $\rho \sim r^{-2}$ as we move from monolithic dark halos to the realm of the nebulae.

Does all this suggest any particular model for the origin of galaxies? The heights of mountains are limited by the strength of the crust of the Earth. Might there be a similar sort of limit that discourages more than about 5×10^{11} Solar masses from accumulating in a sphere of radius 10 kpc? The exclusion principle for a dark (weakly interacting) fermion would do it, but the limit would not scale with radius as we want. If the material were assembled as hydrogen then the proto-halo could tend to blow itself apart when the mass exceeds a critical value. Thus galactic winds or blast waves driven by supernovae in young galaxies would limit growth by accretion.[16] It remains to be seen whether we could rely on such a self-limiting process to build extensive and dilute halos of the sort found around M87.

Fabian describes at this conference accretion flows, which are a sort of reverse of galactic winds. The observed X-ray luminosities of plasma pools like the one around M87 are large enough that in the absence of an energy source the plasma must be setting into the gravitational potential well in a quasi-static way. As the cooling time varies inversely with density the flow accelerates toward the center. Also, the flow is thermally unstable, dense spots tending to lose energy and so shrink faster than the mean. What becomes of the mass flux at the center? Part may serve to power the active nucleus of M87. As Fabian discusses, because the flow is unstable it is natural to expect that part also tends to fragment and form stars.[17] If the proto-halo of M87 consisted entirely of plasma at the present total density and temperature then even at $r = 200$ kpc from the center the cooling time would have been less than the Hubble time (at redshifts z less than about 3), and so all this gas would have been thermally unstable. What is more, as Ostriker noted, there is a natural limit operating here in that if the plasma were much hotter than we have assumed it could not have cooled to form stars in a Hubble time.[18] On the other hand, the Jeans length at $r = 200$ kpc still would have been ~ 200 kpc and I find it hard to imagine that thermal instability could have driven this down to the mass of a star without leaving a hierarchical clustering structure rather than the smooth stellar halo we see. Thus an accretion flow does not seem to me to be a promising way to build up the gross features of halos of galaxies.

The strong central concentration characteristic of a galaxy is produced by the free collapse and violent relaxation of an initially nearly homogeneous cloud of pressureless matter.[19,20] As the matter could be a "gas" of galaxies or of stars or of exotic weakly interacting particles, this free collapse picture might be considered as a model for the formation of clusters of galaxies, or of stellar spheroid components of galaxies, or of dark halos. The collapse produces a mass density run that approximates $\rho \sim r^{-3}$, which is a reasonable match to the distribution of starlight in the stellar spheroid component. However, we have seen that where the mass density run is measured it usually is closer to $\rho \sim r^{-2}$ than $\sim r^{-3}$. The discrepancy could be remedied by properly shaping the radial mass distribution in the original collapsing cloud, but that seems unsatisfactory because we would have to assume a common special shape for the original clouds in order to make $\rho \sim r^{-2}$ halos a common phenomenon.

There is a second more conjectural problem. The free collapse process might be invoked to account for the formation of halos, or the formation of clusters of galaxies, but we cannot expect it to account for the observed continuity of dark halos with the large-scale pattern of the clustering of galaxies. The same applies to the Ostriker-Ikeuchi scenario, where galaxies form through a chain reaction of explosions in the intergalactic medium while clusters of galaxies form by gravitational instability.[21,22] In such pictures the continuity between galaxies and clusters of galaxies would have to be only a coincidence, which certainly is possible, though it would be a shame to have to abandon such an interesting clue.

If continuity were significant it could mean that galaxies and clusters of galaxies formed by scaled versions of the same process, or it could mean one component grew out of the other. An example of the latter would be the pancake scenario,[23] where the first generation are proto-clusters. Galaxies would form by fragmentation of these objects. If this proceeded in a scale-invariant way, in the manner of the energy cascade in turbulence, it could produce the wanted scaling law $M \sim N \sim r^{1.2}$. However, there is the problem that in this picture galaxies are no older than clusters, while the observations indicate that galaxies by and large are old, while superclusters are young, forming only now.[13]

In clustering hierarchy scenarios galaxies and clusters form in a progression of increasing mass by scaled versions of the same process, gravitational instability. One imagines that these objects developed out of small primeval density fluctuations $\delta\rho/\rho$. If $\delta\rho(r)/\rho$ is a random Gaussian process with power spectrum not far from flat then larger objects form later as clusters of smaller objects. If dissipation could be neglected the result would be a clustering hierarchy[15] that resembles one of Mandelbrot's fractals.[24]

The currently popular CDM verison of the clustering hierarchy scenario assumes the primeval density fluctuations are adiabatic, and it assumes that the present mass of the inverse is dominated by exotic matter that is weakly interacting and has low primeval pressure (cold dark matter, or CDM). A further standard assumption is that the pertrubations to space curvature by the mass fluctuations diverge only as the logarithm of the wave number k. Under all these assumptions the spectrum of density fluctuations out of which galaxies and clusters of galaxies would have to form is fixed up to one free parameter, the amplitude.[25] The result is not unpromising as a way to produce objects with nominal masses and radii comparable to those of galaxies and clusters of galaxies.[26]

Can the CDM scenario account for the halo phenomena? In the absence of dissipation the mass distribution develops into a clustering hierarchy that at least roughly approximates the observed galaxy distribution, though to account for large-scale structure -- superclusters and voids -- we must invoke "crosstalk" that promotes galaxy formation in some regions, suppresses it in others.[27] On small scales we want monolithic halos rather than a fractal distribution. We might imagine that that came about because gravitational relaxation or non-gravitational dissipation erased the clustering heirarchy on small scales. This "cannibalism" process[28,29] is generally believed to occur now and to have caused appreciable luminosity evolution of giant elliptical galaxies subsequent to redshift z = 1. Of course, evolution could have been more rapid in the past if we started with more pieces. The shape of the halo that would form by this process would depend on the energy distribution of neighboring galaxies, so we can see how a connection between the masses of dark halos and clusters might develop even if there was none to begin with. In this model the halo stars seen at r ~ 200 kpc from the center of M87 could be the stellar components of the "pregalaxies" from which the galaxy was assembled.

Is this sequence of assumptions and conjectures reasonable? My impression is that it is a useful guide to research but that a good deal of work will have to be done before we can place much weight on it as a theory for the origin of galaxies.

Yet another highly conjectural scenario might be mentioned. Let us suppose that the dark matter consists of neutrinos with a mass m of a few tens of eV. This has the virtue, if we are going to invoke exotic matter, that we at least know neutrinos exist. Also, as Cowsik and McClelland[30] noted, the mass we need yields a beautiful coincidence. The mean number density n of neutrinos thermally produced in the hot Big Bang is known. (It is comparable to the density of primeval fireball photons at T = 2.7 K.) With m ~ 30 eV the product ρ = mn is a mean mass density comparable to what is wanted for a reasonable cosmological model. Second, the mass is just large enough that the wanted number density $\rho(r)/m$ in a dark halo is small enough that it avoids inconsistency with the exclusion principle. A problem is that if we assume adiabatic initial conditions, so that radiation, neutrinos and baryons all have similar space distributions, we find that the primeval thermal motions of the neutrinos erase initial density fluctuations out to uncomfortably large scales.[31,32] To avoid this let us adopt "isocurature" density fluctuations: in the very early universe the baryons (or more properly the density of the absolute value of the baryon number excess) are concentrated in randomly placed clumps of typical mass m_o, and the radiation is distributed so as to cancel these mass concentrations so that space curvature is not perturbed. (The second assumption is needed because if space curvature fluctuations were present the related mass density fluctuations would grow and eventually dominate as adiabatic perturbations.) At high redshifts the neutrinos are relativistic and so would preserve a smooth distribution out to the horizon around each lump of baryons. As the universe expands and the neutrino peculiar velocities decrease the neutrinos would tend to accrete around the mass lumps. As long as only one mass lump is in the horizon the accretion is accurately spherically symmetric. It is an easy exercise[33] to see that the resulting neutrino mass distribution that accumulates around a mass clump varies as $\rho \propto r^{-3/2}$, which implies a rotation curve $v_c(r) \propto r^{1/4}$, which is not unreasonable for dark halos of galaxies.

The centers of objects that formed this way would be dominated by the baryons we put there to begin with. To get circular velocities of 200 to 300 km s^{-1} in dark halos we would need to assume that the baryon clumps had masses m_o on the order of 10^9 Solar masses,[33] which is comparable to what is seen. The randomly placed clumps gravitate and so would move into a clustering hierarchy on large scales. Because the sources for the development of halos and clusters of halos both are the baryon lumps we know the density run from dark halos to clusters of halos would be at least roughly continuous.

A virtue of this scheme is that it provides a direct picture for the formation of the remarkably symmetric and extended dark halos seen around objects like M87 and isolated giant spirals (though the stellar component is more problematic). It is advertised that relaxation processes might produce such structures in clustering hierarchy scenarios; a convincing demonstration over so many orders of magnitude of density is not going to be easy. The great problem with the isocurvature scheme is that so much rests on the lumps of baryons. They are not entirely ad hoc, for there have been discussions of models for baryosynthesis in which the baryon number is a function of position.[34] If the mass distribution prior to baryosynthesis were accurately homogeneous then the perturbations due to inhomogeneous baryon productions would be isocurvature, as wanted. But just as for the CDM scenario, a good deal of careful analysis will have to be done before we can decide whether this isocurvature scenario really is a sensible model for the origin of the halo phenomena.

I am grateful to Steve Boughn, Earl Spillar and Dave Wilkinson for helpful discussions. This research was supported in part by the U.S. National Science Foundation.

REFERENCES

1. Bosma, A., A. J. 86, 1791 (1981).
2. Rubin, V. C., Science 220, 1339 (1983).
3. Milgrom, M., Ap. J. 270, 365 (1983).
4. Felten, J. E., Ap. J. 286, 3 (1984).
5. Fabricant, D. and Gorenstein, P., Ap. J. 267, 535 (1983).
6. de Vaucouleurs, G. and Nieto, J.-L., Ap. J. 220, 449 (1978).
7. Boughn, S. P. and Saulson, P. R., Ap. J. 265, L55 (1983).
8. Oemler, A., Ap. J. 209, 693 (1976).
9. Arp, H. C. and Bertola, F., Astrophysical Letters 4, 23 (1969).
10. Pagels, H. R., in the proceedings of the Eleventh Texas Symposium on Relativistic Astrophysics, ed. D. S. Evans, Ann. N.Y. Acad. Sci. 422, 15 (1984).
11. Tinsley, B. M. and Gunn, J. E., Ap. J. 203, 52 (1976).
12. Davis, M. and Peebles, P. J. E., Ap. J. 267, 465 (1983).
13. Peebles, P. J. E., Science 224, 1385 (1984).
14. Bean, A. J., Efstathiou, G., Ellis, R. S., Peterson, B. A. and Shanks, T., M.N.R.A.S. 205, 605 (1983).
15. Peebles, P. J. E., "The Large-Scale Strucute of the Universe," Princeton: Princeton University Press (1980).
16. Bookbinder, J., Cowie, L. L., Krolik, J. H., Ostriker, J. P., and Rees, M., Ap. J. 237, 647 (1980).

17. Sarazin, C. L. and O'Connell, R. W., Ap. J. 268, 552 (1983).
18. Rees, M. J. and Ostriker, J. P., M.N.R.A.S. 179, 541 (1977).
19. Peebles, P. J. E., A. J. 75, 13 (1970).
20. Gott, J. R., Ann. Rev. Astron. Astrophys. 15, 235 (1977).
21. Ostriker, J. P. and Cowie, L. L., Ap. J. 243, L127 (1981).
22. Ikeuchi, S., Publ. Astron. Soc. Japan 33, 211 (1981).
23. Zel'dovich, Ya. B., Einasto, J. and Shandarin, S. F., Nature 300, 407 (1982).
24. Mandelbrot, B. B., "The Fractal Geometry of Nature," San Francisco: Freeman (1982).
25. Peebles, P. J. E., Ap. J. 263, L1 (1982).
26. Blumenthal, G. R., Faber, S. M., Primack, J. R. and Rees, M. J., Nature 311, 517 (1984).
27. Rees, M. J., J. Astrophys. Astr. 5, 331 (1984).
28. Ostriker, J. P. and Tremaine, S., Ap. J. 202, L113 (1975).
29. McGlynn, T. A. and Ostriker, J. P., Ap. J. 241, 915 (1980).
30. Cowsik, R. and McClelland, J., Ap. J. 180, 7 (1973).
31. Peebles, P. J. E., Ap. J. 258, 415 (1982).
32. White, S. D. M., Frenk, C. S. and Davis, M., Ap. J. 274, L1 (1983).
33. Peebles, P. J. E. in "The Origin and Evolution of Galaxies," eds. B. J. T. Jones and J. E. Jones, Dordrecht: Reidel, p. 143 (1983).
34. Mohanty, A. K. and Stecker, F. W., Physics Letters 143B, 351 (1984).

OBSERVATIONAL AND THEORETICAL ASPECTS
OF RELATIVISTIC ASTROPHYSICS AND COSMOLOGY, pp. 269-310
edited by J. L. Sanz and L. J. Goicoechea
© 1985 by World Scientific Publishing Co.

THE MICROWAVE BACKGROUND RADIATION

Juan M. Uson

Joseph Henry Laboratories, Physics Department, Princeton University, and National Radio Astronomy Observatory*

ABSTRACT

The microwave background radiation is a remnant of the early Universe. It provides us with information on physical processes and (limits on) initial conditions that have determined the formation of the galaxies and clusters of galaxies that we see today. Its spectrum matches that of the radiation from a blackbody at $T \sim 2.75$ K with no statistically significant distortions measured at any frequency, which is strong evidence of a hot Big Bang model for the evolution of the Universe. It is highly isotropic: $\Delta T/T < 10^{-4}$ over all angular scales except for a dipole component which can be attributed to our motion with respect to the preferred reference frame in which the radiation would appear isotropic. This lack of anisotropies provides important constraints on key cosmological parameters such as Hubble's constant and the density parameter for the different theoretical scenarios for the Universe and its contents. A recent (small-scale) anisotropy is produced as the radiation interacts with the ionized gas in dense clusters of galaxies; this is the Sunyaev-Zel'dovich effect which might eventually become an important cosmological tool. These lectures discuss briefly these topics but are not a review of the subject. Instead, I concentrate mainly on recent measurements by the Princeton group, with special emphasis on the experimental difficulties involved and future (likely) improvements.

---

* Operated by Associated Universities, Inc., under contract with the National Science Foundation.

1. INTRODUCTION

Modern cosmology is based on the Cosmological Principle[1]: "Matter is distributed in space in a homogeneous and isotropic way," and relies on two experimental facts:

A) The recession of galaxies. This is Hubble's law[2] which states that "galaxies recede from us (and from each other) with speeds that are proportional to their distance from us (from each other)":

$$v_{12} = H_o r_{12} \qquad (1)$$

with Hubble's constant, H_o, in the range (50-125) km sec^{-1} Mpc^{-1}. (1 Mpc = 3.086 x 10^{22}m.)

From Hubble's law, the standard model yields an age for the Universe of the order of $H_o^{-1} \sim 1.5 \times 10^{10}$ years. Hubble's first determination of H_o was 500 km sec^{-1} Mpc^{-1} which produced an estimate of only 2 x 10^9 years for the age of the Universe; shorter than the estimated ages of stars and of the earth. Therefore, Hubble's law motivated two opposed theories:

i) The steady-state model, which tried to avoid the experimental (erroneous) conclusion that the Universe was younger than its contents. It was based on an extension of the cosmological principle by Bondi and Gold[3] who proposed that the Universe was also homogenous in time, with all the relevant cosmological parameters like its density (ρ) and the distance scale factor (H_o) being constant. In order to make this requirement compatible with the observed expansion, this required continuous creation of matter, but at a level which is below experimental limits and so, in principle, allowed. (The required creation rate was about one atom per cubic kilometer per year.)

ii) The Big-Bang model which establishes that the Universe is expanding from a state of extremely high density, a big explosion that occurred about 1.5 x 10^{10} years ago.

B) The second major experimental base of modern cosmology is the cosmic microwave background which bathes the earth in a sea of radio waves that reach us from everywhere and that where emitted during the Big-Bang. Its most important contribution corresponds to wavelengths in the range of 10^{-4}m to 1m.

2. THE SPECTRUM OF THE MICROWAVE BACKGROUND

The microwave background is blackbody radiation. It was predicted in 1948 by Gamow[4] as a necessary component of the Big-Bang theory. From the observed abundances of helium and deuterium, Alpher and Herman[5] deduced that its temperature should be about 5K (a lower temperature would correspond to a higher helium abundance and conversely). The sensitivity of the available equipment was not enough at that time to detect such a weak radiation and the notion was forgotten. (Dicke and collaborators[6] had set an upper limit of T < 20K on the temperature of a possible background radiation, but this went largely unnoticed and was generally forgotten, even by Dicke himself).

In 1964, Dicke and Peebles reinvented the notion of a microwave background as a necessary ingredient for a cyclic model of the universe. Here, photons were needed to ensure thermal equilibrium in each successive contraction, at a high enough temperature to dissociate heavy nuclei in order to start each cycle with mainly protons, neutrons and electrons. Their calculation prompted Roll and Wilkinson to set out to measure such a background; but before they obtained their measurement, Penzias and Wilson found an unwanted source of noise while carefully calibrating the Bell Labs horn antenna in order to make an absolute map of the galactic background radiation. As soon as both groups discussed their work, they agreed that this noise was indeed the primeval background radiation[7].

Measuring such a small and diffuse signal is difficult and this is the reason why it was not done before 1965. The experimental problems are due to the atmosphere and ground that contribute radiation that has to be carefully measured and subtracted; galactic and extragalactic radio sources also contribute to the signal and have to be eliminated, which is particularly hard at low frequency (less than a few GHz). Measurements at high frequency (above ∼ 100 GHz) have to be done above the atmosphere and are normally performed with balloon-borne equipment. Finally, noise is contributed by the mixer and amplifier, as well as by other components of the system such as waveguide; these have to be carefully estimated or measured and kept as constant as possible.

The traditional way of dealing with these problems follows a procedure invented by Dicke[8], which amounts to a rapid comparison of the wanted signal with a stable reference source. This "Dicke switching" filters out low frequency components of the noise, which are normally the worst ones, and allows very small signals to be measured in the presence of much larger noise by synchronous detection of the receiver output. The measurements are normally not limited by statistical fluctuations but by systematic effects, and the experiments have to be carefully designed to deal with them. Figure 1 shows a sketch of a typical setup that addresses the experimental problems successfully. The Dicke switch compares the radiation entering the main horn with that seen through the sky horn which, having a much broader beam, acts as a reference. One calibrates in Kelvin by placing absorbers at known temperature over the main horn. The radiation coming from the sky through the main horn consists of whatever background radiation there is, plus the contribution due to the atmosphere, which is modeled as a plane slab with a secant-law contribution that is measured by tilting the large aluminum reflector. The contribution of the reflector can be accurately measured and subtracted. The various horns have low side and back-lobes and ground screens help to minimize the signal picked up from the earth which is hot (∼300K!). Finally, an absolute measurement is achieved by placing a cold load (absorber dipped in liquid Helium) in front of the main horn. Notice that the design shown in figure 1 does not require moving the instrument while doing all the required measurements. This avoids systematic effects due to variable losses in the instrument; these vary as stresses change in ways that are hard to predict and avoid. This design allowed Wilkinson and collaborators to achieve accuracies of about 0.2K over a wide range of frequencies as early as 1967, with variations in the atmosphere being the limiting factor. (See ref. 10 for details.) Measurements on the Rayleigh-Jeans region fit a blackbody spectrum with T = 2.72K±0.08K.

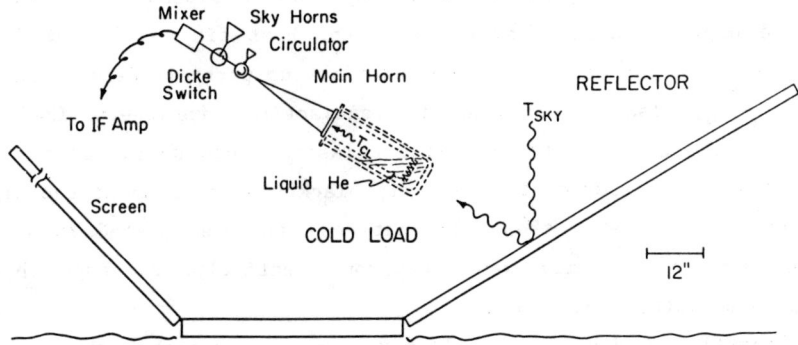

Figure 1. A successful design to measure T_{BG}.

Measurements near the peak of the blackbody spectrum or at even higher frequencies are hard to make as one has to get "above" the atmosphere. This is normally done from balloons with the added complication of loosing some control over the apparatus; the different environmental conditions might induce systematic problems different from the ones at ground level, which have to be foreseen in order to add the necessary devices to measure them in flight.

Therefore, early measurements showed disagreements with the measurements at the Rayleigh-Jeans region[10]. A careful measurement of this region was done by Woody and Richards[11] with a balloon-borne Fourier transform spectrometer with a bolometric detector. Their best fit blackbody temperature was 2.96K and the results showed systematic distortions. It was pointed out by Weiss (ref. 10) that an overall calibration error could account for the discrepancy in the (mean) blackbody temperature, with no statistically significant distortions left. This had been noted by Woody and Richards who argued that the calibration error necessary (∼30%) for this explanation was too large; Weiss argued that even though there was no specific reason to produce such a large calibration error, it was not excluded by the calibration procedure[10].

This "distortion" caused a flurry of theoretical models[12]. It was attributed to Compton scattering by ionized plasma (which would cause a high frequency distortion of the best fit to the low frequency measurements); or to out-of-equilibrium particle decays at t ∼ 1 year$^{+100}_{-0.99}$; also to emission of pregalactic stars thermalized by intergalactic dust; or to free-free emission (this would impose low frequency distortions on the best fit blackbody spectrum at the high frequencies). Some of these alternatives have been pushed in order to account for the microwave background entirely, although these models seem rather contrived.

Recently, Richards and collaborators have repeated their measurement with a different setup, consistent of a set of broad-band filters and a bolometer with an internal cold reference. This measurement now agrees with a 2.75K blackbody[13] and does not show any significant distortion; there is some hint of possible deviations in the direction of the ones previously found, although not at a significant level.

An indirect method of measuring the background radiation uses cyanogen (also CH and CH^+) as a thermometer (Figure 2). Consider an interstellar cyanogen cloud and assume that it is (nearly) in equilibrium with the background radiation. Choose such a cloud on the line of sight to a continuum background source, like a star, and observe the absorption lines in the spectrum of the background source. The relative intensity of the absorption lines yields the relative populations of the lower energy levels in the molecules through $(n_1/n_2) = \exp\{-E/KT\}$. The transitions between these levels correspond to the excitation of the rotational degrees of freedom of the molecule and are assumed to be (mainly) induced by the background radiation. One can then derive T for various frequencies, $\nu = E/h$. This provides upper limits on the microwave background temperature, as some excitation could be due to other mechanisms like collisions with other molecules. A recent measurement was done by Meier and Jura[14], who estimate the collisional excitation of the absorbing molecules and derive $T_{BG} \lesssim (2.73\pm0.04)$K at $\lambda = 2.64$ mm ($\nu \sim 113$ GHz) and $T_{BG} \lesssim (2.8\pm0.3)$K at $\lambda = 1.32$ mm ($\nu \sim 227$ GHz). It is not clear that the collisional processes have been completely accounted for and the errors could have been underestimated, but their results are nevertheless impressive.

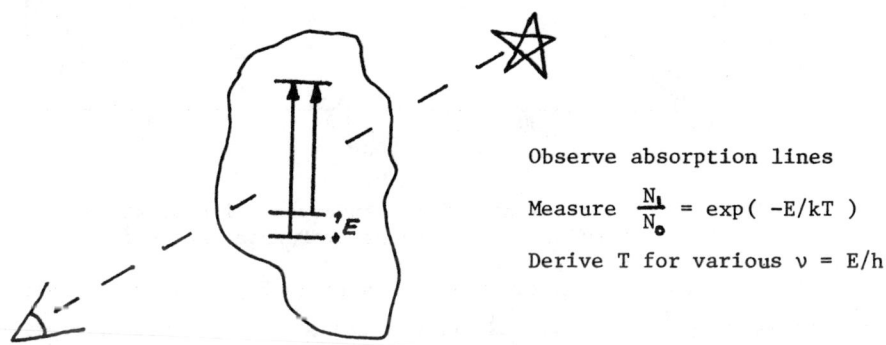

Observe absorption lines

Measure $\frac{N_1}{N_0} = \exp(-E/kT)$

Derive T for various $\nu = E/h$

Figure 2. Indirect measurements using interstellar CN.

Figure 3 shows the different measurements of the absolute (thermodynamic) temperature of the background radiation. They agree with a blackbody spectrum with $T_{BG} = 2.75$K with no statistically significant distortions over a frequency range from about 400 MHz to over 500 GHz[9,10,13-15]. (A weighted fit of all measurements to a single thermodynamic temperature yields $T = (2.747\pm0.024)$K with a Chi-square of 22 for 32 degrees of freedom.) Nevertheless, low level broad-band distortions are not excluded and could be present at the

±0.2K level; narrow-band distortions could be present with even higher amplitudes.

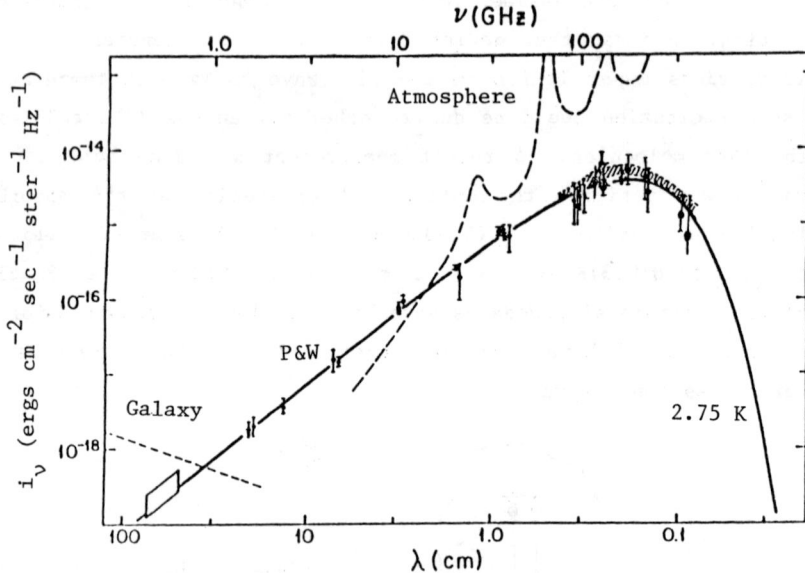

Figure 3. Measurements of the spectrum. The point P&W is the first measurement by Penzias and Wilson.

As the cross-section for the interaction of photons with neutral matter is very small, the radiation travels almost interaction-free if the universe is not ionized. Therefore, blackbody radiation in an expanding non-ionized universe preserves its spectrum and only changes its temperature as $T(t) = T_i(a_i/a(t))$ as shown by Tolman[16] (a is the expansion parameter). Therefore, the microwave background tells us that:

1) There was once thermal equilibrium in the universe, as the expansion does not affect the spectrum and only "cools" the radiation.

2) The universe has <u>evolved</u> from a hot initial state, which therefore rules out the steady-state theory in favor of the Big-Bang model.

3) The absence of spectral distortions limits some cosmological processes (the ones that were invoked to produce it when it was thought to be seen), although these limits are not yet very strong.

4) Its isotropy (see below) lends support to the cosmological principle and limits possible anisotropies in the expansion of the universe. There is perhaps the problem of it being too isotropic!

5) The microwave background could (maybe) fix the initial conditions from which the structure seen in the Universe has evolved and places some limits on cosmological parameters like Hubble's constant and the density parameter.

3. SMALL-SCALE ISOTROPY

Soon after the discovery of the microwave background, Peebles pointed out that observations of its small-scale structure could provide a direct observational probe of the power spectrum and characteristic lengths and masses of the initial density fluctuations that eventually became galaxies and groups of galaxies[17]. This picture of the early universe could be blurred by subsequent scattering of the microwave photons by ionized matter at a much later epoch[18].

A number of observers have published upper limits on small-scale anisotropy of the microwave background at various angular scales and frequencies[19,20], but no small-scale structure has been detected.

The expected anisotropy can be deduced from the present-day structure of the Universe, and a number of authors have studied this relation in detail[21-26]. So far, ways have been found to lower theoretical estimates of the "minimal" anisotropy of the microwave background whenever they have conflicted with the observations.

The connection is sketched in Figure 4 for the case of a universe in which baryons provide the dominant component to its mass. During the initial phase, before the universe cools below $T \sim 4000K$, the universe is optically thick: matter is mostly ionized and in close interaction with the background photons. Matter and radiation can be treated as a single fluid with pressure mainly due to the photons. This pressure provides a restoring force that opposes gravity and matter tends to oscillate instead of piling-up.

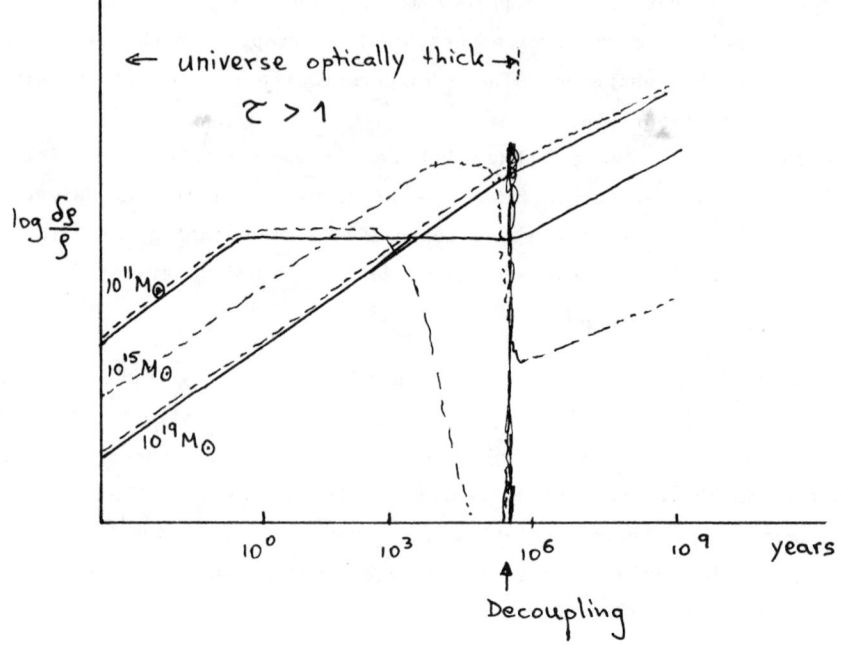

Figure 4. Sketch of the evolution of density perturbations in a baryon dominated universe. The masses of the perturbations are indicated and the dashed lines correspond to adiabatic perturbations with the full lines describing isothermal perturbations.

The fate of a fluctuation is determined by its size as it enters the horizon, i.e., as its various components become causally connected. At any given time there is a characteristic mass scale, the Jeans mass, such that perturbations of higher mass will tend to grow whereas perturbations of lower mass are unstable to oscillations. The Jeans mass grows with time during the first thousand years since the explosion and then levels off at about $3 \times 10^{14} \, \Omega^{-2} \, h^{-4} \, M_\odot$ (h is Hubble's constant in units of 100 km sec^{-1} Mpc^{-1}, Ω is the density parameter:ratio of the actual density to that of an Einstein-de Sitter universe; and $M_\odot$ is the mass of the sun, 2×10^{33} g).[27] Perturbations of larger mass grow as $t^{2/3}$. An important mass scale is that below which photon diffusion and damping are important[28]. This mass grows with the expansion as the mean free path for photons grows. Its value at decoupling is the so-called "Silk mass", $M_s \sim 10^{12} h^{-5/2} \Omega^{-5/4} M_\odot$. Adiabatic fluctuations of lower masses will be severely damped and essentially erased by this time, whereas isothermal fluctuations are not affected by this mechanism.

When the expansion has lowered the temperature to $T \sim 4000K$, after a time of about 10^5 years, protons, α-particles and electrons combine to form mostly Hydrogen and Helium and the Universe becomes optically thin. The Jeans mass drops to $\sim 10^5 M_\odot$ and, after this epoch, matter fluctuations are free to evolve under gravity and gas dynamics independently of the background photons. This decoupling occurs at a redshift $Z_{dec} \sim 1000$.

It is fairly straightforward to estimate that density fluctuations ($\delta\rho/\rho$) have grown at most by a factor $(1 + Z_{dec})^{-1}$ since the decoupling time. Detailed calculations show that for the adiabatic scenario this requires fluctuations in the background radiation of $\delta T/T \sim (1/3) (\delta\rho/\rho) > 10^{-4}$ in order to account for present-day structures. In the isothermal scenario, density fluctuations do not induce fluctuations in the radiation directly; nevertheless, smaller fluctuations are produced as a result of Doppler shifts at the last scattering of each photon. Such effects should have produced fluctuations of the order $\delta T/T \sim v/c \gtrsim 10^{-5}$. In order to fix the scale, notice that $10^{15} M_\odot$ subtend about 5' (this depends mainly on the density parameter Ω).

Matter could become ionized again at a much later epoch, associated with galaxy formation. The initial burst of star formation could release enough energy to achieve a significant degree of ionization and sufficient optical thickness[18]. This could erase fluctuations on scales causally connected at this time, although the process would enhance fluctuations on the scale of the ionized patches; besides, some degree of polarization would be induced by protogalactic magnetic fields[29]. There are some arguments against this possibility that suggest that such a phase would have not affected the angular structure of the microwave background imprinted at the decoupling time[30].

It is hard to measure small-scale anisotropy in the microwave background at the level of 10^{-5}. These measurements require the use of large telescopes in order to obtain the wanted narrow beams. The main experimental problems are again atmospheric emission at $\lambda < 1$ cm, ground radiation at all wavelengths, interference and variable instrumental effects, radio source confusion at $\lambda > 1$ cm and (perhaps) galactic dust emission at $\lambda < 1$ mm. Besides these sources of systematic errors, one again has to deal with Johnson noise from the receiver. This noise determines the accuracy with which a measurement can be made due to statistical fluctuations in the receiver output as:

$$\Delta T_{rms} \sim \frac{2 T_{sys}}{e_{ff}(\Delta \nu \cdot t)^{\frac{1}{2}}} \qquad (2)$$

where T_{sys} is the effective noise temperature of the receiving system, e_{ff} is an efficiency factor, generally of the order of 0.5, $\Delta \nu$ is the pre-detection bandwidth of the system, t is the effective integration time, and the factor 2 assumes a 50% switching cycle with square-law modulation.

For example, the 140-foot telescope of the NRAO in Green Bank (West Virginia) is equipped with a maser amplifier that at a frequency of 19.5 GHz features $T_{sys} \sim 40K$, $\Delta \nu \sim 400$ MHz, $e_{ff} \sim 0.55$ and a 66% useful duty cycle when beam-switching at 3.33 Hz is used; so that in 1 sec, $\Delta T_{rms} \sim 9mK$, in one hour one could reach $\Delta T_{rms} \sim 0.15mK$ and 30 hours would be needed to reach $\Delta T/T \sim 10^{-5}$ on one spot. One could reach this sensitivity on a dozen spots in about two weeks (of perfect weather!).

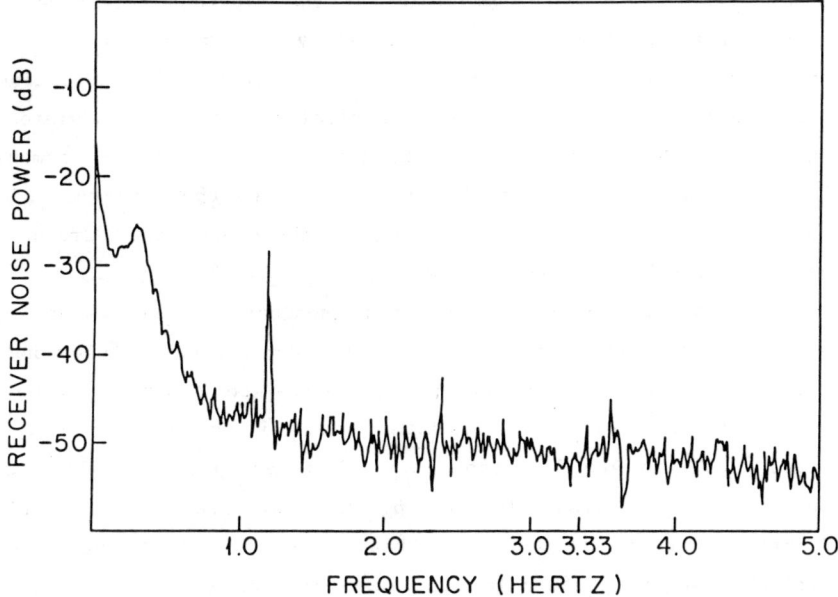

Figure 5. Power spectrum of the NRAO K-band maser receiver at the 140-foot telescope. The maser is tuned at 19.5 GHz, with a bandwidth of 400 MHz. The spectrum shows a 1/f characteristic with narrow lines that are due to a mechanical refrigerator.

But, in high sensitivity radiometry, a crucial parameter is the receiver power spectrum near the Dicke-switch frequency. In general, electronic interference or non-linearities in some of the various components of the receiver, IF amplifier and synchronous detector might add noise to their intrinsic (1/f) spectrum. Some of these contributions could even be in the form of narrow spikes which could conceivably drift in time. All this will at best increase the rms noise above the theoretical level; if any of these spikes drifted in time, it could even mimic a sky signal by contributing power at the Dicke-switching frequency. Therefore, the receiver power spectrum has to be monitored and kept as clean and stable as possible. The optimum Dicke-switching frequency should be determined by this power spectrum. Figure 5 shows the current power spectrum of the NRAO K-band (18-25 GHz) maser amplifier at the 140-foot telescope. The narrow spike at $\nu \sim 1.2$ Hz is induced by a refrigerator that keeps the temperature in the Dewar at 4.2K. The small (mK) modulation of this temperature induced by the mechanical refrigerator modulates the maser gain and produces this spike together with several harmonics. This frequency is known to be stable although the amplitude of the spikes changes with time. One should therefore avoid frequencies close to 1.2n Hz for the Dicke-switching.

One then has to decide on how to conduct the experiment. This decision cannot be made before using the corresponding telescope, as the systematic problems associated mainly to ground pick-up and scattered radiation that is intercepted by the sidelobes of the antenna pattern are different for each telescope. One also has to deal with limitations imposed by the hardware, like restricted pointing ability; the details of beam separation, position angle and available switch frequencies if beam-switching is used; or spurious signals induced by nearby structures like service buildings. All these effects are hard to estimate "a priori"; the best strategy is to measure them and decide on the observing strategy in order to eliminate the worst problems. In order to illustrate some of these problems and how to go about them, I discuss next one such experiment in which Dave Wilkinson and myself have collaborated over the last four years[31].

We have used the NRAO 140-foot telescope, which is sketched in Figure 6. The Dicke-switching is achieved by nutating the Cassegrain subreflector (at $\nu = 3.33$ Hz), which switches the main beam between two positions on the sky (separated 4.5 minutes of arc in our measurements). Here, the dominant source of systematic errors is due to residual ground pick-up introduced by the variations in the diffraction pattern of the telescope as the subreflector nutates. This residual ground pick-up varies with telescope position. One minimizes this variation by going to high declination (in our case, dec $\sim$

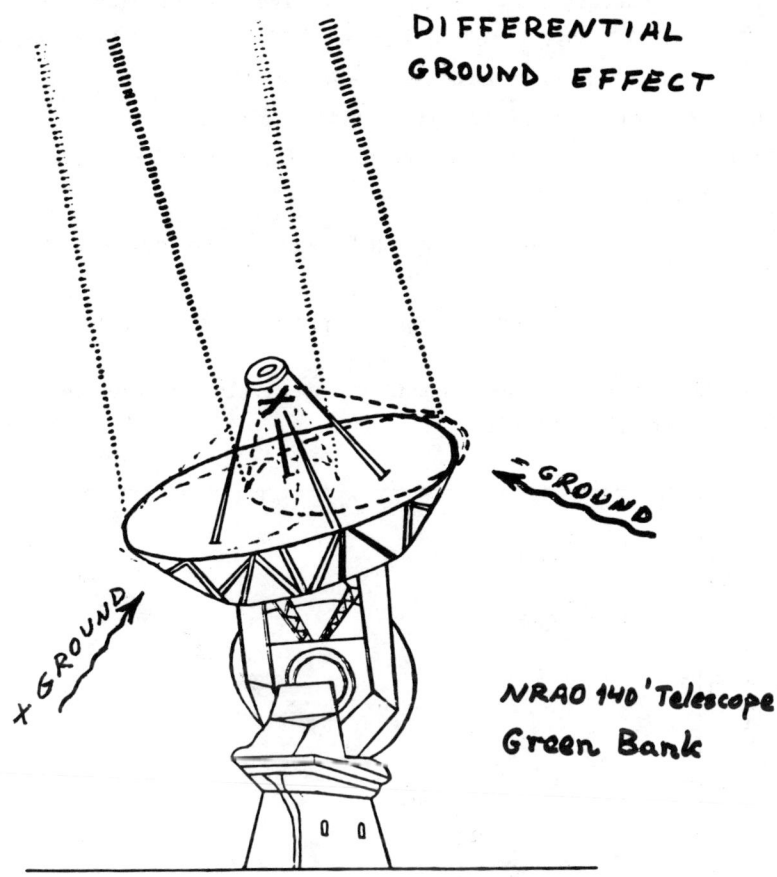

Figure 6. Sketch of the observing setup. The ground pickup is about 1.5 K. The nutation of the subreflector does not cancel it out, but produces a residual ground pick-up that depends on the beam throw and changes with telescope position. For the measurements discussed here, the drift is about 55 mK/hour. A double subtraction scheme is needed to cancel this systematic effect.

87°), so that the motion of the telescope with respect to the ground as it follows a position on the sky is slow and the systematic variation of the residual ground signal is only about 55 mK/hour

(Figure 7). Furthermore, the effect is cancelled by a double subtraction technique. To each of twelve fields (at RA: 1^h30^m, 3^h30^m....23^h30^m, Dec = 86°51'; epoch 1950), we associate two reference fields on opposite sides and separated 4.5 minutes of arc from each main field, which is alternatively placed in the "+" beam (ON scans) and the "-" beam (OFF scans). This provides for each group two sets of data:

$$\delta T_{ON} = T_{FIELD} - T_{REF2} + T_{OFFSET-ON} \qquad (3a)$$

and

$$\delta T_{OFF} = T_{REF1} - T_{FIELD} + T_{OFFSET-OFF} \qquad (3b)$$

where REF1 and REF2 correspond respectively to the westerly and easterly reference fields.

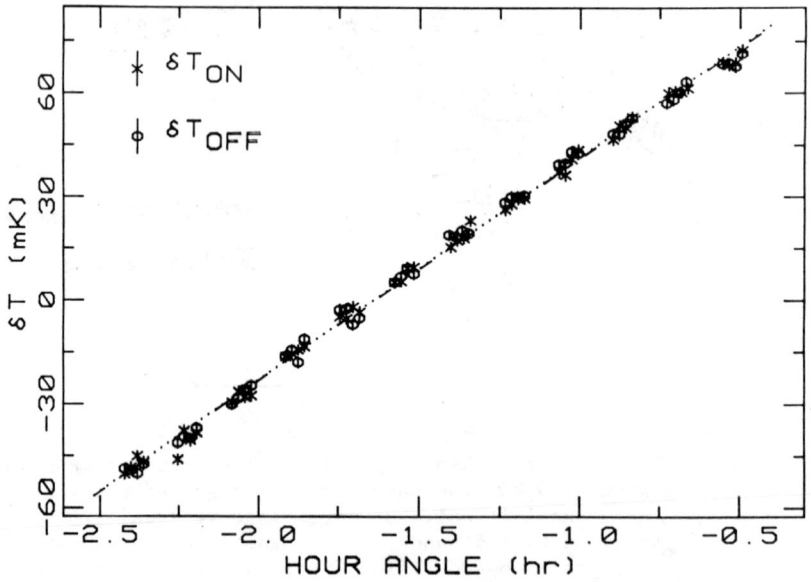

Figure 7. Drift of δT_{ON} and δT_{OFF} (eq. [3]) with hour angle for a 2 hour run. Each point is averaged over 72 seconds, giving four points per scan. The linear drift is due to changing ground and atmospheric pickup by the antenna sidelobes. The second difference ΔT_{FIELD} (eq. [4]) shows no significant drift with hour angle, as expected.

The quantity measured for each field,

$$\Delta T_{FIELD} = (\delta T_{ON} - \delta T_{OFF})/2 = T_{FIELD} - \frac{T_{REF1}+T_{REF2}}{2} + \Delta T_{OFFSET} \quad (4)$$

compares each field to the average of its associated reference fields. One tries to eliminate ΔT_{OFFSET} by careful timing of the OFF-ON sequence such that both scans are taken with the telescope going through the same range of positions with respect to the ground. (The OFF and ON scans last 4 min 48 sec each). This achieves the best cancellation of the ground signal.

The data shown in Figure 8 were taken during 1982 December 27 to 31, 1983 May 16 to 20, and 1984 February 25 to March 5. These runs

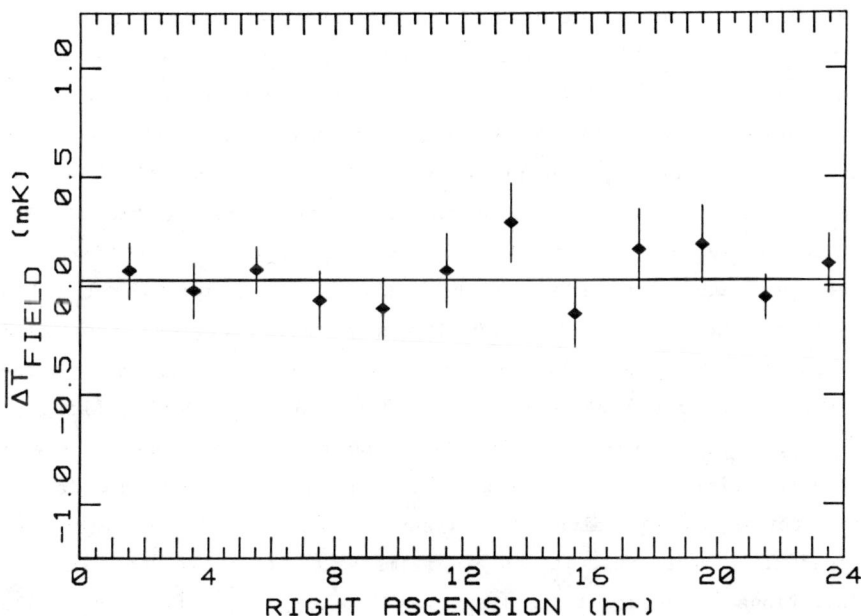

Figure 8. Final averages for the observed fields. The solid line indicates the overall weighted average, $\Delta T_{AVE} = (26\pm39)$ μK. ΔT_{FIELD} is the difference in thermodynamic temperature between the field and the average of two reference positions (see eq. [4]).

gave a total of 174 hours of useful data. After correcting the data for the beam efficiency of the antenna and converting them to thermodynamic temperatures, we compute for each day (i) the mean values ΔT_{FIELD}^i and associated standard deviations σ_{FIELD}^i calculated from the scatter of the individual 10 minute OFF-ON measurements. A typical value for data taken with clear sky is $\sigma_{FIELD}^i \sim 0.25$ mK whereas one expects 0.14 mK from system noise alone. The excess noise at this stage is due to the 1/f component in the maser noise spectrum and in the variable components of the ground and atmospheric contributions to δT_{ON} and δT_{OFF}.

The final values, $\overline{\Delta T}_{FIELD}$, for each one of the fields were obtained from the different two-hour observations of that field as weighted averages, using the standard σ^{-2} weights. The errors associated to each point have to be accurately estimated: an underestimate could lead to a spurious detection as the scatter of the measured values would be blamed on a true sky signal, whereas an overestimate could bury a sky signal as the scatter of the points would no longer be statistically significant. In principle, one associates to each point an error that corresponds to the standard deviation in the mean estimated from the measurement statistics in the usual way. In order to check that this is actually the best estimate, we test whether the two-hour values deviate from their corresponding mean value as expected from gaussian statistics. Figure 9 shows a histogram of the 87 weighted residuals ($\Delta T_{FIELD}^i - \overline{\Delta T}_{FIELD})/\sigma_{FIELD}^i$ (i = 1 to 87 and FIELD denotes any one of the 12 observed fields). The histogram is indeed reasonably gaussian and the corresponding value of Chi-square is 78 with 75 degrees of freedom. Therefore, it is legitimate to estimate the standard deviations in Figure 8 as $\overline{\sigma}_{FIELD} = \{\sum_i (\sigma_{FIELD}^i)^{-2}\}^{-\frac{1}{2}}$ for each field. The variation of $\overline{\sigma}_{FIELD}$ from field to field is mainly due to the different number of days over which each field was observed. The weighted mean of the points in Figure 8 is $\Delta T_{AVE} = (26\pm39)$ μK.

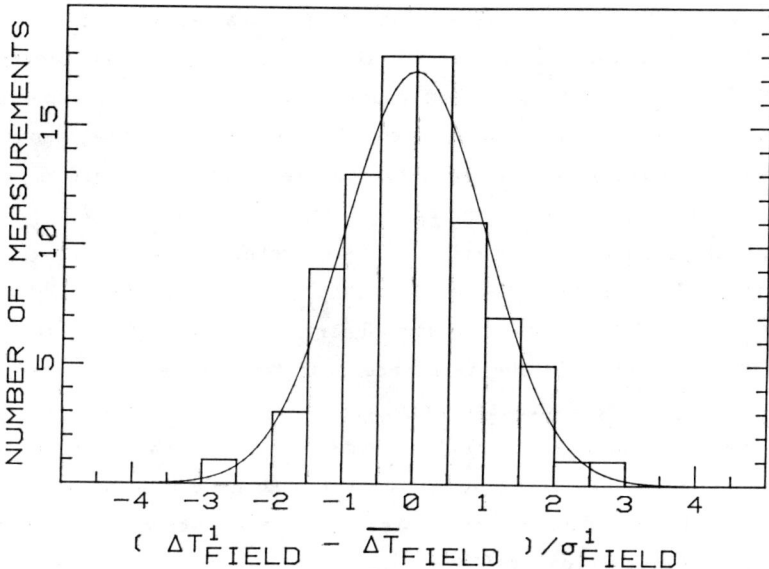

Figure 9. The day-to-day scatter in measurements averaged over ∼ 2 hours. As the residuals from the field averages are used, the possible sky signals have been removed. If there were no extraneous day-to-day effects, the histogram would fit the gaussian curve shown; it does remarkably well.

The main systematic effect in this experiment is the variable ground pick-up and its effect has been minimized by the observing procedure. Nevertheless, any residual offset could still affect the various two-hour measurements. We searched for other systematic effects in the data like a possible correlation of ΔT_i with system temperature, which varies with ambient temperature and water content in the atmosphere; besides the system could drift in time and induce a systematic drift of ΔT_i or the diurnal cycle could induce variations with a 24h or a 12h period. No such effects were found at a statistically significant level.

As none of the 12 points in Figure 8 differs significantly from their average value, ΔT_{AVE}, no anisotropy has been detected at a level of $\sim 2\sigma_{FIELD}$. A more stringent limit can be set using all the data in a statistical argument, although this requires adopting a model for the anisotropy. The underlying sky signal is expected to be gaussian distributed between the 12 (triplets of) field positions (separated $\sim 1.6°$) and can be characterized by its standard deviation σ_{SKY}. Assuming there is no correlation among the signals arriving from each given field and its associated reference positions (4.5 arc minute scale), the rms residual in ΔT_{FIELD} due to a true sky signal is $\sigma_{rms} = (1.5)^{\frac{1}{2}}\sigma_{SKY}$ because the observing procedure used is sensitive to a "signal" in the reference positions as well. The ratio of σ_{rms} to σ_{SKY} depends on the angular autocorrelation that is assumed for the possible sky signal (in this case it vanishes) at the beam-throw scale and has to be evaluated for each different model.

The Neyman-Pearson lemma prescribes the optimal statistic for testing a hypothetical value of σ_{SKY}. In our case, the best statistic is a weighted Chi-square[32], which yields

$$\sigma_{SKY}/T < 2.4 \times 10^{-5} \text{ at the 95\% confidence level.} \qquad (5)$$

Limits on anisotropy from different experiments are difficult to compare. Various scanning patterns and statistical methods have been used, so careful statements of limits on a particular model should examine each measurement separately. Figure 10 shows current experimental limits, most of which assume that the underlying anisotropy is uncorrelated gaussian noise[33]. For this assumption, the result of eqn. 5 applies at an angular scale of about 2 minutes of arc (beam size between 1/e points).

A different way to interpret our result is to assume that all the anisotropy is the result of an incoherent superposition of (spatially) monochromatic sine waves of a certain scale on the sky. This is shown by the dashed curve on figure 10. The strongest limit corresponds to a scale (half wavelength) of 4 minutes of arc and is

$$\Delta T_{rms}/T < 2.1 \times 10^{-5} \text{ (at the 95\% confidence level)} \qquad (6)$$

which is more stringent than the one in eqn. 5 because the sinusoidal fluctuations are anticorrelated on the scale of our beam-throw which increases the efficiency of our observing procedure for this kind of fluctuations.

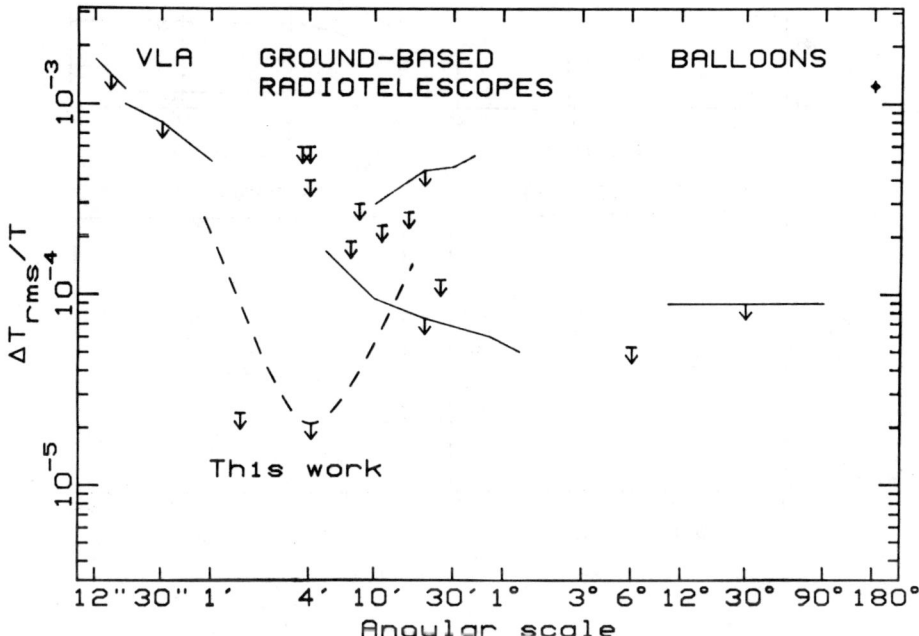

Figure 10. Upper limits, at the 95% confidence level, on anisotropy of the microwave background at various angular scales. Some of the points between 1' and 1° have been revised (Ref. 19) from the values given in the original references. The limits at sub-arcminute scale should be improved soon by about one order of magnitude.

A physical interpretation of the results requires a model for the nature of the underlying fluctuations. For simplicity, most authors assume fluctuations with a spectrum of the form $(\delta\rho/\rho) \propto k^n$ for the dominant contribution to the mass of the Universe at the time of decoupling, and $n = 1$ is preferred as the most natural spectrum. The

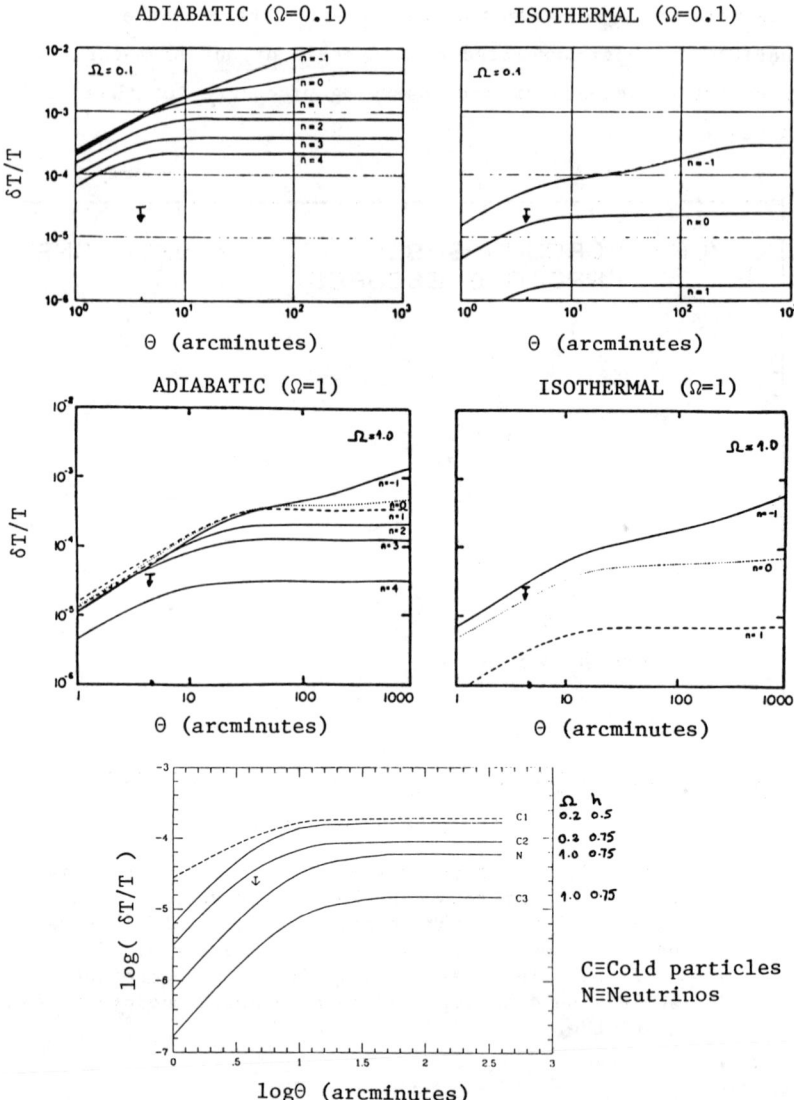

Figure 11. Predicted small-scale anisotropy of the microwave background for various cosmological models. The indicated arrows correspond to the upper limit discussed above. The predictions are statistical and the confrontation between them and the measurement relies on a number of assumptions; the most important one is that the fluctuations are gaussian distributed.

amplitudes of the fluctuations are fixed by matching the evolved autocorrelation function of the matter to the available data from studies of galaxy clustering. The details of the interaction between the mass components and the background radiation determine the anisotropy required by each particular model. Figure 11 shows a sample of such predictions for a number of cosmological models[24-26]. The limit of equation (5) conflicts with those models in which the initial fluctuations are adiabatic and dominated by baryonic matter. Baryon universes in which the initial perturbations are isothermal are currently out of fashion but only somewhat limited by the measurement; for $n < 0$ they need to satisfy $0.1 < \Omega < 0.3$, whereas for $n > 0$ no restriction is imposed. Models dominated by massive neutrinos require $\Omega \sim 1$ although they suffer other problems. More exotic possibilities are models in which the dominant component is in the form of still undiscovered cold dark matter such as axions, photinos or the like. These models would like that $\Omega \cdot h > 0.2$.

I should like to point out that these conclusions are only indicative of possible constraints. The data analysis rests on the assumption that the underlying fluctuations are gaussian distributed; if not, Chebyshev's theorem would allow the limits to be up to three times the value in eqn. (5), (for rather pathological distributions). Besides, the theoretical predictions are probably uncertain by factors of 2. Although it is interesting to note that there is a confrontation between theoretical models and the experimental data, the fate of the various models is far from settled and baryon dominated universes are only in trouble, but not ruled out.

4. LARGE SCALE ISOTROPY

The microwave background is isotropic to better than one part in 10^4 on all angular scales, except for the dipole component discussed below, as is shown in Figure 10, and this isotropy supports the cosmological principle. It also helped to rule out alternative explanations of the radiation in terms of a superposition of emission due to different populations of radio sources or as starlight thermalized by dust. Indeed, it was realized a long time ago that it might be too isotropic as regions that had apparently never been in causal contact before showed the same intensity of the background radiation to such an astounding precision. The inflationary models of the very early Universe provide a possible explanation of this isotropy[34].

However, one does expect large scale anisotropies to occur at some level; although the most important such anisotropy is not intrinsic to the radiation. Soon after its discovery, Peebles pointed out that an observer in motion with respect to another one that sees the radiation isotropic, would see it "hotter" in the direction towards which he is moving and "colder" on the opposite side. This is the "aether drift" effect and it is straightforward to show[35] that an observer moving at speed v, relative to the reference frame in which the radiation appears to be isotropic, would see a blackbody temperature

$$T(\theta) \simeq T_{BG} (1 + \frac{v}{c} \cos \theta), \quad v/c \ll 1 \qquad (7)$$

from a direction at an angle θ from the direction of his (her) motion. The spectrum of the radiation is still that of a blackbody in all directions with a modulation of the temperature that corresponds to a dipole anisotropy.

One also expects an intrinsic large-scale anisotropy due to the large-scale distribution of mass in the universe. As explained by Peebles[36], this is best understood in terms of the Sachs-Wolfe effect[37] as due to large-scale fluctuations in the gravitational potential at the decoupling time. Photons are gravitationally redshifted as they climb out of whatever potential well they are emitted from. Roughly speaking, the mass inside a sphere of radius r is

$$M(r) \propto r^3 \qquad (8)$$

and fluctuations on that scale are of the order

$$\delta M(r) \propto r^{3/2} \tag{9}$$

which induces fluctuations in the gravitational potential and acceleration given by:

$$\delta\phi(r) \propto \frac{\delta M}{r} \propto r^{\frac{1}{2}} \tag{10a}$$

and

$$\delta g(r) \propto \frac{\delta\phi(r)}{r} \propto r^{-\frac{1}{2}} \tag{10b}$$

Notice that whereas δg vanishes at large distances, as it should, $\delta\phi(r)$ formally diverges. This induces large-scale fluctuations in the background radiation

$$\frac{\delta T}{T} \sim \delta\phi \propto \psi^{\frac{1}{2}} \tag{11}$$

that should roughly grow as the square-root of the angular separation between two lines of sight. Notice that this effect should contribute at some level to the dipole anisotropy discussed above. Silk has estimated this contribution as well as the contribution to a quadrupole component[30]. The estimated values of $\Delta T/T$ for an Einstein-de Sitter ($\Omega = 1$) universe are 1.5×10^{-4} (dipole) and $(0.7\text{--}2.1) \times 10^{-4}$ (quadrupole) for a baryon-dominated universe, $(1\text{--}5) \times 10^{-3}$ and 3×10^{-5} if the dominant mass is provided by neutrinos and $(1\text{--}2) \times 10^{-3}$ and 10^{-5} if the universe is dominated by cold, dark matter. The observed dipole is about 10^{-3} although most of this could be motional, whereas current limits on the quadrupole amplitude are below 8×10^{-5}.

Intermediate scale anisotropy could have been produced by the reionization mechanism discussed above, which could produce a signal on scales of 2°-10°. Here, limits from balloon-borne bolometers are $\Delta T/T < 5.5 \times 10^{-5}$ at the 95% confidence level at $\theta = 6°$[36].

Experiments to measure large-scale anisotropy have been done since the pioneering work of Partridge and Wilkinson[37]. Atmospheric inhomogeneities are a serious source of systematic errors and have forced observers to give up on ground-based measurements. The most

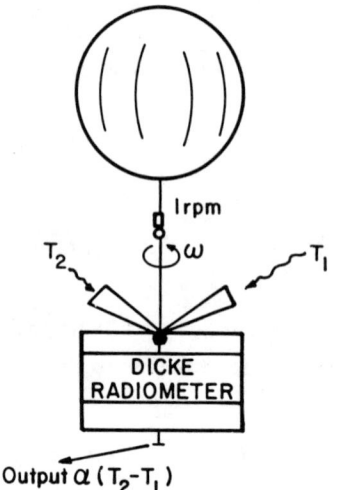
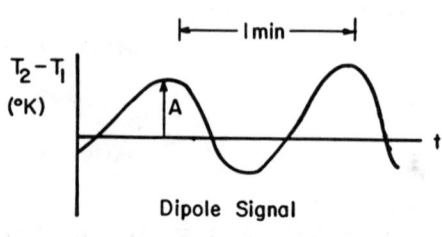
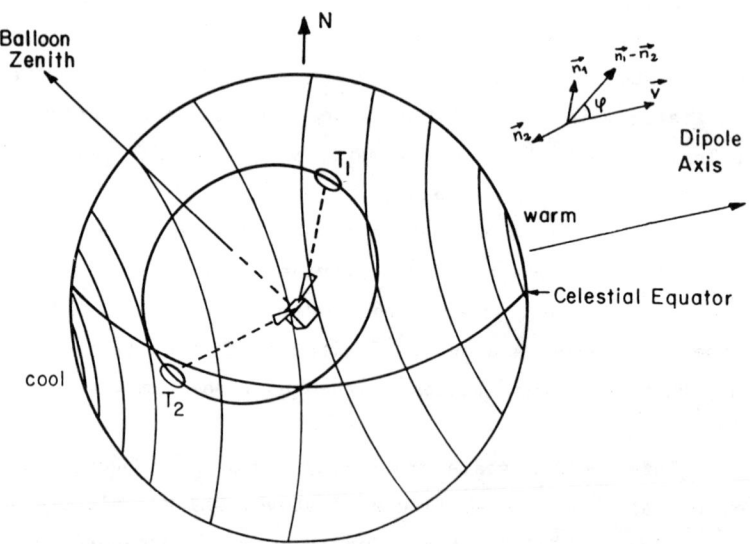

Figure 12. Large-scale structure has to be measured from balloon altitudes to avoid variable atmospheric emission. The package is set into rotation (typically at 1 rpm) and earth's rotation is used to scan the sky. The observed signal corresponds to a dipole anistropy.

successful measurements have been made with balloon-borne radiometers, an approach first used by Henry[38]. Several different experimental schemes are used; one of them, used by the Princeton group, features two horns separated 45° from the balloon zenith on opposite sides (Figure 12). The radiometer is Dicke-switched between both horns and compares the radiation coming from two directions n_1 and n_2, 90° apart. Furthermore, the package is made to revolve at one turn per minute around its symmetry axis and the earth's rotation is used to scan the sky. The response of such procedure to a dipole component is

$$\frac{T_1 - T_2}{T_{BG}} = (\vec{n}_1 - \vec{n}_2) \cdot \frac{\vec{v}}{c} = |\vec{n}_1 - \vec{n}_2| \frac{v}{c} \cos \psi \qquad (12)$$

where ψ is the angle between $(\vec{n}_1 - \vec{n}_2)$ and $\vec{v}$ (see Fig. 12). One does not expect a sinusoidal modulation as ψ is not a linear function of time because $\vec{n}_1$, $\vec{n}_2$ and $\vec{v}$ are generally not in the same plane. The amplitude of the modulated signal depends on the position of the balloon zenith. Current sensitivities are such that this signal is indeed seen in real time[39]. Typically, one such balloon flight scans about 35% of the sky. Figure 13 shows the coverage obtained from two of the Berkeley measurements.

The first convincing detection of the dipole anisotropy was published by Smoot, Gorenstein and Muller[40]; several groups have since detected this signal[41]. The most accurate measurements have been made by groups at Princeton[42] and Berkeley[39,43]. The Princeton radiometer has a system temperature of only 30K, at 24.5 GHz and has achieved the best statistical errors, whereas the Berkeley group, working at a higher frequency (90 GHz) is less subject to contamination due to galactic radiation. Each radiometer has been flown as well from the southern hemisphere, which is necessary in order to separate the polar quadrupole component from the third component of the dipole. After removing galactic contamination, the dipole components found by both groups[42,43] have amplitudes of (3.18±0.17)

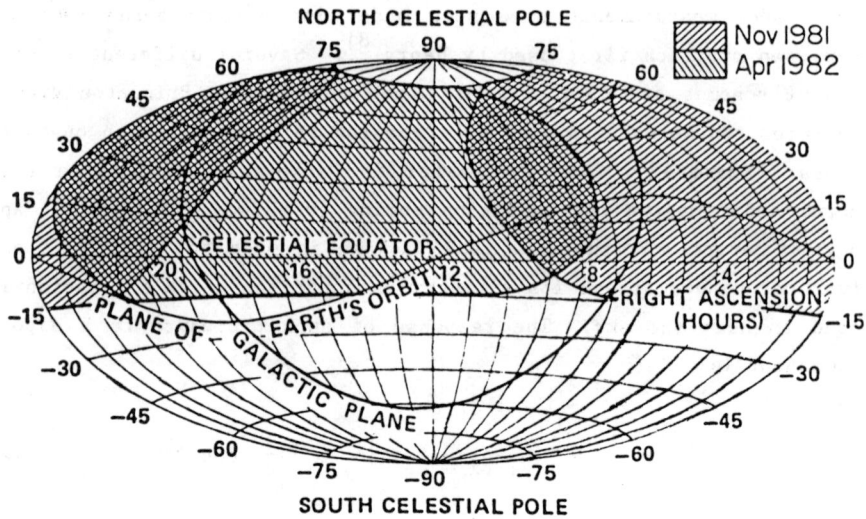

Figure 13. On a typical balloon flight about 35% of the sky is covered (unevenly). The coverage shown corresponds to two flights by the Berkeley group (ref. 39).

mK (Princeton) and (3.46±0.17) mK (Berkeley); with the maxima in the directions RA = (11.18±0.05) h, Dec = (-8.0±0.7)° (Princeton), and (11.3±0.1) h, (-6.0±1.4)° (Berkeley). The discrepancy in the amplitudes is small and most likely due to (systematic) calibration errors; the agreement of both experiments is (surprisingly) excellent. This is an important check as any intrinsic anisotropy as well as the motional effect should be frequency independent[41]. Serious discrepancies would indicate distortions in the blackbody spectrum or the contribution of a local effect. Nevertheless, systematic effects are important and have to be excluded before deriving cosmological consequences.

Assuming that all this dipole component is motional implies that the sun is moving towards RA = (11.2±0.1) h, Dec = (-7.0±1.0)° at a speed of (372±25) km/sec; which implies that the local group of galaxies that includes the Milky Way is moving at (610±50) km/sec towards RA = (10.5±0.4) h and Dec = (-26±5)°. This direction is 49°

away from the direction towards the Virgo cluster, but only 17° away from the direction of motion of the local group with respect to a distant shell of 78 Sbc galaxies (recession velocities between 1000 and 5500 km/sec)[44].

Two of the Princeton flights were made at times when the earth's orbital motion added to, and opposed, the solar motion. This alignment allowed them to detect the earth's motion at the 7-σ level[42].

The announcements of the detection of a quadrupole component have not been confirmed and it is assumed that (again) systematic errors, mainly due to galactic emission, did produce it. Current limits on the amplitude of quadrupole components are of the order $\Delta T/T < 8 \times 10^{-5}$ at the 95% confidence level. Quadrupole components measured are about ±0.1 mK with errors smaller than 0.07 mK.

Recently, a Soviet group has announced preliminary results from an experiment done at 37 GHz with a radiometer carried on board of a satellite in a highly eccentric orbit[45]. Their results correspond to six months of observations and show the dipole effect with a smaller amplitude and roughly the same direction but no errors are quoted. If the errors are accurately evaluated and if systematic errors do not turn out to be a limiting factor, this experiment would produce interesting results that could improve on present limits, as their tentative limits on a quadrupole component are already claimed to be as good as those from the balloon experiments.

5. THE SUNYAEV-ZEL'DOVICH EFFECT

The interaction of the background photons with ionized plasma is an important source of distortions of the radiation. Dense clusters of galaxies are strong emitters of X-rays, which are believed to be produced by thermal bremsstrahlung by intergalactic gas[46]. The X-ray spectra show emission lines due to Fe^{24+} and Fe^{25+} and yield gas temperatures as high as 10^8K. Several mechanisms have been proposed, like supersonic shock waves due to the passage of nearby galaxies. In any case, it is assumed that such heating would have occurred during an early phase of the cluster history. As the gas is fairly diffuse, it cannot cool easily and therefore remains hot. It came as a surprise that the gas constitutes a component as massive as the galaxies and that the abundance of metals is of the same order of magnitude as that of the sun. This shows that the gas is made of heavily processed material that was perhaps ejected from the galaxies through collisions with other galaxies, or through ram pressure due to the hot intragalactic gas.

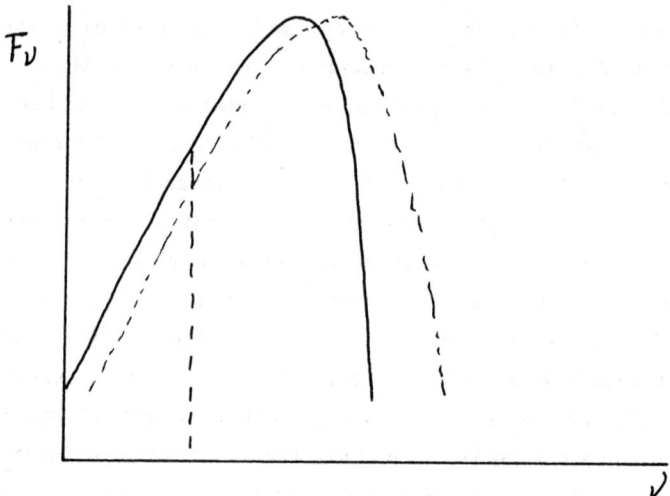

Figure 14. Some microwave photons are shifted to higher frequencies by inverse Compton scattering as they go through hot ionized gas in dense clusters of galaxies. Photon number is conserved but the blackbody spectrum is distorted. In the Rayleigh-Jeans region the spectral energy density of the radiation is decreased by this "Sunyaev-Zel'dovich" mechanism.

In 1972, Sunyaev and Zel'dovich[47] predicted that the background photons would interact with this gas through inverse Compton scattering, as the gas is much hotter than the radiation. This allows the gas to cool but, more important, distorts the spectrum of the radiation. The interaction conserves the number of photons but the scattered photons emerge with higher energy than the incoming ones.

The radiation becomes "cooler" on the Rayleigh-Jeans region and "warmer" on the Wien side, with no effect at a frequency slightly higher than the one at which the undistorted spectrum peaks (Figure 14). At any given frequency, the change in the spectral energy density, $F(\nu)$, is

$$\Delta F(\nu) = F_{after}(\nu) - F_{before}(\nu) = X\nu \frac{d}{d\nu} \{\nu^4 \frac{d}{d\nu} (\nu^{-3} F_{before}(\nu))\} \quad (13)$$

where $X = \frac{kT_e}{m_e c^2} \tau$, with T_e being the gas temperature and me, k and c the mass of the electron, Boltzmann's constant and the speed of light; τ is the line integral

$$\tau = \int N_e \sigma_T d\ell \quad (14)$$

across the cloud, with N_e being the electron density and the cross-section $\sigma_T = 6.65 \times 10^{-25} \text{ cm}^2$.

In the Rayleigh-Jeans region, $F(\nu) \propto \nu^2$ and

$$\Delta F_\nu = -2X F_{before}(\nu) \quad (15)$$

If one measures the background intensity at a single frequency, the Sunyaev-Zel'dovich effect appears as a small-scale anisotropy when one compares the intensity of the radiation in the direction of the cluster of galaxies with that from (nearby) "blank" sky. In the Rayleigh-Jeans region one expects

$$\frac{\Delta T}{T} = -2X \sim -1 \text{ mK} \quad (16)$$

for the densest and hottest clusters, based on the estimated gas densities and temperatures; one could hope to measure this with current sensitivities.

A convincing measurement of the Sunyaev-Zel'dovich effect would provide indirect evidence of the cosmological nature of the background radiation as well as confirm the "hot gas" explanation of the origin of the X-ray emission of clusters of galaxies. Besides, the X-ray emission, being due to bremsstrahlung, depends on cluster parameters through the combination $\exp\{-h\nu/kT_e\}\, T_e^{-\frac{1}{2}}\, N_e^2$ whereas the Sunyaev-Zel'dovich effect is linear in the product $N_e T_e$ in the Rayleigh-Jeans region. A measurement of both effects could allow an independent determination of both parameters and therefore allow the computation of the absolute X-ray luminosity of the cluster. A comparison with the measured X-ray luminosity would then yield a direct determination of the luminosity distance to the cluster, and from this a determination of Hubble's constant without the systematic problems due to the stepwise determination of the distance scale[48].

A number of observers have sought to detect this effect since 1972 but many claims have been disputed by other observers whose claims were negated as well[19,49,50]. The problem is once again that the small magnitude of the effect requires pushing the equipment to its limits and long accumulations of data are needed in order to reduce the statistical errors; the measurements are then normally dominated by systematic errors that are hard to identify and eliminate.

The choice of frequency is important. At low frequencies, confusion due to radio sources is a severe problem and one can also pick up a contribution due to bremsstrahlung from a colder gas component in the cluster itself[51,52]. At higher frequencies one runs into problems due to the atmosphere of the same kind as the ones discussed in section 3. This makes observations near the peak of the blackbody spectrum as well as on the Wien side extremely difficult. Ground radiation is an important problem at all wavelengths.

Dave Wilkinson and myself have worked on such a measurement during the last few years using the NRAO 140-foot telescope with the maser receiver. The setup and main systematic problems are as discussed in section 3. However, here one does not have the option of minimizing systematic problems by going to high declination; one has to observe the clusters wherever they happen to be! As a result, the ground effect is bigger and not linear with hour angle anymore; the diffraction pattern is not merely rotating with respect to the ground, which is the case if one looks somewhere close to the north pole, but moves appreciably with respect to the environment (Figure 15). Nevertheless, one again hopes that careful timing of the off-on

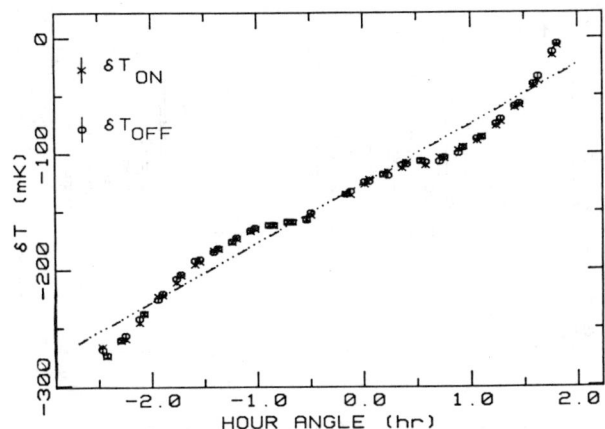

Figure 15. Drift of δT_{ON} and δT_{OFF} with hour angle for a five hour run. Each point is averaged over one minute. The data correspond to the cluster Abell 665 (dec = 66°). As the motion of the telescope with respect to the ground is more severe than for the anisotropy measurements (figure 7), the drift does not follow a linear model; the effect is repeatable for each declination but hard to model.

sequence will still produce a cancellation of systematic effects. Indeed, it seems that whatever the residual offset might be, at least it stays constant over about five hours in hour angle (Figure 16). This residual effect has to be measured, which one hopes to accomplish by observing "blank sky." Besides, the observations could be contaminated by faint radio sources contributing to any of the sky positions observed. Ultimately, one could hope to separate the contribution of radio sources through multifrequency observations, but this requires achieving comparable sensitivities at several somewhat spaced frequencies which is hard and out of the question on a first approach; one has to work hard enough at one frequency to try to measure something and thinks about other frequencies as something to be done later on.

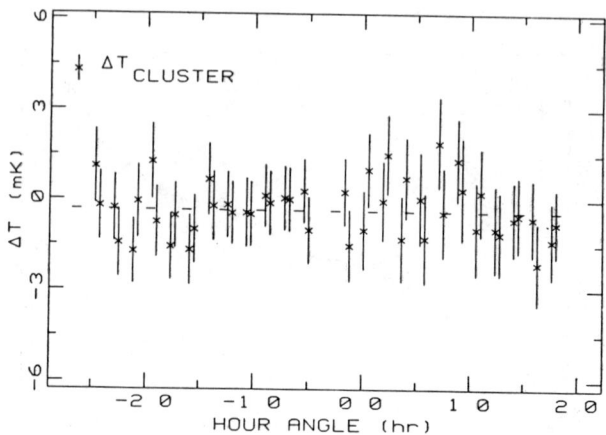

Figure 16. Accurate timing of the scans allows the drift shown in figure 15 to cancel, so that ΔT shows no significant drift with hour angle.

We decided on a complex observing procedure in order to tackle both problems (Figure 17). We chose a beam-throw of 8 minutes of arc and selected two blank fields at the same declination as each cluster, together with their pair of reference fields. Therefore we observe a total of nine positions on the sky. The separations between the blank fields and the cluster are chosen such that we make the six observations in a set with the telescope moving through the same range of positions with respect to the ground. Therefore we can combine the observations in several different ways and hope that the ground effect is subtracted out from each combination (Figure 17). The OFF-ON procedure discussed in section 3 can be applied to each "central" position and associated reference positions, giving:

$$\Delta T_{CLUST} = T_{CLUSTER} - \frac{T_{REF1} + T_{REF2}}{2} \qquad (17a)$$

$$\Delta T_{BF1} = T_{BF1} - \frac{T_{REF3} + T_{REF4}}{2} \qquad (17b)$$

and

$$\Delta T_{BF2} = T_{BF2} - \frac{T_{REF5} + T_{REF6}}{2} \qquad (17c)$$

which gives a measurement of the residual offset, assuming that the blank fields contribute no signal. One can then adopt as the best estimate of the signal produced by the cluster:

$$\Delta T_{CLUSTER} = \Delta T_{CLUST} \frac{\Delta T_{BF1} + \Delta T_{BF2}}{2} \qquad (18)$$

and derive $\sigma_{\Delta T_{CLUSTER}}$ from the measurement statistics.

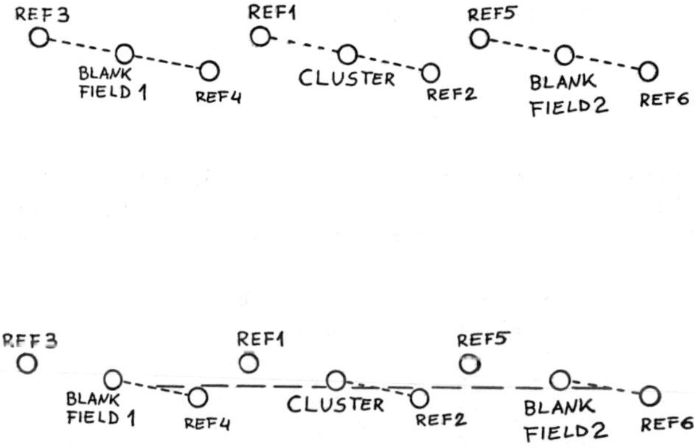

Figure 17. The observing scheme is designed to allow a cancelation of the ground effect and a calibration of the residual offset through observations of blank sky (upper sketch). Combination of the ON scans (or the OFF scans) allows a test for radio sources in the reference beams (lower sketch, see text).

If the spacing of the blank fields and the integration time are properly chosen, one can combine the six measurements in different ways. For example, combining the ON and OFF measurements separately as

$$\Delta T_{ON} = \delta T_{ON,CLUSTER} - \frac{\delta T_{ON,BF1} + \delta T_{ON,BF2}}{2} \qquad (19a)$$

and

$$\Delta T_{OFF} = \delta T_{OFF,CLUSTER} - \frac{\delta T_{OFF,BF1} + \delta T_{OFF,BF2}}{2} \qquad (19b)$$

can be used to check that the signal in $\Delta T_{CLUSTER}$ is contributed equally by the ON and OFF differences (i.e., one expects $\Delta T_{ON} \sim -\Delta T_{OFF}$). This could uncover a radio source in either of the reference fields although it would not be sensitive to contamination in the cluster or "blank field" positions.

The ON-OFF procedure is also used on "blank sky" positions at the same declination as the clusters but further away as the clusters could be part of superclusters and draw an overabundance of radio sources in their neighborhoods.

TABLE 1

Recent measurements of the Sunyaev-Zel'dovich effect (mK) in the clusters of galaxies observed by Uson and Wilkinson.

Cluster	Birkinshaw, Gull et al.	Schallwich et al.	Lasenby and Davies	Uson and Wilkinson
0016+16	-1.58±0.26 -0.37±0.16 -0.72±0.18	-0.9±0.9	--	-0.74±0.22
Abell 401	--	--	--	-0.88±0.41
Abell 545	--	--	--	+0.51±0.43
Abell 665	-0.50±0.14 -0.55±0.13 +0.03±0.25	--	--	-0.50±0.33
Abell 1763	--	--	--	-0.36±0.25
Abell 2218	-0.31±0.13 -0.38±0.19 -0.51±0.14	-1.84±0.33*	0.23±0.77	-0.29±0.24

* This result has apparently been withdrawn.

We observed a number of clusters from 1984, January 3 to 13; and February 25 to March 5. Only data obtained with clear, cold sky have been used, which produced good data for only six clusters. The results[53] are shown in Table 1, which also shows other recent measurements of the effect[19,54,55,56]. Notice that our numbers are in reasonable agreement with those of Birkinshaw, Gull and collaborators. There is more of a problem within their own measurements as they disagree with themselves by up to 5 standard deviations, using the same telescope and setup (?). I suspect that the problem is that they underestimate their errors, which are statistical only; besides, all data were used, no matter how bad the weather was, and this might have introduced some systematic contamination. Scaling their errors by about 2.5 would allow internal agreement of their results while preserving the overall agreement with our measurements, but this is only an indication and not the correct approach. A more careful analysis of their data might show the problems and allow them to be eliminated. Nevertheless, the overall agreement is encouraging.

As indicated above, one could try to combine these measurements with X-ray observations to try to derive a value for Hubble's constant. A quick-and-dirty estimate for the cluster 0016+16 yields 30 km sec^{-1} Mpc^{-1} $\lesssim H_o \lesssim$ 180 km sec^{-1} Mpc^{-1} which is encouraging as it comes from such a different way of estimating it. The main problems are that the clumpiness of the gas in the clusters is unknown but important (as the X-ray emission is proportional to N_e^2 and the Sunyaev-Zel'dovich effect is linear in N_e). Besides, in general the X-ray data only provide a lower limit to T_e and the radio data have poor resolution, so one has to assume that the cluster is smooth.

Among the solutions, determining the cluster dynamics by measuring a fair amount of redshifts for the galaxies will certainly help: one can get an alternative estimate of T with the usual isothermal model for the clusters and detailed dynamical studies could constrain the form of the gravitational potential and provide reasonable limits of the clumpiness of the gas. Besides, current sensitivities allow "mapping" the Sunyaev-Zel'dovich effect, i.e. measuring a few positions within a cluster. In the future, improved X-ray data will give T_e through good spectra, and provide slightly better resolution. This should allow determining H_o to within a factor of 2, but through a totally independent way than the usual one, which will allow an interesting consistency check. A determination of H_o to ±10 km sec^{-1} Mpc^{-1} would seem to require plenty of observing time, hard work and luck.

6. FUTURE LIKELY IMPROVEMENTS

The next few years should see definite improvements in almost all the measurements that I have discussed here. Measurements of the absolute blackbody temperature might reach errors of only ±0.05K in the Rayleigh-Jeans region and possibly of 0.1K near the peak and in the Wien region (at any measured frequency). Small-scale anisotropy searches might reach errors in $\Delta T/T$ of $\sigma \sim 4 \times 10^{-6}$, which might yield the long sought-for detections or at least upper limits below 10^{-5} on scales up to 10 arc minutes and, more important, will have better sky coverage. These sensitivities might also be reached on scales of 30 arc minutes to 1 degree. The VLA should soon produce limits on very-small-scale anisotropy of about 10^{-4}. Large-scale measurements should improve the sensitivity to a quadrupole component by about a factor of 2, galaxy contamination should be reduced and calibrations should improve; this could provide good constraints on spectral distortions. Measurements of the Sunyaev-Zel'dovich effect should reach sensitivities of $\sigma_{\Delta T} \sim 0.1$ mK in a few days which will allow "mapping" a few positions per cluster and show how the effect vanishes when moving away from the cluster center. Measurements at higher frequency are going to help as well. Overall, as more observers refine their procedures, an interesting consistency check will be available.

I am indebted to Dave Wilkinson whose contribution to any of the measurements that I have described above is enormous. We have been fortunate to work with Chuck Brockway and Rick Fisher on the Green

Bank measurements, which would not have reached the results described above without them. I thank Dave and Phil Lubin for supplying many of the figures and for very interesting discussions.

REFERENCES

1. Milne, E.A., "Relativity, Gravitation and World Structure." Oxford: Clarendon Press (1935).
2. Hubble, E., Proc. Nat. Acad. Sci. 15, 168 (1929).
3. Bondi, H. and Gold, T., M.N.R.A.S. 108, 252 (1948).
4. Gamow, G., Phys. Rev. 74, 505 (1948).
5. Alpher, R.A. and Herman, R.C., Nature 162, 774 (1948).
6. Dicke, R.H., Berynger, R., Kyhl, R.L. and Vane, H.V., Phys.Rev. 70, 340 (1946).
7. Dicke, R.H., Peebles, P.J.E., Roll, P.G. and Wilkinson, D.T., Ap.J. 142, 414 (1965). Penzias, A.A. and Wilson, R.W., Ap.J. 142, 419 (1965). See also Wilson, R.W., Rev. Mod. Phys. 51, 433 (1979).
8. Dicke, R.H., Rev. Sci. Instr. 17, 268 (1946).
9. Wilkinson, D.T., Phys. Scripta 21, 606 (1980).
10. Weiss, R., Ann. Rev. Astr. Ap. 18, 489 (1980).
11. Woody, D.P. and Richards, P.L., Phys. Rev. Lett. 42, 925 (1979).
12. Jones, B.J.T., Phys. Scripta 21, 732 (1980).
13. Peterson, J.B., Richards, P.L. and Timusk, T., submitted to Phys. Rev. Lett.
14. Meyer, D.M. and Jura, M., Ap.J.Lett. 276, 1 (1984).
15. Smoot, G.F., De Amici, G., Friedmann, S.D., Witebsky, C., Mandolesi, N., Partridge, R.B., Sironi, G., Danese, L., and De Zotti, G., Phys. Rev. Lett. 51, 1099 (1983).
16. Tolman, R.C. "Relativity, Thermodynamics and Cosmology." Oxford Univ. Press 1934 (§171).
17. Peebles, P.J.E., Lectures in Applied Mathematics 8, 274. Ed. J. Ehlers (Providence: American Mathematical Society, 1967).
18. Dautcourt, G., M.N.R.A.S. 144, 255 (1969).
19. Lasenby, A.N. and Davies, R.D., M.N.R.A.S. 203, 1137 (1983).

20. Partridge, R.B., in "The origin and evolution of galaxies". VIIth course of the International School of Cosmology and Gravitation. Eds. B.J.T. Jones and J.E. Jones (Reidel: 1983), p. 121. See also: Fomalont, E.B., Kellermann, K.I. and Wall, J.V., Ap.J.Lett. 277, 23 (1984) and Knoke, J.E., Partridge, R.B., Ratner, M.I. and Shapiro, I.I., Ap.J.Lett. 284, 479 (1984).
21. Peebles, P.J.E. and Yu, J.T., Ap.J. 162, 815 (1970).
22. Sunyaev, R.A. and Zel'dovich, Ya.B., Ap. Space Sci. 6, 358 (1970).
23. Davis, M. and Boynton, P.E., Ap.J. 237, 365 (1980).
24. Wilson, M.L. and Silk, J., Ap.J. 243, 14 (1981). See also Wilson, M.L., Ap.J.
25. Vittorio, N. and Silk, J., Ap.J.Lett. 285, 39 (1984).
26. Bond, J.R. and Efstathiou, G., Ap.J.Lett. 285, 45 (1984).
27. Rees, M.J. "Growth and fate of inhomogeneities in a Big Bang cosmology" in "Observational Cosmology". Eds. A. Maeder, L. Martinet and G. Tammann. (Geneva Observatory, SAAS-FEE: 1978), p. 259.
28. Silk, J. I., Ap.J. 151, 459 (1968).
29. Hogan, C.J., Kaiser, N. and Rees, M.J., Phil. Trans. R. Soc. London A307, 97 (1982).
30. Silk, J., Proceedings of the "Inner Space/Outer Space" Workshop. Univ. of Chicago Press (1985).
31. Uson, J.M. and Wilkinson, D.T., Ap.J. 283, 471 (1984).
32. Boynton, P.E. and Partridge, R.B., Ap.J. 181, 243 (1973).
33. Uson, J.M. and Wilkinson, D.T., Nature 312, 427 (1984), and references therein.
34. See, for example, the Proceedings of the "Inner Space/Outer Space" Workshop. Univ. of Chicago Press (1985).
35. Peebles, P.J.E. and Wilkinson, D.T., Phys. Rev. 174, 2168 (1968).

36. Melchiorri, F., Melchiorri, B.O., Ceccarelli, C. and Pietraneva, L., Ap.J.Lett. 250, 1 (1981).
37. Partridge, R.B. and Wilkinson, D.T., Phys. Rev. Lett. 18, 557 (1967).
38. Henry, P.S., Nature 231, 516 (1971).
39. Lubin, P.M., Epstein, G.L. and Smoot, G.F., Phys. Rev. Lett. 50, 616 (1983).
40. Smoot, G.F., Gorenstein, M.V., and Muller, R., Phys.. Rev. Lett. 39, 898 (1977).
41. Wilkinson, D.T. Proceedings of the "Inner Space/Outer Space" Workshop. Univ. of Chicago Press (1985).
42. Fixsen, D.J., Cheng, E.S. and Wilkinson, D.T., Phys. Rev. Lett. 50, 620 (1983).
43. Lubin, P.M. and Neto, T.V. (1984, preprint).
44. Hart, L. and Davies, R.D., Nature 297, 191 (1982).
45. Strukov, I.A., Sagdeev, R., Kardashev, N., Skulachev, D., and Eysmont, N. (1984, preprint).
46. Forman, W. and Jones, C., Ann. Rev. Astr. Ap. 20, 547 (1982). See also Sarazin, C.L. "X-ray emission from clusters of galaxies" (1984: preprint).
47. Sunyaev, R.A., and Zel'dovich, Ya.B., Comm. Astrophys. Sp. Phys. 4, 173 (1972).
48. Gunn, J.E., "The Friedmann models and optical observations in Cosmology" in "Observational Cosmology", eds.: A. Maeder, L. Martinet and G. Tamman (Geneva Observatory, SAAS-FEE: 1978), p. 1.
49. Lake, G., "Scattering of the microwave background in clusters of galaxies" in "Objects of High Redshift." Proceedings of I.A.U. Symposium #92. Eds. G.O. Abell and P.J.E. Peebles. (Reidel: 1980), p. 305.
50. Birkinshaw, M., "Measurements of the scattering of the microwave background radiation in clusters of galaxies" in "Objects of

High Redshift". Proceedings of I.A.U. Symposium #92. Eds. G. O. Abell and P.J.E. Peebles (Reidel: 1980), p. 313.
51. Rudnick, L., Ap.J. 223, 37 (1978).
52. Tarter, J.C., Ap.J. 220, 749 (1978).
53. Uson, J.M. and Wilkinson, D.T., submitted to Phys. Rev. Lett. (1985).
54. Shallwich, D., Ph.D. Thesis (Univ. of Bochum, Fed. Rep. Germany: 1982).
55. Andernach, H., Schallwich, D., Sholomitski, G.B. and Wielebinski, R., Astron. Astrophys. 124, 326 (1983).
56. Birkinshaw, M., Gull, S.F., M.N.R.A.S. 206, 359 (1984). See also Birkinshaw, M., Gull, S.F., and Hardebeck, H., Nature 309, 34 (1984) and Birkinshaw, M., Gull, S.F. and Moffet, A.T. Ap.J.Lett. 251, 69 (1981).

SOLITONS AND THE GENERATION OF NEW COSMOLOGICAL SOLUTIONS

Enric Verdaguer

Departament de Física Teòrica
Universitat Autònoma de Barcelona
Bellaterra (Barcelona, Spain)

SUMMARY.- A review is made of the soliton solutions to Einstein's equations with some cosmological interest. All these solutions admit an Abelian two parameter group of isometries. Most of them can be considered as vacuum inhomogeneous cosmological models. Special emphasis is made of the cosmological solutions containing solitons evolving towards gravitational waves, since they give a possible mechanism for the production of a gravitational wave background. Soliton solutions with matter and with an electromagnetic field are also considered.

CONTENTS

I. Introduction
II. Inhomogeneous cosmologies
III. The soliton technique
IV. Soliton solutions on Bianchi type - I
V. Real-pole trajectories
    Va. Diagonal metrics
        A. From Kasner to other Bianchi's
        B. Pulse waves on Kasner backgrounds
    Vb. Nondiagonal metrics
VI. Complex-pole trajectories
    VIa. Diagonal metrics
        A. n-solitons with $\sigma^{(-)}$
            Ai. The real-pole limit
            Aii. Degenerate complex poles
        B. Cosmologies with gravisolitons
    VIb. Nondiagonal metrics
        A. Two-soliton solutions
        B. n-soliton solutions
VII. Soliton solutions on Bianchi type - II
VIII. Soliton solutions on Bianchi type - V
IX. Soliton solutions with matter
X. Soliton solutions in Kaluza-Klein

I. INTRODUCTION

The aim of these lectures is to review and summarize the soliton solutions of the Einstein's equations which have some cosmological interest. When the spacetime admits an Abelian two-parameter group of isometries, new vacuum solutions of the Einstein equations can be generated from particular "seed" solutions by means of the soliton technique . Since, in particular, the anisotropic and homogeneous Bianchi models from type-I to type-VII admit one such a subgroup of isometries, the soliton technique provides a method for generalizing these models. Taking one of these homogeneous models as a seed, the new solutions break the homogeneity in one direction (the propagation direction). Not all the "soliton solutions" generated that way have a cosmological interpretation. The type of solution one gets depends on the seed used but also on the particular election of parameters allowed by the soliton technique.

Among the solutions with cosmological interest some can be interpreted as inhomogeneous generalizations of Bianchi models, however, others are just homogeneous Bianchi models but of different type than the seed.

Of special interest are the inhomogeneous solutions that can be interpreted as soliton perturbations propagating on a cosmological homogeneous background. These solutions have generally the property that the solitons evolve towards gravitational waves. They are an example of production of gravitational waves of cosmological origin and they may be relevant to the present, hypothetical, cosmological gravitational wave background.

The review begins, in section II, presenting some arguments for the study of inhomogeneous cosmologies. We emphasize the relevance that the detection of a gravitational wave background can have and comment on some of the efforts made in this direction.

In section III a summary is made of the soliton technique of Belinskii and Zakharov (1979) with the practical purpose of making these lectures reasonably selfcontained.

Sections IV, V and VI all deal with the Kasner cosmological metric as the seed metric. Section V is devoted to the case in which only real parameters are allowed. The classification of the solutions is based in that presented in Carr and Verdaguer (1983). If one wishes to interpret these solutions in terms of the Kasner background they are, in general, metrics having discontinuities in their first derivatives. When the solutions are diagonal metrics, one finds that some can be related to other solutions obtained by different methods. For example, some solutions are the family of Ellis and MacCallum (1969) of Bianchi Type III, V and VI, and the inhomogeneous generalization of Wainwright et al. (1979). Others are the pulse wave solutions of Carmeli and Charach (1980).

In section VI complex parameters are allowed and all the possible soliton solutions are classified and analysed. When the metric is diagonal special simplifications occur and the analysis of all the solutions can be carried out explicitly. Some of these solutions have singularities in the direction of inhomogeneity, so they have no clear cosmological interpretation, others become flat in that direction. All of them have the cosmological singularity. The most interesting solutions are those which can be interpreted as gravitational solitons (gravisolitons) propagating on the Kasner background. The gravisolitons show features similar to those of classical solitons (such as in hydrodynamics). The analysis of these metrics is based on the properties of the Riemann tensor and its curvature scalar invariants. When the metrics are not diagonal their properties can be inferred from the properties of the diagonal metric and from the simplest of the nondiagonal solutions.

In section VII, a soliton solution obtained with a Bianchi type -II seed (Belisnkii and Francaviglia, 1982) is presented. It generalizes one of the soliton solutions with a Kasner seed.

Section VIII is devoted to soliton solutions that can be interpreted as gravisolitons propagating on a Milne Universe (Ibañez and Verdaguer, 1984b). The Milne Universe approaches, at late times, an open Friedman-Robertson-Walker (FRW) Universe and the soliton solutions are relevant for a model of FRW with gravitational waves.

A review of soliton solutions in the presence of matter is given in section IX. Such solutions can be obtained only when the matter is a perfect fluid such that the speed of sound equals the speed of light (stiff matter) (Belinskii, 1979a) or when it is an anisotropic fluid for which this speed equality is true in the direction of propagation of the solitons (Letelier, 1982).

Finally, the soliton technique is extended to Einstein vacuum equations in an arbitrary number of dimensions, providing that the metric depends on two coordinates only. When this is done in five dimensions it can be applied to find soliton solutions of a four-dimensional metric coupled to an electromagnetic field or, also, coupled to an electromagnetic and a scalar fields.

II. INHOMOGENEOUS COSMOLOGIES

The present Universe in its large structure seems to be isotropic and spatially homogeneous. Physical cosmology is based on the relativistic Friedman -Robertson-Walker (FRW) models which describe the Universe as completely homogeneous and isotropic in all its evolution (Peebles (1971), Weinberg (1972), Zeldovich and Novikov (1983)).

Cosmological models which are not FRW have also been studied in the past and, particularly, in recent years. Partially, this study was motivated by the need to understand better Einstein's theory of gravity (the theory in which all cosmological models are described) and partially because the FRW cannot explain several features and enigmas of the present Universe. One of them, for instance, is to explain why the Universe is FRW today which motivated the "chaotic cosmology" program (Misner, 1969). But also to explain the fact that if one perturbes the FRW models encounters decaying modes which have been much more important in the past suggesting a finite deviation from FRW at early epochs, or that statistical fluctuations in FRW models cannot collapse fast enough to form the observed galaxies (see MacCallum (1979) for a review of these and other reasons).

The study of non-FRW models started, basically, with the anisotropic and spatially-homogeneous models in which a three dimensional isometry group G_3 acts simply transitively on the hypersurfaces of homogeneity (Ellis and MacCallum 1969, 1970). These models are included in the Bianchi classification and some of them have the interesting property that they evolve towards isotropy (see MacCallum (1973) and Ryan and Shepley (1975) for reviews).

Spatially inhomogeneous cosmologies have been limited mainly to the case of spacetimes admitting an Abelian two-parameter group, G_2, of isometries. They were initiated with the closed models of Gowdy (1974) and, recently, there has been an increased interest in such cosmologies (Wainwright (1979a-b, 1981),Wainwright and Marshman (1979), Carmeli and Charach (1980), Adams et al. (1982)). See Carmeli et al. (1981) and MacCallum (1984) for a review of the main work.

Part of this interest has been motivated by the possibility that there could exist a background of gravitational waves (Rosi and Zimmerman (1976), Carr (1980)), present cosmological constraints to such background are very small (Zimmerman and Hellings, 1980), and a lot of effort has been devoted to find means for detecting it. If these waves have a cosmological origin they would generally have a longer period than the waves generated at the present epoch and could be observable by new detection techniques for example, by the Doppler tracking of interplanetary spacecraft (Hellings (1979), Bertotti and Carr (1980), Mashhoon and Grishchuk (1980)) by scrutinizing the timing noise in pulsars (Mashhoon (1982), Bertotti et al. (1983), Hellings and Downs (1983), Romani and Taylor (1983)) or by monitoring perturbations to planetary orbits (Bertotti (1973), Turner (1979), Mashhoon et al. (1981)).

These gravitational waves may reflect an irregularity in the initial structure of the Universe. Finding exact cosmological solutions which evolve towards homogeneous cosmological models with a background of cosmological gravitational waves, is of interest as a classical mechanism for the creation of such a background.

Inhomogeneous cosmological solutions admitting a G_2 provide exemples of such a mechanism although, of course, they produce a very correlated gravitational wave background (all waves travelling in the same direction) rather than a stochastic background like the one probably present.

Some of such solutions can be considered as generalizations of homogeneous Bianchi models in which the homogeneity is broken in one direction. In fact, all Bianchi models from type I to type VII admit an Abelian subgroup G_2 of isometries and can be written (Ryan and Shepley 1975) as

$$ds^2 = f(t)(dz^2 - dt^2) + \eta_{cd}(t) e_a^c(z) e_b^d(z) dx^a dx^b \qquad (2.1)$$

$$(a,b,c,d = 1,2)$$

where the functions $e_a^c(z)$ depend only on the particular Bianchi type. One may break the homogeneity in the z-direction by assuming that f and η_{cd} are also functions of z: $f(t,z)$ and $\eta_{cd}(t,z)$.

Adams et al. (1982) have studied solutions which describe gravitational waves in Bianchi backgrounds, particularly of type I, by solving the Einstein equations when the above assumption is made.

The soliton technique of Belinskii and Zakharov (1979) gives a systematic way for solving vacuum Einstein equations when the metrics admit an Abelian G_2 group of isometries and can be written (Gowdy 1974) as

$$ds^2 = f(t,z)(dz^2-dt^2) + g_{ab}(t,z)dx^a dx^b ,$$

where we have assumed two spacelike commuting Killing vectors: ∂_{x^1} and ∂_{x^2}. They include (2.1) and their inhomogeneous generalizations. The soliton technique allows to generate new solutions once a particular "seed" solution is given. In the search for solutions with cosmological interes it is only natural to start with the Bianchi metrics (2.1) as seed metrics. Most of the soliton solutions that we shall discuss have been generated using this guide. The persistance of the solitons that are found, suggests that these solutions may be singled out as modeling the most likely form of irregularity in the early Universe and we would expect many of their features to be valid in the more general case.

III. THE SOLITON TECHNIQUE

The so-called "solitonic methods" were developed in the sixties to solve certain kind of non-linear differential equations in fluid dynamics and quantum mechanics. A method known as "inverse scattering technique" was applied to solve in a systematic way the Korteweg de Vries (KdV) equation for shallow water waves (Gardner et al. 1967), the nonlinear Schrödinger equation and the sine-Gordon equation (see Scott et al. 1973 for a review). A peculiarity of these solutions is that generally they present "solitons". These are solutions of the field equations which share a number of common properties with classical particles. The solitons are the result of a balance between non-linear effects and wave dispersion and therefore they can appear only in a very special class of non-linear equations.

Belinskii and Zakharov (1979) were able to generalize and adapt the inverse scattering technique to solve certain subclass if vacuum Einstein's equations, namely when the spacetime admits an orthogonally transitive two-parameter group of isometries.

In this section we shall summarize the soliton technique of Belinskii and Zakharov with a view towards its cosmological applications. The details of the technique can be found in the well written original papers of Belinskii and Zakharov (1979, 1980) and short summaries for application purposes can be found in Belinskii and Francaviglia (1981) or Carr and Verdaguer (1983).

Another soliton technique based in the theory of Bäcklund transformations (Harrison (1978), Neugebauer (1979,1980)) has been proved by Cosgrove (1980) to be equivalent, in some sense, to the presented here.

We shall write the metric, which admits two commuting Killing vectors, in adapted coordinates such that the two ignorable coordinates do not appear,

$$ds^2 = f(dz^2-dt^2) + g_{ab}dx^a dx^b \qquad (a,b=1,2) , \qquad (3.1)$$

where the metric coefficient f and the two-dimensional matrix g are functions of t and z alone. As remarked in the last section, this metric includes the homogeneous Bianchi models of type I to VII (see (2.1)) which have two commuting Killing vectors, as well as their generalizations to models with the homogeneity broken in the z direction.

We shall impose to metric (3.1) that

$$\det g = t^2. \tag{3.2}$$

This does not involve a loss of generality because we can take any other value for det g, provided its square root is a solution of the wave equation in t and z (as required by the Einstein equations), by performing a coordinate transformation

$$\begin{aligned} z' &= a(z+t)+b(z-t) \\ t' &= a(z+t)-b(z-t) \end{aligned} \tag{3.3}$$

with arbitrary functions a and b. This leaves the metric (2.1) invariant while changing the coefficient f according to

$$f(dz^2-dt^2) = \frac{f(dz'^2-dt'^2)}{4(a,_z+a,_t)(b,_z-b,_t)}$$

Thus, for a given det g, different from (3.2), all the results that follow can be generalized by performing a transformation of the form (3.3).

The Einstein equations in vacuum $R_{\mu\nu}=0$ split into the following two groups:

$$U,_t - V,_z = 0 \tag{3.4}$$

$$\begin{cases} (\ln f),_t = -\frac{1}{t} + \frac{1}{4t} \operatorname{Tr}(U^2+V^2) \\ (\ln f),_z = \frac{1}{2t} \operatorname{Tr}(U.V) \end{cases} \tag{3.5}$$

where

$$U \equiv tg,_t g^{-1} \qquad V \equiv tg,_z g^{-1} \tag{3.6}$$

Equations (3.4) are just $R_a^b = 0$ and form a non-linear system of equations in the three unknowns g_{ab}. The Belinskii-Zakharov soliton technique is based on the inverse scattering transform and provides a procedure for generating soliton solutions to the non-linear system (3.4) when a particular "seed" solution g_0 is known. Once a solution g is found f can be calculated by solving equations (3.5) by quadratures.

The main idea of the method consists of the following. One looks for a linear eigenvalue problem having the non-linear equations (3.4) as integrability conditions. Solving the linear problem will produce, by an appropriate procedure, solutions to the original non-linear equations.

The linear eigenvalue problem that one associates to (3.4) is:

$$\begin{aligned} (\partial_z - \frac{2\lambda^2}{\lambda^2-t^2} \partial_\lambda)\psi &= -\frac{(tV+\lambda U)}{\lambda^2-t^2} \psi \\ (\partial_t - \frac{2\lambda t}{\lambda^2-t^2} \partial_\lambda)\psi &= -\frac{(tU+\lambda V)}{\lambda^2-t^2} \psi \end{aligned} \tag{3.7}$$

where λ is a complex "spectral" parameter and $\psi(\lambda,t,z)$ is a two-dimensional matrix that satisfies

$$g(t,z) = \psi(0,t,z) \tag{3.8}$$

Taking into account (3.6) and (3.8) one can re-interpret equations (3.7) as a genuine differential system for the unknown ψ. The equations (3.4) are integrability conditions for these. Once ψ has been calculated by solving equations (3.7), g is given by (3.8).

Let us suppose that a particular solution $g_0(t,z)$ of the system (3.4) is given. Equations (3.7) can be integrated to find the corresponding solution $\psi_0(\lambda,t,z)$. This integration can be done easily for diagonal metrics, as shown by Jantzen (1980), and has also been done for some non-diagonal metrics (Belinskii and Francaviglia, 1982). In some cases this may not be a trivial step.

Once ψ_0 has been found, a new solution ψ can be generated by purely algebraic operations if one assumes that ψ is the product of a two-dimensional matrix, with n simple (nondegenerate) poles in the complex λ plane, and ψ_0. Equation (3.8) shows that an n-soliton solution for the two-dimensional matrix g(t,z) can then be found.

The explicit procedure is as follows.

1.- One starts by choosing the number n and by specifying whether the "pole trajectories", defined by

$$\mu_k = u_k - z \pm [(u_k-z)^2 - t^2]^{1/2} \qquad (k=1,\ldots,n) \tag{3.9a}$$

are real or complex. Here the μ_k are solutions of the equations

$$\mu_{k,z} = \frac{2\mu_k^2}{t^2-\mu_k^2} \qquad\qquad \mu_{k,t} = \frac{2t\mu_k}{t^2-\mu_k^2} \tag{3.9b}$$

and the u_k are arbitrary (real or complex) constants. From the ψ_0 matrix associated with a given seed metric g_0, one then constructs the vectors

$$m_a^{(k)} = (m_0)_c^{(k)} [\psi_0^{-1}(\mu_k,t,z)]_{ca} , \tag{3.10a}$$

where $(m_0)_c^{(k)}$ are arbitrary real or complex parameters. If one starts with real-pole trajectories, then the parameters $(m)_c^{(k)}$ also have to be real. If one starts with a complex trajectory μ_k, its complex conjugate is also a trajectory. Thus, complex trajectories always go in pairs, and we can put $\mu_{k+n/2} = \bar{\mu}_k$. The complex parameters $(m_0)_c^{(k)}$ will then satisfy

$$(m_0)_c^{(k+n/2)} = (\bar{m}_0)_c^{(k)} .$$

2.- The next step is the construction of the n×n matrix

$$\Gamma_{kl} = \frac{m_c^{(k)}(g_0)_{cb}m_b^{(l)}}{\mu_k\mu_l - t^2} \tag{3.10b}$$

and its inverse, $D_{kl} = (\Gamma_{kl})^{-1}$, from which one can derive a matrix g':

$$g'_{ab} = (g_0)_{ab} - \sum_{k,l}^{n} \frac{D_{kl} m_c^{(k)}(g_0)_{ca} m_d^{(l)}(g_0)_{db}}{\mu_k \mu_l} \qquad (3.11)$$

This is a solution of equation (3.4) and, while it does not satisfy condition (3.2), the matrix

$$g = t^{-n} \left(\prod_{k=1}^{n} \mu_k \right) g' \qquad (3.12a)$$

does and so is the required n-soliton solution of the Einstein equations. We see now that for n large one may not be able to find explicit expressions for g since matrix Γ_{kl} becomes impractical to solve. This will be the case for nondiagonal metrics, fortunatelly the study of one and two-soliton solutions will be enough the obtain a qualitative picture of the general case. If we restrict ourselves to find diagonal metrics, explicit expression can be given for the matrix elements as we shall see later.

3.- After the solution for g has been found, we can calculate the metric coefficient f, defined by equations (3.5). It is a remarkable fact that these solutions can be integrated explicitly for the n-soliton solution (3.12a). This has been proved by Belinskii and Zakharov (1980) for axisymmetric metrics, but it is also true in the cosmological context. The proof is inductive and uses the fact that the n-soliton solution can be obtained step by step, starting with the one-soliton solution. The final result is

$$f = f_0 t^{-n^2/2} \frac{\left(\prod_{k=1}^{n} \mu_k \right)^{n+1}}{\prod_{\substack{k,l=1 \\ k>l}}^{n} (\mu_k - \mu_l)^2} \det \Gamma_{kl} \qquad (3.12b)$$

where the term in the denominator is 1 for n=1 and f_0 is the value of f for the seed metric.

Although the solutions given by (3.12) have been obtained for non-degenerate poles, solutions with multiple poles can be obtained from them by a limiting procedure in which $u_k \to u_1$. Thus, in the axisymmetric version of metric (3.1), taking the Minkowski metric as seed and using 2n real poles, the n-Kerr-NUT metric is generated; by approaching the same pole trajectory (giving n double poles), the Tomimatsu-Sato (δ=n) solution is generated (Alekseev and Belinskii (1981), Tomimatsu and Sato (1981)). Some cosmological exemples of double pole solutions will be discussed later. Francaviglia and Sgarra (1983) have shown that the soliton technique can be extended also to degenerate poles without need of this limiting procedure.

In the cosmological context, the technique has been applied to generate vacuum soliton solutions from Bianchi type-I metrics by Belinskii and Zakharov (1979), Belinskii and Fargion (1980) and Carr and Verdaguer (1983); from type-II metrics by Belinskii and Francaviglia (1982, 1983) and from type-V metrics by Ibañez and Verdaguer (1984b). We shall review these solutions and also some non-vacuum solutions in the next sections.

IV. SOLITON SOLUTIONS ON BIANCHI TYPE I

As we have seen in the last section the soliton solutions are characterized by two main ingredients. The first is the seed metric, i.e. particular solutions of Einstein's equations which are used as starting points for the generation of the new solutions. The second ingredient is the number and kind (real or complex) of pole trajectories. This may be used as a means to classify the soliton solutions. Moreover, when the seed metric chosen is diagonal, one can take the arbitrary parameters in such a way that the corresponding soliton solution be also diagonal, in which case a number of simplifications occur.

As we remarked in section II on the search for soliton solutions of cosmological interest it is natural to start with the homogeneous and anisotropic Bianchi types I-VII as seeds. The corresponding soliton solutions will break the homogeneity of these models in one direction.

The simplest of the Bianchi models is the Kasner metric, which is a solution of vacuum Einstein equations belonging to the Bianchi type-I. The soliton solutions which have been studied in more detail have this metric as seed. In the form (3.1) it is

$$ds^2 = t^{(\delta^2-1)/2}(dz^2 - dt^2) + t^{1+\delta}dx^2 + t^{1-\delta}dy^2 , \qquad (4.1)$$

which verifies the condition (3.2). It is related to the standard Kasner form (Kramer et al. 1980)

$$ds^2 = -dT^2 + T^{2p_1}dx^2 + T^{2p_2}dy^2 + T^{2p_3}dz^2 \qquad (4.2)$$

$$(p_1 + p_2 + p_3 = p_1^2 + p_2^2 + p_3^2 = 1)$$

by a time transformation, $T = t^{(\delta^2+3)/4}$, and

$$p_1 = \frac{2(1+\delta)}{\delta^2+3} , \quad p_2 = \frac{2(1-\delta)}{\delta^2+3} , \quad p_3 = \frac{\delta^2-1}{\delta^2+3} . \qquad (4.3)$$

The real parameter δ is arbitrary, but we shall assume that we always have $\delta > 0$ or $\delta < 0$ since one can obtain one from the other by interchanging x by y. The axisymmetric Kasner solution is given by $\delta = 0$, while $|\delta| = 1$ corresponds to Minkowski space. The z axis is expanding for $|\delta| > 1$ and contracting for $|\delta| < 1$, this will lead to very different behaviour in the corresponding soliton solutions.

V. REAL-POLE TRAJECTORIES

Real-pole trajectories are obtained from (3.9) with $u_k = z_k^0$ (real). To see the general features of the soliton solutions with real poles we can start with the simplest case of the one-soliton solution. The pole will be

$$\mu_1^{(\pm)} = z_1^0 - z \pm [(z_1^0 - z)^2 - t^2]^{1/2} \qquad (5.1)$$

with either sign being allowed. We shall call z_1^0 the "origin" of the soliton. Given a seed metric, equations (3.12) generate a solution only in the region $(z_1^0-z)^2 \geqslant t^2$, where the pole is real. In $(z_1^0-z)^2 < t^2$, the solution remains the seed metric (Belinskii and Zakharov 1979). The spacetime is thus divided into two regions: inside the light cone $(z_1^0-z)^2 = t^2$, the solution is unperturbed; outside the light cone, one has an inhomogeneous one-soliton solution depending on just two parameters. The overall metric is continuous, but it has discontinuities in its first derivatives on the light cone itself. Plane-symmetric cosmological metrics with discontinuous first derivatives have been discussed by Wairwright (1979) and Carmeli et al. (1981) and we shall come to this later. Belinskii and Francaviglia (1982) describe these solutions as gravitational shock waves. Note that in the axisymmetric context this problem does not arises because the square root in (5.1) is then $[(z_1^0-z)^2+\rho^2]^{1/2}$ where ρ is the radial coordinate. Therefore one-soliton solutions can be found (Verdaguer, 1982) without discontinuities in their first derivatives.

We can now imagine the following sequence of solutions. The two-soliton solution generated from the one-soliton solution will contain two light cones: $(z_2^0-z)^2=t^2$ and $(z_1^0-z)^2=t^2$. In the region contained in both light cones, the solution will be the seed metric. Inside each light cone, excluding the intersection region, it will be of the "one-soliton" form, in the remaining regions, it will be of the "two-soliton" form and be described by four parameters. The overall metric will have discontinuous first derivatives along both light cones. From this one can generate the three-soliton solution and so on.

The common feature of these metrics, for an observer at finite z, is that they start very complicated but evolve towards the seed metric as the inhomogeneities propagate outwards at the speed of light. If the soliton origins $z=(z_1^0, z_2^0, \ldots, z_n^0)$ are equally spaced, the inhomogeneity will be reduced in steps at regular intervals. The light cones separating the different regions have been represented in Figure 1.

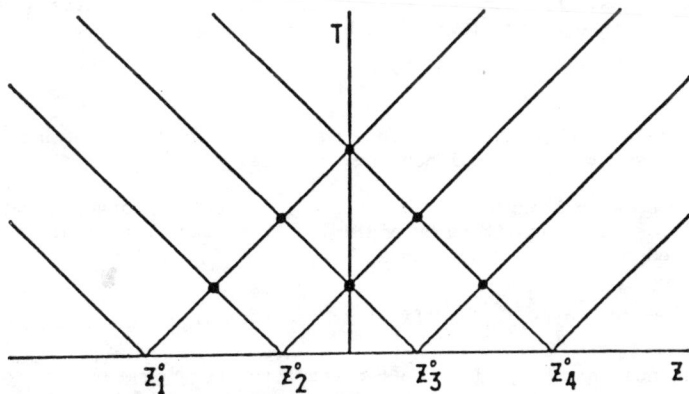

FIG. 1.- This represents the light cones for 4 real-pole soliton solutions with equally spaced origins. The spacetime is different as we cross any of these lines becoming homogeneous when we cross upwards.

After this very general overview of the soliton solutions of cosmological interest with real poles we should look at them with more detail since they still store some surprises.

It is convenient now, to write down the main asymptotic values for $\mu_k^{(-)}$ and $\mu_k^{(+)}$. We should first note that

$$\mu_k^{(+)} = t^2/\mu_k^{(-)} \tag{5.2}$$

Thus we shall consider only $\mu_k^{(-)}$. We may distinguish four asymptotic regions: (i) the future timelike infinity or "causal region" ($|z_k| \ll t \to \infty$ for all k, with $z_k \equiv z_k^0 - z$), which is contained in the intersection of the set of light cones with origins at $z=(z_1^0, z_2^0, \ldots, z_n^0)$, (ii) the future null infinity ($|z_k| \sim |z| = t \to \infty$), (iii) the spacelike infinity ($t \ll |z_k| \sim |z| \to \infty$), and (iv) the "initial region" ($|z_k|, |z_1| \gg t \to 0$). In the regions $z_k^2 \gg t^2$ the limiting values of $\mu_k^{(-)}$ are $\mu_k^{(-)}/t \to 1$ at the future null infinity and

$$\frac{\mu_k^{(-)}}{t} = \frac{t}{2|z_k|}\left[1 + 0\left(\frac{1}{|z_k|}\right)\right] \quad \text{(spacelike infinity)}$$

$$\frac{\mu_k^{(-)}}{t} = \frac{t}{2|z_k|}\left[1 + 0(t^2)\right] \quad \text{(initial region)} \tag{5.3}$$

Belinskii and Zakharov (1979) have shown that one recovers the seed solution in the limit $\mu_k/t \to 1$, thus the soliton solution in $z_k^2 \gg t^2$ matches the seed solution in the region where $t^2 > z_k^2$.

Va. Diagonal metrics

Starting with a diagonal metric as seed we can generate diagonal soliton solutions. This can be done by taking one of the arbitrary parameters $(m_0)_c^{(k)}$ in equation (3.10a) to be zero. We shall assume

$$(m_0)_1^{(k)} = 0 . \tag{5.4}$$

Therefore the soliton solutions with diagonal metrics can be seen as a particular case of the general nondiagonal case.

The general expression for the metric coefficients g_{11} and g_{22} can be obtained for n-solitons by adding solitons one at a time. For any diagonal seed metric the result is

$$g_{11} = \prod_{k=1}^{n}(\mu_k/t)(g_0)_{11}, \quad g_{22} = t^2/g_{11} \tag{5.5a}$$

When g_0 depends on t only, the coefficient f is best found by integrating equations (3.5) directly, rather than using equation (3.12b). For the Kasner seed we get (Carr and Verdaguer, 1983).

$$f = f_0 t^{n(4-n)/2}\left[\prod_{k=1}^{n}\left(\frac{\mu_k}{t}\right)\right]^{(2+\delta-n)} \prod_{\substack{k,l=1 \\ k>l}}^{n}(\mu_k-\mu_l)^2 \prod_{k=1}^{n}(\mu_k^2-t^2)^{-1} \tag{5.5b}$$

Since the metric coefficient f is determined once g is known, the essential of the metric features can be seen from g. Consequently, we can concentrate on the metric elements g_{11} and g_{22} from (5.5a).

From the asymptotic limits (5.3) and the relation (5.2), it is obvious that at spacelike infinity the metric will deviate from the seed metric, becoming in general singular, unless we take the same number of poles $\mu_k^{(-)}$ as $\mu_i^{(+)}$. We can also take degenerate poles. For exemple, by taking the limit $z_k^0 \to z_1^0$ for all k, we get the new solution

$$g_{11} = \left(\frac{\mu_1}{t}\right)^n (g_0)_{11} , \quad g_{22} = t^2/g_{11} \qquad (5.6)$$

This is also a solution of Einstein equations when <u>n is an arbitrary real number</u>. Another exemple, which will be used later, is the case of two degenerate poles $z_k^0 \to z_1^0$ for k=3...m and $z_k^0 \to z_2^0$ for k=m+1,...,n taking different prescriptions for μ_1 and μ_2, say $\mu_1^{(+)}$ and $\mu_2^{(-)}$, the new soliton solution is then

$$g_{11} = \left(\frac{\mu_1^{(+)}}{t}\right)^m \left(\frac{t}{\mu_2^{(+)}}\right)^{n-m} (g_0)_{11} , \quad g_{22} = t^2/g_{11} \qquad (5.7)$$

This is also a solution of Einstein equations when <u>n and m are arbitrary real numbers</u>.

Therefore from the soliton solutions (5.5) new solutions can be generated by a trivial limiting procedure. We can relate these solutions with other solutions found in the literature (Verdaguer, 1984).

A. FROM KASNER TO OTHER BIANCHI'S

Diagonal solutions of type (3.1) have been studied intensively, the reason being that Einstein equations become linear differential equations. This is seen easily when we write metric (3.1) in the standard form

$$ds^2 = f(dz^2 - dt^2) + t(e^\phi dx^2 + e^{-\phi} dy^2) , \qquad (5.8)$$

and the Einstein equations (3.4) become

$$\frac{\partial^2 \phi}{\partial z^2} + \frac{1}{z}\frac{\partial \phi}{\partial z} - \frac{\partial^2 \phi}{\partial t^2} = 0 , \qquad (5.9)$$

which is linear in the "potential" ϕ. Therefore a linear superposition of solutions for ϕ is also a solution; thus the product in (5.5a) can be interpreted as linear superpositions of the corresponding potential, ϕ, for each soliton.

As a solution of equation (5.9) we may take (Lamb 1906, Rosen 1954)

$$\phi = \alpha \ln t + \beta \int_0^{z-t} \frac{g(\lambda) d\lambda}{[(z-\lambda)^2 - t^2]^{1/2}} \qquad (5.10)$$

where $g(\lambda)$ is an arbitrary bounded function and α and β are arbitrary real parameters. The last term of (5.10) represents a travelling pulse wave.

The particular case $g(\lambda)=1$ for $0<\lambda<\infty$ leads to the solution of Wainwright et al. (1979),

$$\phi = \alpha \ln t + \beta \ln\left[\frac{z}{t} + \sqrt{\frac{z^2}{t^2}-1}\right], \qquad (5.11)$$

which corresponds to our solution (5.6) with the Kasner seed $(g_0)_{11}=t^{1+\delta}$, taking $z_1^0=0$ and the prescription $\mu_1^{(+)}$. The parameters α and β are related to our δ and n by

$$\alpha = \delta, \quad \beta = n. \qquad (5.12)$$

These solutions have curvature singularities at t=0, but also at $|z|\to\infty$ with the only exception of

$$n = -(3+\delta^2)^{1/2} \qquad (5.13)$$

which has only the cosmological singularity at t=0 (Carmeli and Charach 1980). But this solution is just the Ellis and MacCallum family of vacuum Bianchi models. In fact, introducing new coordinates T, Z.

$$t = e^{-2aZ}\sinh 2aT, \quad z = e^{-2aZ}\cosh 2aT \qquad (5.14)$$

with a an arbitrary positive parameter, we obtain the Ellis and MacCallum (1969) metric,

$$ds^2 = (\sinh 2aT)^{1+\delta^2}(\tanh aT)^{\delta(3+\delta^2)^{1/2}}(dZ^2-dT^2) +$$
$$+ (\sinh 2aT)e^{-2aZ}\left[(\sinh 2aT)^{\delta}(\tanh aT)^{(3+\delta^2)^{1/2}} dx^2 + \right. \qquad (5.15)$$
$$\left. +(\sinh 2aT)^{-\delta}(\tanh aT)^{-(3+\delta^2)^{1/2}} dy^2\right].$$

The particular case $\delta=0$ (axisymmetric seed) corresponds to the Bianchi V model, whereas $\delta^2=1$ (Minkowski seed) leads to the Kantowski-Sachs solutions of Bianchi type-III. Other values of δ correspond to Bianchi type VI$_h$ (Bbi case). All these solutions have the cosmological singularity at T=0 (MacCallum, 1971) and are of type I in the Petrov classification with the exception of $\delta^2=1$ which is of type D; note that the Kasner metrics are of Petrov type I with the exception of $\delta=0$ which is type D and $\delta^2=1$ (Minkowski).

Thus our soliton solutions (5.6), obtained from the Kasner seed and limited to the region $t^2 \leqslant z^2$, contain the inhomogeneous Wainwright et al. solutions (with limited cosmological interest due to their space singularities) and the homogeneous Bianchi models of type III, V and VI, of Ellis and MacCallum. Similar results have been obtained by D. Kitchingham (1984).

B. PULSE WAVES ON KASNER BACKGROUNDS

The possibility of finding soliton solutions having only the cosmological singularity at $t\to 0$ and becoming the regular Kasner solution at $|z|\to\infty$, is imbeded in the solution (5.7) if we take m=n/2. This can be easily seen using the asymptotic values (5.3).

If we write $z_1^0=0$, m=n/2, the Kasner seed $(g_0)_{11}=t^{1+\delta}$ and use the notation (5.8) we have

$$\phi = \delta \ln t + \frac{n}{2} \{ \ln[z+(z^2-t^2)^{1/2}] - \ln[z-z_2^0+[(z-z_2^0)^2-t^2]^{1/2}] \} \quad (5.16a)$$

which is valid in the region $z \geqslant t+z_2^0$ ($z > 0$) and it is regular at $z \to \infty$. This can be combined with a solution of type (5.6), that is,

$$\phi = \delta \ln t + \frac{n}{2} \ln\left[\frac{z}{t} + \sqrt{\frac{z^2}{t^2} - 1}\right] \quad (5.16b)$$

in the region $t < z < t + z_2^0$.

The solution (5.16) is the metric of Carmeli and Charach (1980), These metrics are regular everywhere, except at the initial hypersurface t=0, they go to the Kasner seed, with δ, at $z \to \infty$ and have a δ-function discontinuity in the derivatives of ϕ across the null hypersurface $z_2^0=z-t$. This can be regarded as an impulsive gravitational wave. Note that if we take $n/2 = -(3+\delta^2)^{1/2}$, this solution (5.16) is equivalent to the Bianchi III, V or VIh models in the region $0 < z-t < z_2^0$, whereas it is completely different in the region $z-t > z_2^0$.

The solutions (5.16) can be considered as impulsive waves propagating on Kasner backgrounds. The discontinuities, in this context, are a consequence of taking real poles, since by writing (5.16b) in the form (5.10)

$$g(\lambda) = \begin{cases} 1 & 0 < \lambda < z_2^0 \\ 0 & \lambda > z_2^0 \end{cases} . \quad (5.17)$$

That is, real poles imply that the function $g(\lambda)$ be discontinuous. One can cure that problem by requiring a smooth $g(\lambda)$ function.

In the soliton context this problem will be avoided when we take complex poles. The reason is that complex poles are defined in the whole (t,z) region. There we shall be able to find also localized wave solutions propagating in Kasner backgrounds and free of discontinuities.

Vb. Nondiagonal Metrics

As we have remarked earlier, the generation of nondiagonal solutions with n > 2 becomes a cumbersome task due to the problem of inverting the n ×n matrix Γ_{kl}, defined by equation (3.10b). Moreover once a nondiagonal metric has been obtained its study is more complicated than the diagonal metrics (the Riemann tensor becomes specially simple for diagonal metrics).

In the soliton context nondiagonal metrics can always be converted to diagonal metrics (if the seed is diagonal) by properly choosing the arbitrary parameters of the solution. Nondiagonal metrics present waves with two polarizations (Adams et al., 1982) whereas diagonal ones have one polarization only. For these reasons many properties of the nondiagonal metrics can be inferred from properties of the diagonal ones.

With real poles the only nondiagonal solution, that I am aware of, was obtained by Belinskii and Zakharov for n = 1,

$$ds^2 = \frac{t^{\delta^2/2}\cosh(\delta r/2+C)}{(z^2-t^2)^{1/2}}(dz^2-dt^2) + \frac{1}{\cosh(\delta r/2+C)}\left\{t^{1+\delta}\cosh[(1+\delta)r/2+C]dx^2 \right.$$
$$\left. + t^{1-\delta}\cosh[(1-\delta)r/2+C]dy^2 - 2\sinh(r/2)dxdy\right\} \tag{5.18}$$

where $r=2\ln(\mu_1/t)$ and C is an arbitrary parameter.

The singularity structure of this solution has not been studied. It has the cosmological singularity at $t \to 0$ and, at least its diagonal limit is singular at $|z| \to \infty$ too. If one considers the metric component g_{11} and determines the position of its extremum, with respect to the spacelike variable z for various fixed instants of time t, it can be seen that the world line of the extremum has the equation $z=t\cosh(r_0/2)$, where r_0 is a constant. So that this maximum cannot be considered to represent the propagation of a physical effect, it is just the evolution of some initial condition.

Taking the limit $C \to \infty$, metric (5.18) becomes diagonal

$$ds^2 = \frac{t^{\delta^2/2}}{(z^2-t^2)^{1/2}}\left(\frac{\mu_1}{t}\right)^\delta (dz^2-dt^2) + t^{1+\delta}\left(\frac{\mu_1}{t}\right)dx^2 + t^{1-\delta}\left(\frac{t}{\mu_1}\right)dy^2, \tag{5.19}$$

which is of type (5.6), with n=1, and it is singular at $t \to 0$ and at $|z| \to \infty$. Therefore its possible cosmological interest is not clear.

VI. COMPLEX-POLE TRAJECTORIES

When we take the parameters u_k in equation (3.9) complex,

$$u_k \equiv z_k^0 - iw_k \tag{6.1}$$

the complex-conjugate poles may be written as

$$\mu_k = \sqrt{\sigma_k}\, t e^{i\phi_k}, \quad \mu_{k+n/2} = \bar{\mu}_k \quad (k=1,2,\ldots,n/2) \tag{6.2}$$

The functions $\sigma_k(t,z)$ and $\phi_k(t,z)$ are given by

$$\cos\phi_k = \frac{2z_k\sqrt{\sigma_k}}{t(1+\sigma_k)}, \quad \sin\phi_k = \frac{2w_k\sqrt{\sigma_k}}{t(1-\sigma_k)} \quad (z_k \equiv z_k^0 - z) \tag{6.3}$$

Two explicit solutions for $\sigma_k(t,z)$ can be found from these equations by solving a fourth-degree equation (Carr and Verdaguer, 1983). This gives two real solutions, $\sigma_k^{(-)}$ and $\sigma_k^{(+)}=1/\sigma_k^{(-)}$, where

$$\sigma_k^{(-)} = \frac{w_k^2+z_k^2}{t^2} + a_k - \sqrt{2}\left[\frac{w_k^2-z_k^2}{t^2} + \frac{(w_k^2-z_k^2)^2}{t^4} + \frac{(w_k^2+z_k^2)}{t^2}a_k\right]^{1/2}$$

$$a_k = \left[1 + \frac{2(w_k^2-z_k^2)}{t^2} + \frac{(w_k^2+z_k^2)^2}{t^4}\right]^{1/2} \tag{6.4}$$

These satisfy $0 < \sigma_k^{(-)} < 1$ and $1 < \sigma_k^{(+)} < \infty$.

As we did for real poles, it is convenient to write down the main asymptotic values for $\sigma_k^{(-)}$. For the four asymptotic regions, defined in section V, we find the values

$$\sigma_k^{(-)} = 1 - \frac{2w_k}{t} + O(t^{-2}) \qquad \text{(future timelike infinity)},$$

$$\sigma_k^{(-)} = 1 - 2\left[\frac{[w_k^2+(z_k^0)^2]^{1/2}-z_k^0}{t}\right]^{1/2} + O(t^{-1}) \qquad \text{(future null infinity)},$$

(6.5)

$$\sigma_k^{(-)} = \frac{t^2}{4z_k^2}\left[1+O(z^{-2})\right] \qquad \text{(spacelike infinity)},$$

$$\sigma_k^{(-)} = \frac{t^2}{4(w_k^2+z_k^2)}\left[1+O(t^2)\right] \qquad \text{(initial region)}.$$

Belinskii and Zakharov (1979) have shown that one recovers the seed solution in the limit $\sigma_k \to 1$. Equations (6.5) show that this is always the case at the timelike and null infinity regions, the perturbations decaying as t^{-1} and $t^{-1/2}$ respectively. Therefore, for an observer at finite z, the n-soliton with complex poles will always evolve towards the seed metric. This situation is different from the real-pole case, however, in that one never gets the exact seed solution and that $\sigma_k^{(-)}$ goes smoothly through the light cone $z_k^2=t^2$. The asymptotic behaviour of the soliton solution at timelike infinity, evolving always towards the seed, can be used as a guide in choosing an appropriate seed metric.

The behaviour of the soliton solutions at spacelike infinity will generally be quite different from the seed metric. Although as we have seen in the real-pole case one can also obtain soliton solutions which tend to the seed metric in that asymptotic region too. In such cases, the soliton solutions can be regarded as localized "perturbations" on the seed background.

The time evolution of $\sigma_k^{(-)}(t,z)$ is represented in Fig. 2 for $w_k=0.2$. The slope of the curves between small and large values of z is governed by the parameter w_k; the smaller w_k, the steeper the slope.

Since the solitons can be associated with the z derivatives of σ_k, the parameter w_k reflects the "width" of the solitons. All the metric-dependent quantities can be obtained from the pole equations (3.9b). In terms of σ_k, these become

$$\sigma_{k,z} = \frac{8z_k\sigma_k^2(1-\sigma_k)}{H_k(1+\sigma_k)t^2}, \qquad \sigma_{k,t} = \frac{2\sigma_k(1-\sigma_k^2)}{H_k t},$$

$$H_k \equiv (1-\sigma_k)^2 + \frac{16w_k^2\sigma_k^2}{(1-\sigma_k)^2 t^2}.$$

(6.6)

These equations give a "recursion" relation, enabling one to find, for instance, the Riemann tensor for the n-soliton with complex poles in terms of $\sigma_k(t,z)$. It is easy to see, using equations (6.5), that $\sigma_{k,z}$ has a maximum at null infinity, indicating that the corresponding soliton solutions contain inhomogeneities propagating at the speed of light as $t \to \infty$.

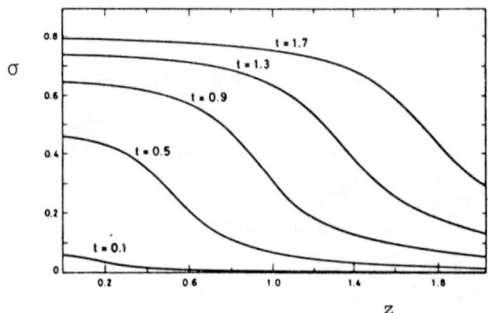

FIG. 2.- This shows the function $\sigma^{(-)}(t,z)$ for different values of t, as defined by equation (6.4). The origin is z=0 and the width parameter is w=0.2. The function is unchanged if one reverses the sign of z. When σ approaches 1, the corresponding soliton solution approaches the seed metric.

In the limit $w_k \to 0$, the two complex conjugate poles become a double real pole, with $\sigma_k=1$ inside the light cone $z_k^2=t^2$ and $\sigma_k=\mu_k^2/t^2$ outside it. Since $\sigma_{k,z}$ has a δ-function discontinuity on the light cone, this suggests a gravitational shock wave propagating at the speed of light. Therefore, double real-pole solutions can be studied as limiting cases of complex-pole solutions.

It is clear that the key point to understand the soliton solutions will be the behaviour of the metric in the spacelike infinity region. Providing the metric approaches the seed metric in that region, the solution can be interpreted as consisting of inhomogeneities moving on the background of the seed metric. We shall now consider all possible soliton solutions that can be obtained using the Kasner metric as seed.

VIa. Diagonal metrics

There are two reasons for studying first the diagonal (one-polarization) metrics: (1) they are simpler than the nondiagonal ones -in particular the Riemann tensor components can be calculated explicitly; and (2) they can be used as a paradigm to understand the general nondiagonal (two-polarization) solutions, since they can be obtained from the latter by a proper election of parameters.

The general expression of the n-soliton solution has been given explicitly in (5.5) for the Kasner seed. Using the notation (6.2) for complex poles we have:

$$g_{11} = t^{1+\delta} \prod_{k=1}^{n/2} \sigma_k \quad , \quad g_{22} = t^{1-\delta} \prod_{k=1}^{n/2} \sigma_k^{-1} \quad \quad (6.7a)$$

and

$$f = \frac{t^{(\delta^2-1-n^2)/2}}{\prod_{k=1}^{n/2} H_k} \prod_{k=1}^{n/2}\left[\frac{\sigma_k^{(\delta-n+4)}}{(1-\sigma_k)^2}\right] \prod_{\substack{k,l=1 \\ k>l}}^{n/2}\left\{\left[(\sigma_k+\sigma_l)t^2 - \frac{8z_k z_l \sigma_k \sigma_l}{(1+\sigma_k)(1+\sigma_l)}\right]^2 - \frac{64 w_k^2 w_l^2 \sigma_k^2 \sigma_l^2}{(1-\sigma_k)^2(1-\sigma_l)^2}\right\}^2$$

(6.7b)

As discussed earlier, the metric at timelike infinity represents a Kasner background on which some inhomogeneities with "amplitude" decreasing as t^{-1} are superposed. This can be seen from (6.7a) and (6.5). The amplitude, on the other hand, decreases at null infinity as $t^{-1/2}$, which is typical of linear gravitational waves on a cosmological background. It remains to study the behaviour of the metric at spacelike infinity; this is the only asymptotic region in which it may deviate appreciably from the seed solution.

We shall distinguish several possibilities: (a) n solitons with $\sigma_k^{(-)}$; (b) n solitons with $\sigma_k^{(+)}$; and (c) r solitons with $\sigma_k^{(-)}$ and n-r solitons with $\sigma_k^{(+)}$. Since $\sigma_k^{(+)}=1/\sigma_k^{(-)}$, case (b) reduces to case (a) if one interchanges x and y in the seed and then interchanges them again in the soliton solution. Similarly, in case (c), we only need consider r>n-r; the asymptotic structure in the far region will then be similar to case (a) with 2r-n solitons, although the structure in the regions $z_k \sim t$ will be very different since we still have n solitons there. From the point of view of soliton perturbations propagating on a homogeneous background, the most interesting case is (c) with r=n/2, since at spacelike infinity

$$g \to g_0\left[1 + 0(z^{-1})\right] \ .$$

These models can therefore be interpreted as n/2 pairs of solitons with origins at $z=(z_1^0, z_2^0, \ldots, z_{n/2}^0)$ propagating on a Kasner background. As $t \to \infty$, they propagate near the two branches of the light cones $z_k^2 = t^2$, evolving towards gravitational waves. They have been studied in detail by Ibañez and Verdaguer (1983, 1984a), we shall call these models "cosmologies with gravisolitons" (gravitational solitons).

For a metric of type (3.1) with $g_{12}=0$ (diagonal) the Riemann tensor has only three independent components. Its components can be given in terms of a null tetrad adapted to the metric,

$$\vec{n} = \frac{1}{\sqrt{2f}}(\partial_t + \partial_z) \qquad \vec{l} = \frac{1}{\sqrt{2f}}(\partial_t - \partial_z)$$

$$\vec{m} = \frac{1}{\sqrt{2}}\left(\frac{1}{\sqrt{g_{11}}}\partial_x + i\frac{1}{\sqrt{g_{22}}}\partial_y\right) \qquad \vec{m}^* = \frac{1}{\sqrt{2}}\left(\frac{1}{\sqrt{g_{11}}}\partial_x - i\frac{1}{\sqrt{g_{22}}}\partial_y\right) \ .$$

(6.8)

The non-vanishing components of the Riemann tensor in this tetrad are

$$\psi_0 = R_{\mu\nu\alpha\beta} n^\mu m^\nu n^\alpha m^\beta$$

$$\psi_2 = \frac{1}{2} R_{\mu\nu\alpha\beta} n^\mu l^\nu (n^\alpha l^\beta - m^\alpha m^{*\beta})$$

$$\psi_4 = R_{\mu\nu\alpha\beta} l^\mu m^{*\nu} l^\alpha m^{*\beta}$$

(6.9)

and they have the following physical meaning: ψ_0 and ψ_4 represent the radiative part of the field whereas ψ_2 contains the Coulomb part; ψ_0 gives the radiative component along the left directed perturbations and ψ_4 along the right directed perturbations. The Riemann components can be calculated explicitly, using (6.6) for the derivatives of σ_k, and equation (3.5) for the derivatives of f. The final result is then expressed in terms of the functions $\sigma_k(t,z)$.

A. n-SOLITONS WITH $\sigma^{(-)}$

We first discuss case (a) for the situation with $\delta < 0$. The asymptotic expression for the coefficient f at spacelike infinity is then

$$f = t^{(\delta+n+1)(\delta+n-1)/2} z^{-n(\delta+n)} \left[1+0(z^{-2})\right] \qquad (6.10a)$$

and the Riemann components are

$$\psi_0 = \frac{1}{8ft}\left[\frac{(\delta+n)(1+\delta+n)(1-\delta-n)}{t} - \frac{3n(\delta+2)(1-\delta-n)}{z} + 0(z^{-2})\right]$$

$$\psi_2 = \frac{1}{8ft}\left[\frac{(1+\delta+n)(1-\delta-n)}{t} + 0(z^{-2})\right] \qquad (6.10b)$$

$$\psi_4 = \frac{1}{8ft}\left[\frac{(\delta+n)(1+\delta+n)(1-\delta-n)}{t} + \frac{3n(\delta+2)(1-\delta-n)}{z} + 0(z^{-2})\right]$$

From this we can define a critical value δ_c

$$\delta_c = -n , \qquad (6.11)$$

for which $\psi_0 \to 0$, $\psi_4 \to 0$ and $\psi_2 \to t^{-3/2}$ as $z \to \infty$. The metric thus becomes Petrov type D as $z \to \infty$ (no radiation). This is very different from the behaviour at timelike infinity, where the solution tends to the corresponding Kasner background. The Kasner solution is Petrov type I unless $\delta=-1$ (flat space) or $\delta=0$ (axisymmetric Kasner), in which case it is also Petrov type D.

For $\delta<\delta_c$, the Riemann tensor satisfies $\psi_0 \to 0$, $\psi_2 \to 0$ and $\psi_4 \to 0$ as $z \to \infty$, so the metric is asymptotically flat at spacelike infinity. The interpretation of these metrics is that they represent inhomogeneities on a flat background which evolve towards the Kasner solution at timelike infinity due to soliton propagation along the z axis.

For $0 \geqslant \delta > \delta_c$ and $n > 2$, the metric is singular at spacelike infinity (ψ's $\to\infty$) and so it has a space singularity as well as a cosmological singularity. The physical interpretation of this is not clear; such metrics may not have cosmological interest.

Ai. The real-pole limit

We now discuss the limit in which $w_k \to 0$, that is, the limit in which the complex poles become real. The n-soliton solution then contains n/2 degenerate double real poles. As an illustrative exemple, we shall consider the n=2 case. The complex two soliton solution is

$$g_{11} = t^{1+\delta}\sigma, \quad g_{22} = t^{1-\delta}\sigma^{-1}, \quad f = t^{(\delta^2-5)/2}\frac{\sigma^{(\delta+2)}}{(1-\sigma)^2 H} \quad (6.12)$$

where we have dropped the index 1. If we now take the limit $w \to 0$ we obtain

$$g_{11} = t^{1+\delta}\left(\frac{\mu_1}{t}\right), \quad g_{22} = t^{1-\delta}\left(\frac{t}{\mu_1}\right)^2, \quad f = \frac{t^{(\delta^2+3)/2}}{(z^2-t^2)^2}\left(\frac{\mu_1}{1}\right)^{2\delta} \quad (6.13)$$

where μ_1 is the real pole (5.1), taking the + or - sign depending on the choice of the solution for σ. This is a double real soliton solution. It can be compared with the solution (5.19).

Aii. Degenerate complex poles

Solutions with degenerate complex poles can be obtained from equation (6.7) by taking $z_k^0 \to z_1^0$, $w_k \to w_1$ for all k,l. Similarly as we did in the real-pole case. This leads to a cosmological version of the Tomimatsu-Sato axisymmetric solutions,

$$g_{11} = t^{1+\delta}\sigma^{n/2}, \quad g_{22} = t^{1-\delta}\sigma^{-n/2}$$

$$f = \frac{t^{(\delta^2-n^2-1)/2}\sigma^{n(\delta+n)/2}}{H^{n/2}(1-\sigma)^{n^2/2}}\left[\frac{z^2(1-\sigma)^2+w^2(1+\sigma)^2}{z^2(1-\sigma)^4+w^2(1+\sigma)^4}\right]^{n(n-2)/4} \quad (6.14)$$

where n is here an <u>arbitrary real parameter</u>. As usual, the metric tends to the Kasner seed at <u>timelike infinity</u>, with the perturbation decaying as t^{-1}. At spacelike infinity f has the asymptotic form given by equation (6.10a). Likewise, the Riemann components asymptotically look like (6.10b). Thus, the behaviour is similar to case (a).

B. COSMOLOGIES WITH GRAVISOLITONS

Those are the solutions of case (c) with r=n/2. As we remarked earlier those solutions go to the Kasner background at spacelike infinity. They have the cosmological singularity only and are interesting as inhomogeneous cosmological models. They represent highly inhomogeneous universes wich evolve towards homogeneous Kasner models with gravitational waves. Thus, they are an example of the creation of a background of gravitational waves as a consequence of initial inhomogeneities, in the line of the Adams et al. (1982) work.

The general case, arbitrary n, has been studied by Ibáñez and Verdaguer (1984a). But the interesting features of the model are already present in the n=4 case (Ibañez and Verdaguer, 1983). For that reason we shall restrict to this case. In Figure 3 we represent the t-z diagram with the two soliton origins and the light cones from them. The four solitons will propagate around these light cones.

An interesting feature of this solution, apparent in the Figure, is that two of the solitons will collide. Thus by studying such a solution the soliton behaviour under collision will be revealed.

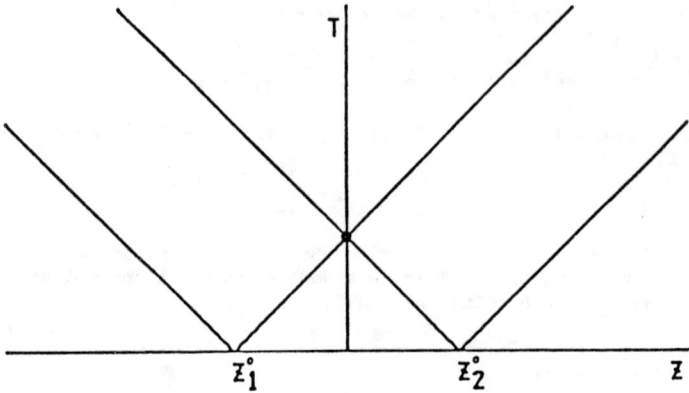

FIG. 3.- The t-z diagram for the 4-soliton solution. The four solitons with origins at z_1^0 and z_2^0 propagate around the light cones represented.

It is appropiate to make here an analogy between the gravitational solitons generated by the soliton technique of Belinskii and Zakharov and the classical solitons such as those found in hydrodynamics. The classical solitons are characterized by their localizability, by their peculiar behaviour under collisions and by carrying energy with some associated velocity of propagation. As we shall see the gravisolitons are very similar to the classical solitons.

We can note that in all the asymptotic regions the metric tends towards the Kasner background. However, the longitudinal expansion, which is given by the coefficient f, is the Kasner background value, say f^k, at spacelike infinity only. At null infinity and timelike infinity the longitudinal expansion is f^k multiplied by a factor depending on the soliton parameters. For instance, for the solutions for which $w_1^2 \sim w_2^2 \sim w^2 \ll 1$, at timelike infinity the asymptotic values are $f \sim f^k/w^4$, while at null infinity they are $f \sim f^k/w^4$ along the light cones of the inner solitons (those which collided, see Figure 3) and $f \sim f^k/w^2$ along the light cones of the outer solitons. Thus it appears that each collision increases the value of the coefficient f relative to its value in the corresponding Kasner background: after each collision the solitons have a greater longitudinal expansion along their direction of propagation.

The soliton structure and the intrinsic properties of the metric can be seen from the Riemann tensor. This can be easily calculated using the tetrad (6.8). Frame-independent soliton properties are seen from the two nonzero curvature scalar invariants:

$$I = \psi_0\psi_4 + 3\psi_2^2 \quad , \quad J = (\psi_0\psi_4 - \psi_2^2)\psi_2 \qquad (6.15)$$

The evolution in time of the ratio I/I^k (I^k is the Kasner value) is shown in Fig. 4 for a model in which $\delta=0$. Before the collision for solitons with small tails are seen. After the collision the outer solitons are clear but the intensities of the inner solitons are very small because of the $1/w^4$ factor of the longitudinal expansion. The same factor makes the value of I at timelike infinity very small too.

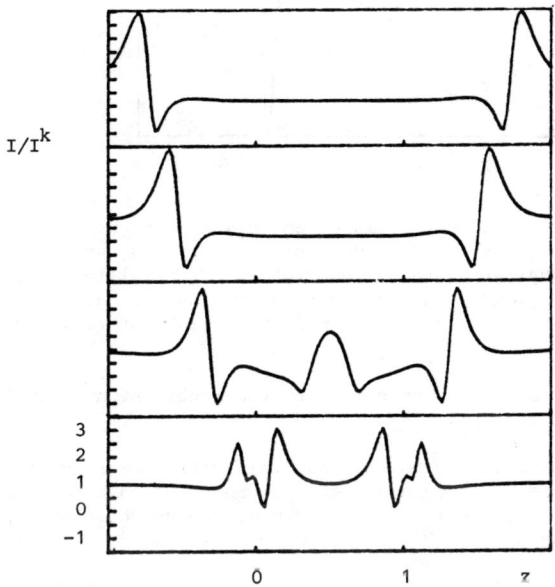

FIG. 4.- Time evolution of the ratio I/I^k for $\delta=0$. The curves are represented against the propagation axis (z axis). The widths and origins of the solitons are $w_1=w_2=0.05$ and $z_1^0=0$, $z_2^0=1$. The different curves from bottom to top of figure represent the time sequence t=0.1, 0.3 (before collision), 0,5 (collision time), and 0,7 (after collision).

We can now represent the components of the Riemann tensor (6.9). In Figure 5 the radiative field ψ_4 is represented, this field induces time-varying tidal forces to test particles on the plane orthogonal to the propagation of the solitons travelling to the right.

The radiative field ψ_0 has a similar behaviour for the solitons travelling to the left and the Coulomb component ψ_2 is specially concentrated near the z=0 and z=1 values.

The asymptotic values of the Riemann components, at null infinity when calculated analytically show that, for z>0, $\psi_0/\psi_4 \to 0$ and $\psi_2/\psi_4 \to 0$ so that the metric in its leading terms becomes of Petrov type N:

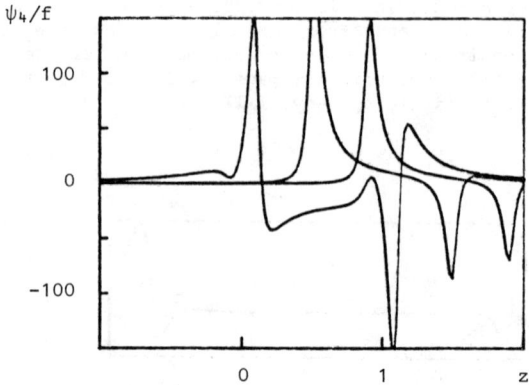

FIG. 5.- Time evolution of the Riemann tensor component ψ_4 (divided by f in order to make the solitons which collided apparent) giving the gravitational strength of the right-ward-travelling solitons. Same model and parameters as in Figure 4. The Kasner background value is $\psi_4^k=0$. The three curves from left to right of figure represent the time sequence t=0.1, t=0.5 and t=0.9.

pure gravitational radiation travelling to the right. At null infinity for z<0 the dominant term is ψ_0.

We can summarize by saying that this metric represents intrinsic inhomogeneities propagating on a Kasner background which behave as classical solitons (i.e. they are localized and have shapes which are not changed by collisions as seen in Figure 5) and evolve towards gravitational waves.

We can still associate a velocity of propagation to these solitons by means of the Bel-Robinson superenergy tensor (Ibáñez and Verdaguer, 1984a). As expected that velocity is small near $t \to 0$ and becomes the velocity of light at null infinity. This indicates that the solitons evolve from quasiparticles to pure gravitational waves.

VIb. Nondiagonal Metrics

In the nondiagonal case, one can no longer give explicit expressions for the n-soliton metric elements. However, information can still be obtained from asymptotic expressions which can be calculated in the limit in which the nondiagonal metric tends towards diagonality. We will first deal with the nondiagonal two-soliton solutions because some of their features apply for general n.

A. TWO SOLITON SOLUTIONS

These solutions are the simplest nondiagonal metrics with complex poles that one can evaluate and their analytical expressions are relatively simple. We shall here study the asymptotic behaviour of the solutions which can be obtained from all Kasner seeds, as well as their diagonal limits. Following the procedure in Section III, the two soliton solution can be shown to be (Belinskii and Fargion 1980, Carr and Verdaguer 1983)

$$g_{11} = \frac{t^{1+\delta}}{D} \{ (\sigma+\sigma^{-1}-2)\sin^2(\phi+\psi) + [L_0^2 \sigma^{-(1+\delta)} + L_0^{-2}\sigma^{(1+\delta)} + 2]\sin^2\phi \}$$

$$g_{22} = \frac{t^{1-\delta}}{D} \{ (\sigma+\sigma^{-1}-2)\sin^2(\phi-\psi) + [L_0^2 \sigma^{(1-\delta)} + L_0^{-2}\sigma^{-(1-\delta)} + 2]\sin^2\phi \}$$

$$\tag{6.16}$$

$$g_{12} = \frac{2w}{D} \{ L_0 \sigma^{-(1+\delta)/2} [\sin(\phi-\psi) + \sigma\sin(\phi+\psi)] + L_0^{-1}\sigma^{-(1+\delta)/2}[\sin(\phi+\psi)+\sigma\sin(\phi-\psi)] \}$$

$$f = Ct^{(\delta^2-5)/2} \sigma^2 DH^{-1}(1-\sigma)^{-2}(\sin\phi)^{-2}$$

where we use the usual notation of section VI (although we drop the index 1) and

$$D \equiv (\sigma+\sigma^{-1}-2)\sin^2\psi + (L_0^2\sigma^{-\delta}+L_0^2\sigma^\delta+2)\sin^2\phi$$

$$\psi \equiv \delta\phi + \psi_0$$

$$\tag{6.17}$$

C, L_0 and ψ_0 are arbitrary real constants.

We can determine the connection with the diagonal (one-polarization) two-soliton solution by taking the limits

$$L_0 \to 0 \qquad CL_0^2 \to C' \text{ (finite)} \tag{6.18a}$$

the metric (6.16) becomes

$$g = g \text{ diag}[1+0(L_0^2)] \quad \text{and} \quad f = f \text{ diag}[1+0(L_0^2)] \tag{6.18b}$$

where (f diag, g diag) is given by (6.12). Thus the two-soliton solution (6.16) can be regarded as a generalization of the diagonal metric (6.12) and the parameter L_0 can be interpreted as a "polarization" parameter.

All the diagonal metrics, whatever the Kasner seed, had the common feature that they evolve towards the seed at the asymptotic timelike and null infinity regions. The same holds for the nondiagonal two-soliton solution. Taking the limits (6.5)

$$g = g_0[1+0(t^{-1})] \quad , \quad f = C\frac{(L_0+L_0^{-1})^2}{16w^2} f_0[1+0(t^{-1})] \tag{6.19}$$

at timelike infinity, and

$$g = g_0[1+0(t^{-1/2})] \quad , \quad f = C\frac{[4\sin^2\psi_0 + (L_0+L_0^{-1})^2]}{32w^2} f_0[1+0(t^{-1/2})] \tag{6.20}$$

at null infinity. One should not deduce from equation (6.20) that the

Riemann tensor behaves the same way at null infinity as in the seed solution, because the z dependence in (6.20) has been hidden in the assumption $|z| \sim t$. That asymptotic region has to be seen as a boundary between the timelike and spacelike infinity regions which contains the propagating solitons.

We can now study the asymptotic behaviour at spacelike infinity. In that limit the behaviour depends crucially on the seed metric. For seeds with $\delta > 3$ all two-soliton metrics become the diagonal solution (6.12)

$$\delta > 3 \qquad g \to g \text{ diag} \qquad \text{(at spacelike infinity)} \qquad (6.21)$$

and for $1 > \delta \geqslant 0$ they become the seed metric

$$1 > \delta \geqslant 0 \qquad g \to g_0 \qquad \text{(at spacelike infinity)} \qquad (6.22)$$

Thus, when the background is contracting in the z direction the two soliton solutions can be interpreted as two perturbations (gravisolitons) propagating along the z axis. Since the solitons have their maximum amplitude on the light cone for large t, they move in oposite directions with a speed asymptotically approaching the speed of light, i.e. they become gravitational waves. The only difference with the gravisolitons in diagonal metrics is that these waves here have two polarizations

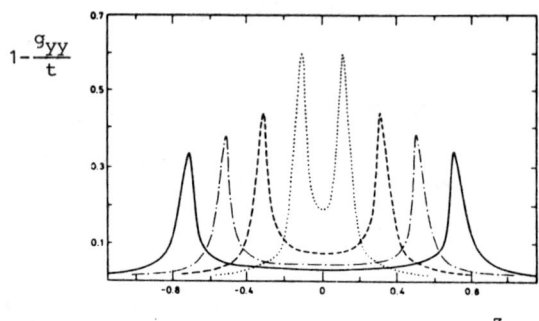

FIG. 6.- This shows the time evolution of $g_{yy}(t,z)$ for the two-soliton solution (6.16) generated from the axisymmetric Kasner seed ($\delta=0$). The Kasner background has been substracted and "normalized" by the factor t; we have also made a $\pi/4$ rotation of the x and y axes. The soliton origin is z=0 and width is w=0.01. The dotted line is t=0.1, the dashed line t=0.3, the dot-dashed line t=0.45, and the continuous line t=0.7.

The evolution of the two-soliton solution when the background is the axisymmetric Kasner ($\delta=0$) has been represented in Figure 6, where it is clear that this soliton tends to Kasner in all asymptotic regions having maximum amplitude near the light cone. We should note, as we did in the diagonal case, that the longitudinal expansion, f, includes a different constant, from the Kasner background, at timelike infinity and null infinity [see equations (6.19) and (6.20)]. Thus, in these asymptotic regions, the existence of solitons modifies the longitudinal expansion with respect to the Kasner background.

For $3 \geqslant \delta \geqslant 1$ the asymptotic behaviour is more complicated (see Carr and Verdaguer, 1983).

B. n-SOLITON SOLUTIONS

Although in general we cannot give explicit expressions for these metrics, we can discuss some of their asymptotic properties. In particular as in all the solutions considered so far, the n-soliton solution evolves towards the Kasner seed, with the perturbation decreasing as t^{-1} at timelike infinity. This is a consequence of the value of σ_k in that region, given by equation (6.5), and it would be true for any seed.

This can be seen explicitly using the soliton technique of section III,

$$g = g_0 \left[1 + (t^{-1}) \right] \qquad \text{(at timelike infinity)} , \qquad (6.23)$$

and similarly

$$g = g_0 \left[1 + 0(t^{-1/2}) \right] \qquad \text{(at null infinity)} , \qquad (6.24)$$

see Carr and Verdaguer (1983) for details.

Again the situation is different at spacelike infinity. However, some of their features can be deduced from the diagonal and non-diagonal two-soliton metrics. They will depend essentially on the seed metric.

For $1 > \delta \geqslant 0$ (z axis contracting), the n-soliton solution will always tend to the seed metric at spacelike infinity. The reason is that one can find the n-soliton solution step by step, using the (n-2)-soliton solution as seed, etc. It is clear that at each step we will recover the seed metric in this limit. Consequently, the general n-soliton solution can be considered as n gravisolitons propagating on a Kasner background which is contracting along the propagation axis. Since the speed of the solitons asymptotically approaches the speed of light, they evolve towards gravitational waves with two polarizations. If we take $|z_k^0 - z_{k-1}^0| = d$ for all k, so that the solitons are equally spaced, the wave period will be d.

As an exemple we show in Figure 7 the g_{yy} coefficient of the four-soliton solution for the axisymmetric Kasner seed ($\delta = 0$). The structure of four solitons propagating on a background is clear. One can also observe the collision of the two inner solitons. The amplitude of the colliding pair is much greater, implying larger curvature, than that of the other pair, but the two solitons leave the collision unmodified (as is typical of classical solitons). There is a "hierarchy" effect, in that at large t the gross features of the four-soliton solution will resemble those of the two-soliton solution.

For seed metrics with $\delta > 3$, the general n-soliton solution still tends asymptotically to the diagonal n-soliton solution (6.7). We have already seen that this is the case in the two-soliton solution, and one can prove the general result by induction (Carr and Verdaguer, 1983). This can be interpreted as the loss of one of the polarizations in the asymptotic regions. Thus the asymptotic results of section VIa(A) for the diagonal metrics apply; some of those metrics, depending on δ and n, will become singular at $z \to \infty$ (without clear cosmological interpretation) others will become flat at $z \to \infty$. Of course in the "near regions" these metrics are very different from the diagonal ones. They have a much reacher fine structure, with two polarizations and many extra parameters.

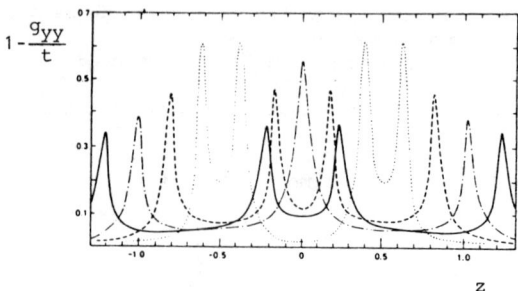

FIG. 7.- This shows the time evolution of the $g_{yy}(t,z)$ component for the four soliton solution generated by the axisymmetric Kasner seed ($\delta=0$). The same conventions as in Figure 6. The soliton widths are $w_1 = w_2 = 0.01$ and the separation of their origins is $|z_1^0 - z_2^0| = 1$. The dot-dashed line corresponds to the time t=0.45 during the collision of the inner solitons. After the collision, they move unperturbed.

The solutions with $1 \leqslant \delta \leqslant 3$, which include the Minkowski seed, have yet to be studied in detail. They are more complicated because they do not tend towards either diagonality or the seed metric as $z \to \infty$.

VII. SOLITON SOLUTIONS ON BIANCHI TYPE-II

Belinskii and Francaviglia (1982) have given the general formalism for calculating soliton solutions in all Bianchi types from I to VII. As we remarked in section III the integration of equation (3.7), to find $\psi_0(\lambda,t,z)$ once a seed $g_0(t,z)$ is given, may not be an easy task when $g_0(t,z)$ is a nondiagonal metric.

Bianchi type II metrics are the first example of nondiagonal metrics to which the solitonic technique has been applied. In their first paper, Belinskii and Francaviglia (1982), found the matrix $\psi_0(\lambda,t,z)$ and, in their second paper (Belinskii and Francaviglia, 1983), they give the corresponding one-soliton solution.

The seed metric coefficient f_0 is

$$f_0(t) = \frac{C_0}{4a_0^2} t^{\frac{\delta^2-1}{2}} (1+p) \quad , \quad p = 4\chi^2 a_0^2 (\delta+1)^{-2} t^{2(\delta+1)} \qquad (7.1a)$$

where C_0, a_0, δ and χ are arbitrary real parameters. For $\chi=0$ we recover the Kasner metric (4.1).

The matrix g_0 is

$$g_0 = \tilde{1} \, \gamma_0 \, 1 \tag{7.1b}$$

where γ and 1 are 2×2 matrices ($\tilde{1}$ is the transpose matrix) defined by

$$\gamma_0 = \begin{pmatrix} a^2(t) & 0 \\ 0 & b^2(t) \end{pmatrix} \quad \text{and} \quad 1 = \begin{pmatrix} 1 & \chi z \\ 0 & 1 \end{pmatrix} \tag{7.1c}$$

with $a^2(t) = 4a_0^2 (1+p)^{-1} t^{\delta+1}$ and $b^2(t) = t^2 a^{-2}(t)$.

Note that here $\det g_0 = t^2$ as in equation (3.2), hence the coordinates are adapted to the notation of section III.

To find the one-soliton solution we must choose a real-pole trajectory as in (5.1). Here, we have again the problem of discontinuities across the light cone $(z_1^0 - z)^2 = t^2$. Only in the region $(z_1^0 - z)^2 \geqslant t^2$ the soliton solution has meaning. Inside the light cone the solution is the homogeneous seed metric (7.1).

The metric coefficient f for the one-soliton solution is

$$f = C_1 \, f_0 \, \frac{\mu^2 Q}{(t^2 - \mu^2) \sqrt{t}} \tag{7.2a}$$

where C_1 is an arbitrary parameter and

$$Q = a^{-2} \Lambda_1^2 + b^{-2} \Lambda_2^2$$

$$\Lambda_1 = a^2 \left[-\chi q_1 \mu^{\frac{\delta+1}{2}} + q_2 \mu^{-\frac{\delta+1}{2}} \right]$$

$$\Lambda_2 = t^2 \left[\chi q_1 \mu^{\frac{\delta-1}{2}} p^{-1} + q_2 \mu^{-\frac{\delta+3}{2}} p \right]$$

with q_1, q_2 arbitrary parameters.

The matrix g of the one-soliton solution is

$$g = \tilde{1} \, \gamma \, 1 \tag{7.2b}$$

with the matrix γ defined by

$$\gamma_{ab} = \frac{|\mu|}{t} (\gamma_0)_{ab} + \frac{t^2 - \mu^2}{t \, |\mu|} \frac{\Lambda_a \Lambda_b}{Q} \quad (a,b=1,2) \;.$$

When we take $\chi=0$ the soliton solution (7.2) reduces to the one-soliton solution on a Kasner background with real-poles (5.18). This can be considered a generalization of (5.18).

This solution has not only the cosmological singularity at $t=0$, but it has also a space singularity when $z \to \infty$. Besides, it shares the problems of the real-pole solutions with Kasner seeds. Its cosmological interest, therefore, is not clear.

The space singularity could be eliminated, in the two-soliton solutions, by taking different prescriptions with the two poles, say $\mu_1^{(-)}$ and $\mu_2^{(+)}$. Such solutions would have a cosmological interpretation as pulse waves on Bianchi II backgrounds. The problem of discontinuities could be solved by taking complex-poles.

VIII. SOLITON SOLUTIONS ON BIANCHI TYPE V

The present Universe seems to fit very well with a Friedman-Robertson-Walker (FRW) model. It would be of interest to find new exact soliton solutions, representing inhomogeneous generalizations of the FRW models, which evolve towards the standard model containing gravitational waves. However the soliton technique can be applied only to vacuum or to a perfect fluid of stiff matter. Belinskii (1979a) has derived solutions with stiff matter (see next section).

Here we shall study the soliton solutions of the Milne Universe (Ibañez and Verdaguer, 1984b). The Milne (1934) model, which is a region of flat space, can be interpreted as a vacuum open FRW, since the open FRW models evolve towards it when the influence of matter can be neglected. The Milne model has τ=const. hypersurfaces of constant negative curvature and belongs to the Bianchi type V class (Siklos, 1984).

It can be written as

$$ds^2 = -d\tau^2 + \tau^2 dl^2$$
$$dl^2 = d\chi^2 + \sinh^2\chi(\sin^2\theta d\phi^2 + d\theta^2) \tag{8.1}$$

where $0 \leq \tau \leq \infty$, $0 \leq \chi \leq \infty$, $0 \leq \theta \leq \pi$ and $0 \leq \phi \leq 2\pi$. The lines χ=constant, radiating from the origin, represent the world lines of physical observers with proper time τ who see a Universe expanding from the origin at τ=0.

Metric (8.1) is not written in the form (3.1), appropiate to the soliton technique. We can make a coordinate change (Belinskii et al., 1971) to cylindrical coordinates,

$$\sinh\rho = \sinh\chi \sin\theta$$
$$\cosh\rho \sinh z = \sinh\chi \cos\theta$$

where now $0 \leq \rho \leq \infty$ and $-\infty \leq z \leq \infty$, and dl^2 becomes

$$dl^2 = d\rho^2 + \sinh^2\rho d\phi^2 + \cosh^2\rho dz^2 = d\rho^2 + dl_1^2 \ . \tag{8.2a}$$

Often, instead of using Milne's time τ we shall use $t = \ln\tau$, where now $-\infty \leq t \leq \infty$. With this time, metric (8.1) reads

$$ds^2 = e^{2t}(d\rho^2 - dt^2) + e^{2t} dl_1^2 \ . \tag{8.2b}$$

Now, the Milne metric (8.2) is written in the appropiate form (3.1). We see that for this seed metric,

$$\det g_0 = e^{4t}\sinh^2\rho\cosh^2\rho \qquad (8.3)$$

does not verify condition (3.2). But it can be put that way by a coordinate change of type (3.3) to (z',t'). When one evaluates the real-pole trajectories (5.1) in terms of the present coordinates t and ρ one finds that they are well defined in the whole range of these coordinates. The reason is that the region $(z')^2 < t'^2$ has no physical significance in Milne's coordinates. Therefore, soliton solutions with real-pole trajectories do not have, here, the problem of discontinuities.

For simplicity we find the diagonal metric with two real-poles with prescriptions $\mu_1^{(-)}$ and $\mu_2^{(+)}$ so that, at $z \to \infty$, the soliton solution will become the Milne metric. And it will be interpreted as gravisolitons on a Milne background.

It can be written (Ibáñez and Verdaguer, 1984b)

$$ds^2 = f(\tau^2 d\rho^2 - d\tau^2) + \tau^2 dl_2^2 \qquad (8.4)$$

where

$$f = C s_1^3 s_2^3 \tau^{-8} \sinh^{-2}\rho \cosh^{-6}\rho (s_1 + \tanh\rho)^{-2} (s_2 + \tanh\rho)^{-2} (s_1^2-1)^{-1}(1-s_2^2)^{-1}(s_1 s_2 -1)^{-2}$$

$$dl_2^2 = \frac{\sinh^2\rho}{s_1 s_2} d\phi^2 + s_1 s_2 \cosh^2\rho \, dz^2$$

with $\quad s_1 \equiv \beta_1/\alpha - [(\beta_1/\alpha)^2 - 1]^{1/2}, \qquad s_2 \equiv \beta_2/\alpha + [(\beta_2/\alpha)^2 - 1]^{1/2}$

$$\beta_a \equiv -\tau_a^2 - \frac{1}{2}\tau^2 \cosh 2\rho \qquad (\tau_1 \neq \tau_2) \qquad \alpha \equiv \frac{1}{2}\tau^2 \sinh 2\rho$$

here C and τ_a are arbitrary real parameters; for convenience we take

$$C = 64 \tau_1^3 \tau_2^3 (\tau_1 - \tau_2)^2 .$$

Metric (8.4) becomes Milne in the asymptotic regions. It turns out that the maximum deviations from the Milne background are concentrated around the future light cone $\rho = \ln\tau$, for large τ, and the past light cone $\rho = -\ln\tau$, for small τ. The asymptotic analysis can be done in terms of the Riemann components (6.9) using the tetrad (6.8). It is found that in the future timelike infinity $\rho << \ln\tau \to \infty$, $\psi_0 \sim \psi_2 \sim \psi_4 \sim 0(\tau^{-2})$, therefore the metric becomes flat i.e. the Milne metric.

A similar behaviour is found at spacelike infinity $\ln\tau << \rho \to \infty$ where $\psi_0 \sim \psi_2 \sim \psi_4 \sim 0(e^{-\rho})$. In the past timelike infinity $\rho << -\ln\tau \to \infty$ ($\tau \to 0$) the Riemann components become constant, thus the metric (8.4) does not have the cosmological singularity.

At future null infinity $\rho \simeq \ln\tau \to \infty$ we find $\psi_4 \sim 0(\tau^{-2})$ but $\psi_0 \sim \psi_2 \sim$ $\sim 0(\tau^{-4})$. Thus, the metric becomes flat there but the outgoing radiative component decreases in amplitude slower than ψ_0 and ψ_2. That means that the metric, in its leading terms, is of type N in the Petrov classification. That is, the soliton perturbations become pure outgoing gravitational radiation. This is similar to what we found with Kasner seeds.

At past null infinity $-\rho \simeq \ln\tau \to -\infty$ ($\tau \to 0$), we have $\psi_4 \sim \psi_2 \sim 0(1)$ and

$\psi_0 \sim O(\tau^{-2})$. Thus the metric, in its leading terms is also of the radiation type. Moreover the incoming radiative component ψ_0 is unbounded, so that the metric is singular in this region: test particles on the plane orthogonal to the direction of propagation of the incoming soliton will find infinite tidal forces at $\tau=0$. However this is not a scalar singularity (Siklos, 1981) since the curvature scalar invariants (6.15) are bounded in that region. As they must be, because the nonradiative components are bounded.

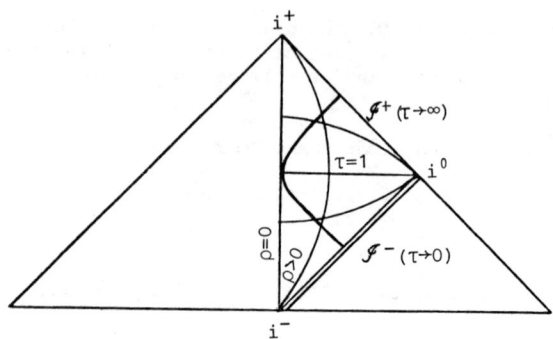

FIG. 8.- Structure of the soliton solution (8.4) on the conformal diagram for the Milne Universe. It is represented for the z=0 plane.

We represent in Figure 8 the structure of the soliton solution (8.4) on the conformal diagram for the Milne Universe (A region of the conformal diagram of Minkowski space). It is represented for the z=0 plane. The "unphysical" region $\rho<0$ is not represented, although it is also allowed by the solution (8.4). The thick line represents the trajectory of the soliton perturbation. The soliton perturbation starts as gravitational radiation of infinite amplitude. It propagates as a cylindrical wave towards the axis, $\rho=0$, where it behaves in a particle-like fashion. Finally the soliton propagates away from the $\rho=0$ axis becoming gravitational radiation again, which propagates with decreasing amplitude on the Milne background.

We should mention now that metric (8.4) can be seen as a paradigm of all diagonal metrics with 2n-solitons constructed with the same number of (+) and (-) prescriptions for the pole trajectories.

IX. SOLITON SOLUTIONS WITH MATTER

The soliton technique of Belinskii and Zakharov is aimed, basically, to the solution of the field equations $R_{ab}=0$, i.e. equations (3.4). When we assume the presence of matter, that is a given energy momentum tensor $T_{\mu\nu}$, the above equations will not be verified because they will contain a non null right hand side. However, when a stiff perfect fluid is assumed (equation of state $\varepsilon=p$), then

$$T_{\mu\nu} = 2\varepsilon u_\mu u_\nu + \varepsilon g_{\mu\nu} \qquad (u^\mu u_\mu = -1) \qquad (9.1)$$

and Einstein equations become

$$R_{\mu\nu} = 2\varepsilon u_\mu u_\nu \qquad (9.2)$$

Since for a metric of the form (3.1) the Ricci components $R_{0a}=R_{3a}=0$, it follows from (9.2) that $u_a=0$ and, consequently, $R_{ab}=0$. Thus the soliton technique can be applied in that case since equations (3.4) still hold.

The remaining Einstein equations (3.5) will be modified now. And only the coefficient f of metric (3.1) will be different from the vacuum case. Wainwright et al. (1979) had shown that for metrics with the form (3.1), solutions with stiff matter could be generated from vacuum solutions. The equation of state for stiff matter implies that the velocity of sound equals the velocity of light and was proposed by Zeldovich (1972) to describe the matter content of the Universe in its early stages (Barrow, 1977).

Belinskii (1979a) applied the soliton technique to that case. Expressing the matter field in terms of a scalar function ϕ (this can be done without loss of generality),

$$\varepsilon = p = -\frac{1}{2}\phi_{,\mu}\phi^{,\mu} \qquad u_\mu = \frac{1}{\sqrt{2\varepsilon}}\phi_{,\mu} \qquad , \qquad (9.3)$$

the Einstein equations (9.2) read

$$R_{\mu\nu} = \phi_{,\mu}\phi_{,\nu} \qquad (9.4a)$$

and the contracted Bianchi identities imply

$$\phi_{;\mu}{}^{;\mu} = 0 \qquad (9.4b)$$

These are the equations for the coupling of the gravitational field with a massless scalar field.

If we write the metric coefficient f in (3.1) as a product $f=f_v \cdot F$ equations (9.4) can be divided into four groups. The first and second of them exactly repeat the Einstein equations in vacuumn (3.4) and (3.5) for a metric (3.1) with f changed by f_v. The third group is just a wave equation for the scalar field ϕ,

$$(t\phi_{,z})_{,z} - (t\phi_{,t})_{,t} = 0 \qquad , \qquad (9.5)$$

and the fourth group determines the factor F which corrects (3.5) for the matter

$$\begin{aligned}(\ln F)_{,t} &= t(\phi_{,z}^2 + \phi_{,t}^2) \\ (\ln F)_{,z} &= 2t\phi_{,z}\phi_{,t}\end{aligned} \qquad (9.6)$$

Hence to solve the problem we must first construct a vacuum soliton solution (f_v, g) as we explain in Section III. After this we must determine the scalar ϕ from equation (9.5) and then find, using equation (9.6), the coefficient F. The final result will be a solution (f,g), with $f=f_v \cdot F$, and the energy density and the components of the velocity of the matter will be determined by the scalar field ϕ according with equations (9.3).

We should observe the parallelism between equations (3.4)-(9.5) and (3.5)-(9.6). In fact, the soliton technique can be applied to equations (9.5) and (9.6) too and one can find soliton solutions for the scalar field ϕ. One can take different number of solitons for the pair (f,g), say n, and for the pair (F,ϕ), say m. Such solutions will be called n-m-soliton solutions.

Belinskii (1979a) found the 1-0-soliton solutions, i.e. the scalar field remains unperturbed, with nondiagonal metrics for the flat, open and closed FRW models. Although there is only one pole and therefore it must be real, there is no problem of discontinuities because such metrics do not verify (3.2) and the physically meaninful range of variables is reduced to the region $(z_1^0-z')^2 \geqslant (t')^2$ (in terms of the new variables z', t' transformed by (3.3) and such that $\det g = (t')^2$).

As an exemple we consider the 1-0-soliton solution in the flat FRW background with stiff matter.

The seed metric is, in this case

$$ds^2 = t(d\rho^2-dt^2)+t\rho^2 d\theta^2+tdz^2$$

$$\phi = \sqrt{3/2} \ln t \qquad \varepsilon = p = \frac{3}{4t^2} \tag{9.7}$$

with $0 \leqslant \rho \leqslant \infty$, $0 \leqslant \theta \leqslant 2\pi$, $-\infty < z < \infty$ and $t \geqslant 0$.

The real-pole trajectory (5.1) with prescriptions (+) now becomes

$$\mu = -\frac{1}{2}(l^2+t^2+\rho^2) + \frac{1}{2}[(l^2+t^2+\rho^2)^2-4t^2\rho^2]^{1/2} \tag{9.8}$$

which is well defined in the whole t-ρ region and where l is an arbitrary real parameter. The 1-0-soliton solution is then

$$f = \frac{l^2 t [s^2 t^2+(t^2+\mu)^2]}{s^2 l^2 t^2+(t^2+\mu)^2}$$

$$g = \frac{t}{[s^2t^2+(t^2+\mu)^2]} \begin{pmatrix} s^2 t^2 \rho^2 + \rho^2(t^2+\mu)^2 + q\rho^2(t^2+\mu) - q^2\mu & qs\mu \\ qs\mu & s^2 t^2 + (t^2+\mu)^2 - q(t^2+\mu) \end{pmatrix} \tag{9.9a}$$

with s an arbitrary parameter and $q \equiv s^2-l^2$. The matter contents is given by

$$\phi = \sqrt{3/2} \ln t$$

$$\varepsilon = p = \frac{3s^2[l^2 t^2+(t^2+\mu)^2]}{l^2 t^2 [s^2 t^2+(t^2+\mu)^2]} \tag{9.9b}$$

This model has the cosmological singularity only. For small t a soliton perturbation is concentrated near the symmetry axis $\rho=0$. After a critical time $t \sim s$ that cylindrical perturbation propagates outward, with decreasing amplitude and at large t it propagates at the speed of light.

This is another exemple of production of gravitational waves. The qualitative behaviour of this solution is very similar to that of the late stages of the soliton solutions on a Milne Universe discussed in section VIII. The 1-0-soliton solutions on the open FRW models are qualitatively similar to (9.9) and in the closed model the soliton perturbation starts decreasing but then it increases again on the final stages.

Letelier (1982) generalized equations (9.5) to the case of an anisotropic fluid described by two-perfect-fluid components. The pressure on the direction of propagation of the wave equals the energy density (stiff matter). He finds the (0-1), (1-0) and 1-1-soliton solutions with real-poles on a Kasner background. In the last case both, the gravitational field and the matter, present a solitonic behaviour.

Note that it seems essential for a solitonic behaviour the existence of a unique velocity defined on the system in the direction of propagation of the wave. In vacuum we have only the gravitational field (speed of light). With stiff matter, the matter and the gravitational field have a coherent coupling: speed of sound = speed of light. One might argue that for a perfect fluid with different speed of sound the solitonic behaviour will not persist and the solitons will be dispersed.

X. SOLITON SOLUTIONS IN KALUZA-KLEIN

Vacuum Einstein equations in N dimension can be written

$$R_{AB} = 0 \quad (A,B = 1,2,\ldots,N). \quad (10.1)$$

Let us assume that the N-dimensional spacetime admits the existence of N-2 commuting Killing vectors, so that we can write the metric tensor in terms of two coordinates. This metric tensor can be written in the form (3.1) where, now, g is an $(N-2)\times(N-2)$ matrix $(a,b = 1,2,\ldots,N-2)$ and Einstein equations (10.1) can be written in the form (3.4)-(3.6) with U and V being $(N-2)\times(N-2)$ matrices.

The soliton technique described in section III can be applied without modification up to equation (3.11). The final n-soliton solution (3.12) suffers a slight modification due to the fact that det g', from (3.11), is different when $N > 4$. The final result is

$$g = t^{\frac{-2n}{N-2}} \left(\prod_{k=1}^{n} \mu_k \right)^{\frac{2}{N-2}} g' \quad (10.2a)$$

$$f = f_0 t^{\frac{-n(n+4-N)}{N-2}} \left(\prod_{k=1}^{n} \mu_k \right)^{\frac{2(n-3+N)}{N-2}} \frac{\left[\prod_{n=1}^{n} (\mu_k^2 - t^2) \right]^{\frac{4-N}{N-2}}}{\prod_{\substack{k,l=1 \\ k>l}}^{n} (\mu_k - \mu_l)^{\frac{4}{N-2}}} \det \Gamma_{kl} \quad (10.2b)$$

Equations (10.2) give the n-soliton solution of Einstein equations in vacuum for a spacetime of dimension N when this depends on two coordinates only.

The interest on Kaluza-Klein theories or theories with dimension greater than four (Kaluza (1921), Klein(1926)) has come recently from the particle physics community, especially from those investigating supergravity (Cremmer and Julia (1978), Scherk and Schwarz (1979)).

The existence of the extradimensions may have interesting applications to cosmology, see for instance Chodos and Detweiler (1980), Alvarez and Belen-Gavela (1983). However, here we shall adopt the point of view of considering extradimensions merely as an auxiliary tool to obtain meaningful solutions in four dimensions.

Restricting ourselves to five dimensions, this point of view will allow us to find soliton solutions of the Einstein-Maxwell equations or even when there is also a scalar field coupled to the electromagnetic and the gravitational fields.

Let us consider the latter case first. Take a five-dimensional metric

$$ds^2 = \gamma_{AB} dx^A dx^B \quad (A,B = 1,2,3,4,5) , \tag{10.3}$$

assume that

$$\gamma_{AB,5} = 0 \tag{10.4}$$

and define the four-dimensional metric

$$g_{\mu\nu} = \gamma_{\mu\nu} - \phi^2 A_\mu A_\nu \quad (\mu,\nu=1,2,3,4) \tag{10.5}$$

where

$$A_\mu \equiv \phi^{-2} \gamma_{5\mu} \quad \text{and} \quad \phi^2 = \gamma_{55} . \tag{10.6}$$

The vaccum Einstein equations in five dimensions, (10.1), for a metric verifying (10.4) can be written

$$R_{\mu\nu} - \frac{1}{2} g_{\mu\nu} R = -\phi^{-1} (\phi_{;\mu;\nu} - g_{\mu\nu} \phi^{;\alpha}_{;\alpha}) - \frac{1}{2} \phi^2 (F_{\alpha\mu} F^\alpha_\nu - \frac{1}{4} g_{\mu\nu} F_{\alpha\beta} F^{\alpha\beta}) \tag{10.7a}$$

$$\phi^{;\mu}_{;\mu} = \frac{1}{4} \phi^3 F_{\alpha\beta} F^{\alpha\beta} \tag{10.7b}$$

$$F^{\mu\alpha}_{;\alpha} = -3\phi^{-1} F^{\mu\alpha} \phi_{,\alpha} \tag{10.7c}$$

where

$$F_{\mu\nu} = A_{\mu,\nu} - A_{\nu,\mu} \tag{10.8}$$

and all tensorial operations are computed with respect to the four-dimensional metric $g_{\mu\nu}$.

These are the equations for the coupling of a scalar field, ϕ, an electromagnetic field, A_μ, and the gravitational field, $g_{\mu\nu}$. It is now apparent that one can obtain soliton solutions of equations (10.7) when the metric $g_{\mu\nu}$ and the other fields admit two commuting Killing vectors. The way to proceed is to start with a seed metric, which could be Minkowski, and use (10.2) to compute the soliton solutions for the five-dimensional metric γ_{AB}. Once this metric has been computed, use (10.5) and (10.6) to find the soliton solutions of the system (10.7).

Belinskii and Ruffini (1980) found the two-soliton solutions with a Minkowski seed in the axisymmetric context. Their solution is asymptotically flat and generalizes the Kerr solution.

The case of the Einstein-Maxwell equations without charges and matter can be seen as a subcase of (10.7). When one assumes $\phi=1$ these equations become

$$R_{\mu\nu} = -\frac{1}{2} F_{\mu\alpha} F^{\alpha}{}_{\nu} \qquad (10.9a)$$

$$F^{\mu\alpha}{}_{;\alpha} = 0 \qquad (10.9b)$$

$$F_{\alpha\beta} F^{\alpha\beta} = 0 \qquad (10.9c)$$

which are the Einstein-Maxwell equations with the constraint (10.9c) (see also Belinskii 1979b).

Therefore soliton solutions of the Einstein-Maxwell equations without charges and matter with the constraint (10.9c) can be found if one assumes that the fields depend on two variables only. The procedure is identical to the one described above when a scalar field is present.

Similarly, soliton solutions in Brans-Dicke theory without ordinary matter can be found; since the Brans-Dicke equations can be obtained from vacuum Einstein equations in five dimensions, after an appropiate identification of the fields (Belinskii and Khalatnikov, 1973).

ACKNOWLEDGEMENTS: It is a pleasure to thank J.L. Sanz and L.J. Goicoechea for their invitation to deliver these lectures and for their kind hospitality and efficient organization. I would like to acknowledge stimulating discussions with B.J. Carr, J. Ibáñez, M.A.H. MacCallum and S.T.C. Siklos and I am also indebted to Mercè Pascual for her help in preparing the manuscript for publication.

Adams, D.J., Hellings, R.W., Zimmerman, R.L., Farhoosh, H., Levine, D.I. and Zeldich, Z., Astrophys. J. 253, 1 (1982).

Alekseev, G.A. and Belinskii, V.A., Sov. Phys. JETP 51, 655 (1981).

Alvarez, E. and Belen-Gavela, M., Phys. Rev. Lett. 51, 931 (1983).

Barrow, J.D., Nature 272, 211 (1977).

Belinskii, V.A., Sov. Phys. JETP 50, 623 (1979a).

Belinskii, V.A., JETP Lett. 30, 29 (1979b).

Belinskii, V.A. and Fargion, D., Nuovo Cimento 59B, 143 (1980).

Belinskii, V.A. and Francaviglia, M., Gen. Rel. Grav. 14, 213 (1982).

Belinskii, V.A. and Francaviglia, M., Gen. Rel. Grav.   (1983).

Belinskii, V.A. and Khalatnikov, I.M., Sov. Phys. JETP 36, 591 (1973).

Belinskii, V.A., Lifshitz, E.M. and Khalatnikov, I.M., Sov. Phys. Uspekhi 13, 745 (1971).

Belinskii, V.A. and Ruffini, R., Phys. Lett. 89B, 195 (1980).

Belinskii, V.A. and Zakharov, V.E., Sov. Phys. JETP 49, 985 (1979).

Belinskii, V.A. and Zakharov, V.E., Sov. Phys. JETP 50, 1 (1980).

Bertotti, B., Astrophys. Lett. 14, 51 (1973).

Bertotti, B. and Carr, B.J., Astrophys. J. 236, 1000 (1980).

Bertotti, B., Carr, B.J. and Rees, M.J., Mon. Not. R. Astron. Soc. 203, 945 (1983).

Carmeli, M., Charach, Ch., Phys. Lett. 75A, 333 (1980).

Carmeli, M., Charach, Ch. and Malin, S., Phys. Rep. 76, 79 (1981).

Carr, B.J., Astron. Astrophys. 89, 6 (1980).

Carr, B.J. and Verdaguer, E., Phys. Rev. D28, 2995 (1983).

Chodos, A. and Detweiler, S., Phys. Rev. D21, 2167 (1980).

Cosgrove, C.M., J. Math. Phys. 21, 2417 (1980).

Cremer, E. and Julia, B., Phys. Lett. 80B, 48 (1978).

Ellis, G.F.R. and MacCallum, M.A.H., Commun. Math. Phys. 12, 108 (1969).

Ellis, G.F.R. and MacCallum, M.A.H., Commun. Math. Phys., 19, 31 (1970).

Francaviglia, M., Journées Relativistes, Grenoble (France, 1981).

Francaviglia, M. and Segarra, C., Nuovo Cimento 80B, 223 (1984).

Gardner, C.S., Greene, J.M., Kruskal, M.D., Miura, R.M., Phys. Rev. Lett. 19, 1095 (1967).

Gowdy, R.H., Ann. Phys. 83, 203 (1974).

Harrison, K.B., Phys. Rev. Lett. 41, 1197 (1978).

Hellings, R.W., Phys. Rev. Lett. 43, 470 (1979).

Hellings, R.W. and Downs, G.S., Astrophys. J. 265, L35 (1983).

Ibáñez, J. and Verdaguer, E., Phys. Rev. Lett. 51, 1313 (1983).

Ibáñez, J. and Verdaguer, E., Phys. Rev. D, in press (1984a).

Ibáñez, J. and Verdaguer, E., to be published (1984b).

Jantzen, R.T., Nuovo Cimento $\underline{59B}$, 287 (1980).

Kaluza, Th., Sitzungsber. Preuss. Akad. Wiss. Phys. Math. Kl. $\underline{LIV}$, 966 (1921).

Klein, O., Z. Phys. $\underline{37}$, 875 (1926).

Kramer, D., Stephani, H., MacCallum, M. and Herlt, E., "Exact solutions of Einstein's field equations", Cambridge University Press (1980).

Kitchingham, D., Queen Mary College preprint (1984), private communication by M.A.H. MacCallum.

Lamb, H., "Hydrodynamics", Cambridge University Press (1906).

Letelier, P.S., Phys. Rev. $\underline{D26}$, 2623 (1982).

MacCallum, M.A.H., Commun. Math. Phys. $\underline{20}$, 57 (1971).

MacCallum, M.A.H., "Cargese Lectures in Physics", ed. E. Schatzman (Gordon and Breach, N.Y., 1973).

MacCallum, M.A.H., "General Relativity, An Einstein Centenary Survey", eds. S.W. Hawking and W. Israel (Cambridge University Press, 1979).

MacCallum, M.A.H., "Lecture Notes in Physics: Retzbach Seminar on exact solutions of Einstein's field equations", eds. W. Dietz and C. Hoenselaers (Springer-Verlag, Berlin 1984).

Mashhoon, B., Mon. Not. R. Astron. Soc. $\underline{199}$, 659 (1982).

Mashhoon, B. and Grishchuk, Astrophys. J., $\underline{236}$, 990 (1980).

Mashhoon, B., Carr, B.J. and Hu, B.L., Astrophys. J. $\underline{246}$, 569 (1981).

Milne, E.A., Quart. J. Math., Oxford $\underline{5}$, 64 (1934).

Misner, C.W., Phys. Rev. Lett. $\underline{22}$, 1071 (1969)

Neugebauer, G., J. Phys. $\underline{A12}$, L67 (1979).

Neugebauer, G., J. Phys. $\underline{A13}$, L19 (1980).

Peebles, P.J.E., "Physical Cosmology", Princeton University Press (1971).

Romani, R.W. and Taylor, J.H., Astrophys. J. $\underline{265}$, L35 (1983).

Rosen, N., Bull. Res. Coun. Israel $\underline{3}$, 328 (1954).

Rosi, L.A. and Zimmerman, R.L., Astrophys. Space Sci. $\underline{45}$, 447 (1976).

Ryan, M.P. and Shepley, L.S., "Homogeneous Relativistic Cosmologies" (Princeton University Press, 1975).

Scherk, J. and Schwarz, J., Nucl. Phys. $\underline{B153}$, 61 (1979).

Scott, A.C., Chu, F.Y.F. and McLaughlin, D.W., Proceedings of the I.E.E.E. $\underline{61}$, 1443 (1973).

Siklos, S.T.C., J. Phys. $\underline{A14}$, 395 (1981).

Siklos, S.T.C., "Relativistic Astrophysics and Cosmology", p.201, eds. X. Fustero and E. Verdaguer (World Scientific Publishing Co., 1984).

Tomimatsu, A. and Sato, H., Prog. Theor. Phys. $\underline{70}$, 215 (1981).

Turner, M.S., Astrophys. J. $\underline{233}$, 685 (1979).

Verdaguer, E., J. Phys. $\underline{A15}$, 1261 (1982).

Verdaguer, E., In preparation (1984)

Wainwright, J., Phys. Rev. $\underline{D20}$, 3031 (1979a).

Wainwright, J., J. Phys. $\underline{A12}$, 2015 (1979b).

Wainwright, J., J. Phys. $\underline{A14}$, 1131 (1981).

Wainwright, J., Ince, W.C.W. and Marshman, B.J., Gen. Rel. Grav. $\underline{10}$, 259 (1979).

Wainwright, J. and Marshman, Phys. Lett. $\underline{72A}$, 275 (1979).

Weinberg, S., "Gravitation and cosmology", John Wiley and Sons (1972).

Zeldovich, Ya.B., Mon. Not. R. Astron. Soc. $\underline{160}$ 1 (1972).

Zeldovich, Ya.B. and Novikow, I.D., "Relativistic Astrophysics: The structure and evolution of the Universe", University of Chicago Press (1983).

Zimmerman, R.L. and Hellings, R.W., Astrophys. J. $\underline{241}$, 478 (1980).

PARTICIPANTS

E. Aguirre (Madrid)
M. Alises (Almería)
M. Araujo (London, U.K.)
A.I. Arauzo (Valladolid)
F. Argueso (Santander)
B. Barberis (Torino, Italy)
X. Barcons (Santander)
O. Bertolami (Cambridge, U.K.)
R. Blanco (Santander)
J.A. Caballero (Badajoz)
J. Calero (Cordoba)
M. Calvo (Madrid)
J. Carot (Palma de Mallorca)
B.J. Carr (Cambridge, U.K.)
M. Castro (Madrid)
A. de Castro (Madrid)
R. Conejo (Madrid)
C. Delgado (Granada)
P. Domingo (Madrid)
R. Dominguez (Madrid)
X. Dong Liu (København, Danmark)
L.J. Echevarría (Santander)
P. Emperador (Madrid)
C. Enrique (Madrid)
A.C. Fabian (Cambridge, U.K.)

J.J. Ferrando (Valencia)
L. Fontes (Madrid)
M.A. Fuente (Cuenca)
M. Gadella (Valladolid)
D. Galletto (Torino, Italy)
J.B. García (Las Palmas)
M.E. García (Madrid)
F. Garzón (Tenerife)
L.J. Goicoechea (Santander)
M. Gonzalez (Santander)
M. Gutierrez (Madrid)
A.J. Heras (Santander)
S. Hernández (Madrid)
J.M. Ibañez (Meudon, France)
E. Jimenez (Madrid)
B.J.T. Jones (København)
R. Lapiedra (Valencia)
F. Lazaro (Tenerife)
E. Lopez-Tello (Granada)
M.A.H. MacCallum (London, U.K.)
J.M. Marcaide (Bonn, Germany)
F. Marques (Barcelona)
M.A. Martín (Valladolid)
J.M. Martín (Salamanca)
A. Martínez (Santander)

E. Martinez (København,Danmark)
C.M. Mascort (Sevilla)
K.P. Mauricio (London,U.K.)
M.A. Mederuelo (Madrid)
M.P. Merino (Santander)
A. Molina (Barcelona)
C. Morales (Madrid)
L.M. Nieto (Valladolid)
J. Oliver (Valencia)
C. del Olmo (Madrid)
V. del Olmo (Valencia)
E. Ojero (Madrid)
M.E. Paez (Tenerife)
P.J.E. Peebles (Princeton,U.S.A.)
F. Pereda (Santander)
A. Pérez (Valencia)
I. Perez (Boon, Germany)
R. Perez-Aloe (Badajoz)
M. Portilla (Valencia)

G. Pujana (Bilbao)
A.M. Rodrigo (Madrid)
E. Ruiz (Salamanca)
B. Ruiz (Santander)
M.D. Sabau (Madrid)
A. Sanchez (Bilbao)
I. Sanchez (Málaga)
F. San José (Madrid)
J.L. Sanz (Santander)
M.J. Senobiain (Salamanca)
J. Sommer-Larsen (København)
R. Sussman (London, U.K.)
I. Tapia (Málaga)
I. Torremocha (Madrid)
J. Uson (Princeton, U.S.A.)
M.C. Velasco (Valladolid)
E. Verdaguer (Barcelona)
F. Vicente (Valladolid)
A. Vidaurre (Valencia)